VISCERAL MYOLOGIK™ 2

Atlas of Human Musculature in Clay

Jon Zahourek

ZOOLOGIK™ SYSTEM KINESTHETIC ANATOMY MANIKEN®

Edited by Kenneth Morgareidge, PhD

Zahourek Systems, Incorporated

ZOOLOGIK Systems N^{o}s:

ATLAS OF HUMAN VISCERAL ANATOMY IN CLAY
ZSP-MA-0002
5 VOLUME ATLAS OF HUMAN ANATOMY IN CLAY
ZSP-MA-0009

ISBN-887087-02-8 (Volume 2)
ISBN-887087-00-1 (5 Volume Set)

Table of Contents

Anatomy is, whether we like it or not, destiny. That is not such a limiting statement, after all, for the patterns and processes of our anatomy reveal the history of the Universe. We are Star Dust and Ocean Sea, the poetry of nature neither at its best nor its worst, but at its miraculous usual. *JZ*

Dedication

The Zoologik™ System has always been dedicated to:

René Dubos, the famed microbiologist who opened my eyes to the patterns of nature, and

Lawrence Angel, the great anthropologist who encouraged me and disciplined me to shape this system.

This atlas is dedicated to:

Jan Robert Zahourek, my son "Jenda", whose birth clarified my life and my life's work, and

Renée Whitman Zahourek, my wife, whose tireless devotion to me and to the ideas of the Zoologik™ System has made her the *mother* of Maniken™.

Thanks to Ken Morgareidge, *PhD,* for editing this Atlas; for his Latin scholarship, his translations of anatomical terminology, and for his encouragement.

Thanks to Cynthia Christy, for her enthusiasm and her helpful suggestions, corrections and proofing.

Jon Zahourek, 1994

Why Study Anatomy?

"If you would become educated, first know thyself." This is the traditional foundation on which to build an education. I believe that learning is mostly patterns-recognition; education is their manipulation. Understanding is realizing the interactivity of patterns, inside and out. Recognize patterns within yourself and you'll soon find yourself looking for the patterns around you.

Our bodies are perfect learning laboratories for the pattern analysis of our worlds, internal and external. Although we are mostly unaware of the process, we are busy in our body-labs, from our earliest fetal twitches, learning the most profound lessons about time, space and gravity; things we know so well we forget ever not knowing. We learned them by doing.

Nature's patterns are the "ways and means" of everyone's life, and can be recognized and *known* "in our bones." Anatomy should be a fundamental subject for everyone, rather than an ultrasubject for the specialist. I also believe that anatomy awareness and understanding should be entry-level human activity practiced lifelong. The most knowable and the most knowing part of yourself is the ancient culture of life-forms and functions we think of as our anatomy.

MYOLOGIK™ & ZOOLOGIK™

Myology is the study of muscles from micro (little) anatomy through gross (big) anatomy. *Zoology* is the general study of *all* animals, including human. The ZOOLOGIK™ System of Kinesthetic Anatomy approaches anatomy through our often under-used psychomotor learning power coupled with the logic of the *building* process. The user forms anatomy in clay and attaches it to uniquely designed skeletal models. One literally builds an understanding as one builds musculoskeletal, neural, vascular or splanchnic ideas in clay.

ZOOLOGIK™ models divide into evolutionary subsystems for clarity of understanding and, coincidentally, for ease of building clay anatomy: Axial, Visceral (face, jaws and throat); and Pectoral and Pelvic Appendicular subsystems.

I have found that *comparative anatomy* study of zoologic patterns is an exciting and vital way to clarify understanding of the logic of form and function. The ZOOLOGIK™ System also features an EQUIKEN™ Series for horse anatomy.

Other models in the series were designed to demonstrate typical human movement. Each model "freezes" a moment of motion, providing one the opportunity to consider and reconsider the problems and solutions within human kinesiology.

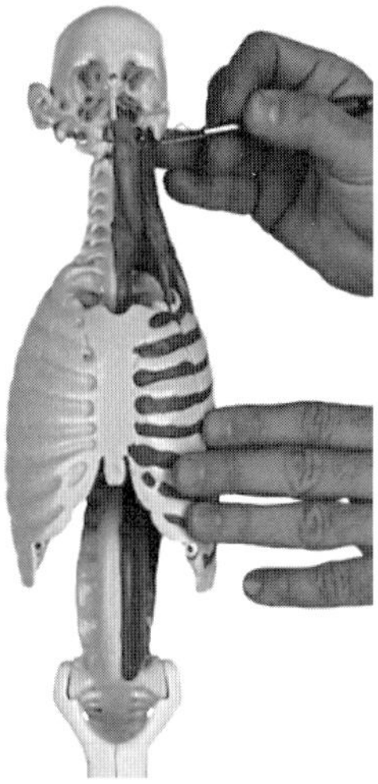

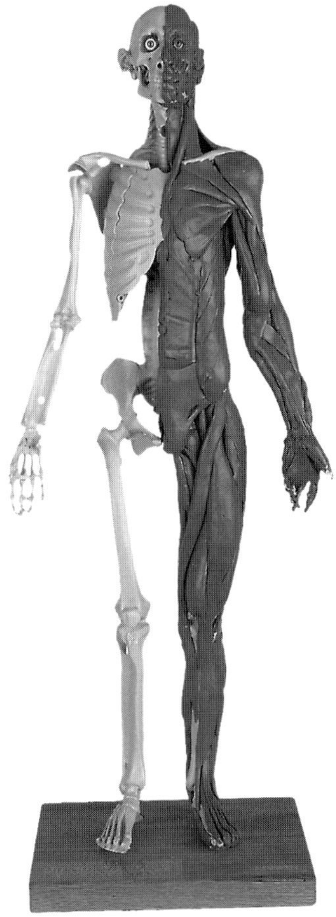

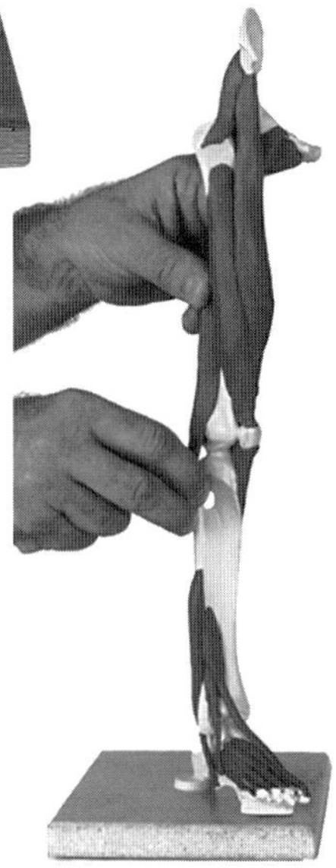

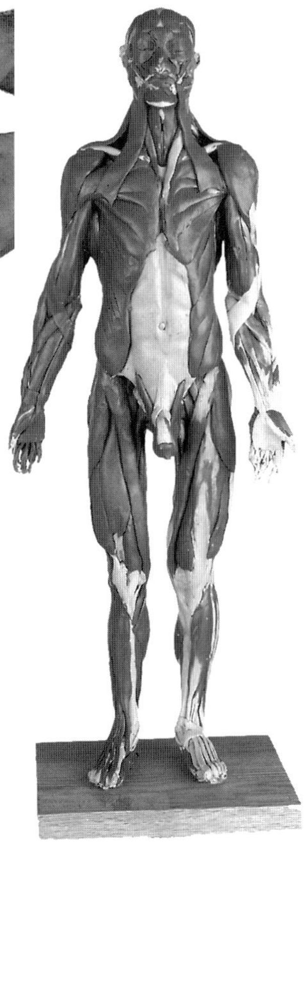

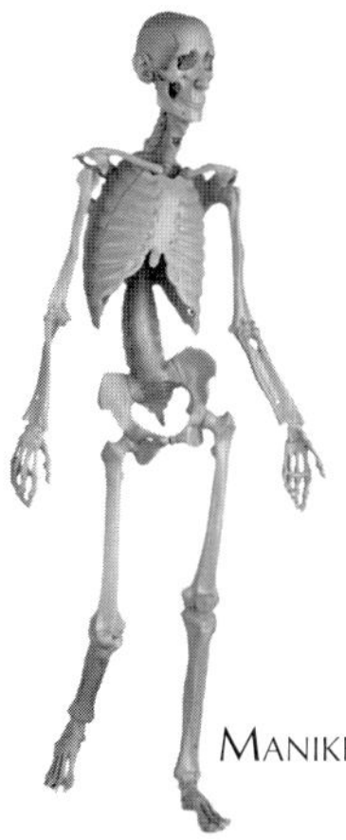

MANIKEN™ II PRO

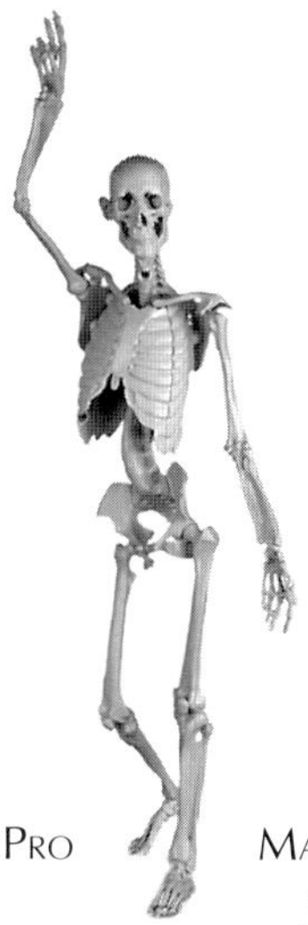

MANIKEN™ II PRO
(shown with optional alternative right arm)

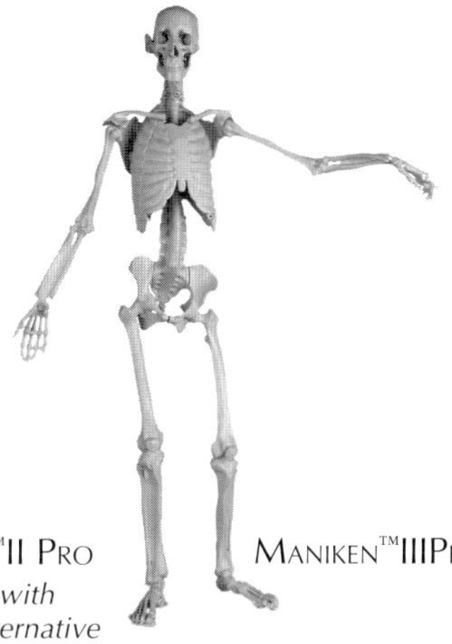

MANIKEN™ III PRO

EQUIKEN™ VI

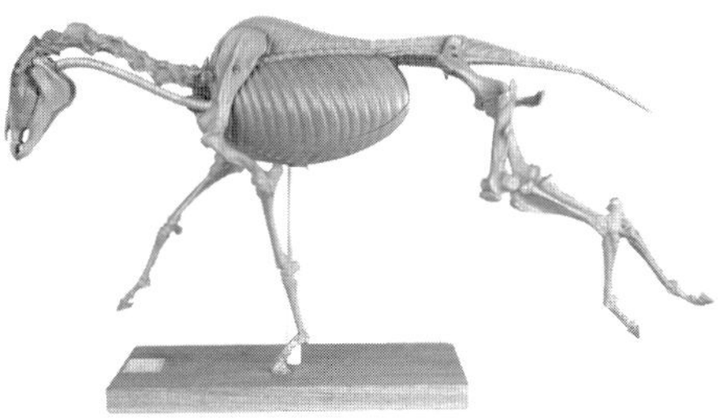

EQUIKEN™ V

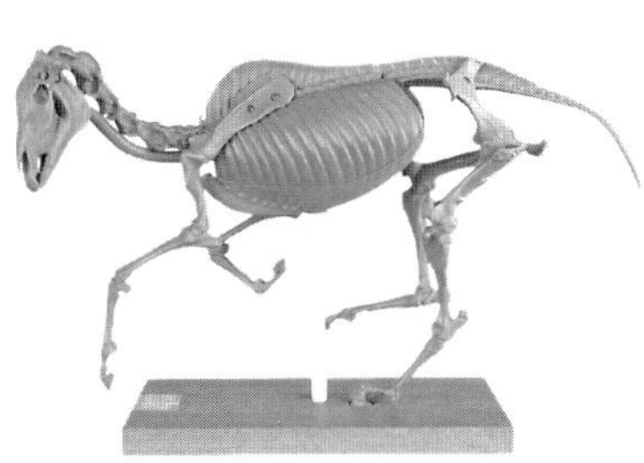

EQUIKEN™ I

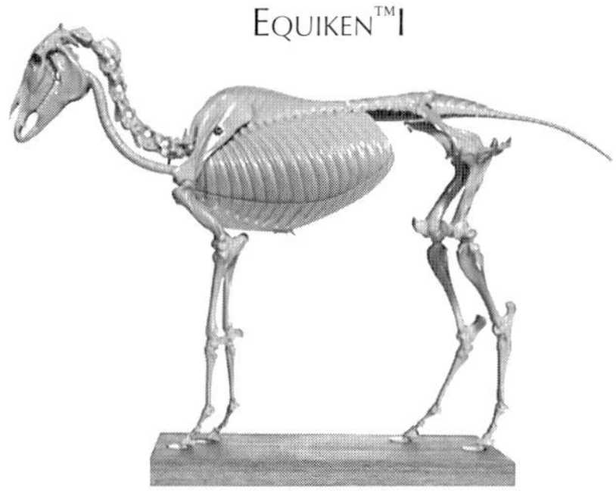

MANIKEN™ I PROFESSIONAL

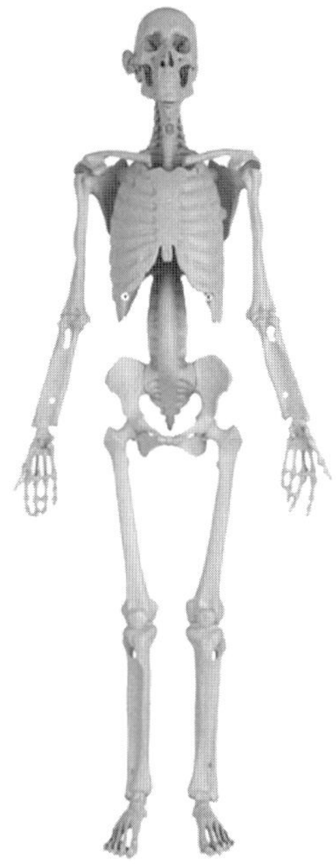

Focusing on human anatomy, the MANIKEN™ PROFESSIONAL Series of ZOOLOGIK™ models is produced in several versions. Models used in this atlas stand in the traditional, international standard Anatomical Position. MANIKEN™ SCHOLASTIC is also used in *this* atlas, because it divides this model into left and right, symmetrical halves, to provide medial views.

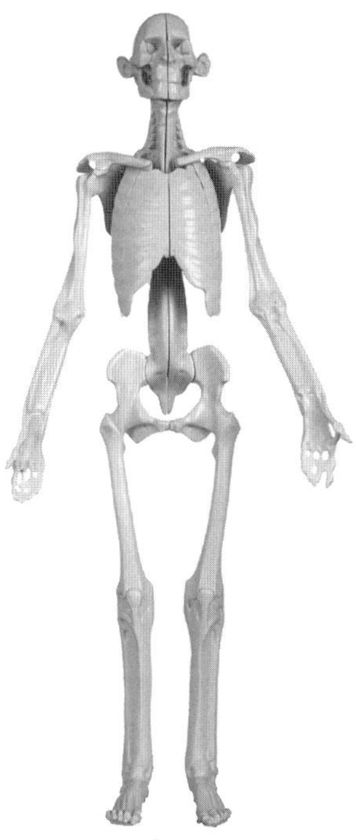

MANIKEN™ SCHOLASTIC

Anatomy Illustration

Anatomy books are necessarily two-dimensional illusions of three-dimensional material. An illustrator must draw a flat shape to *represent* any form, even round ones. Regardless of the artist's understanding of anatomy, or his or her mastery of illusion, every drawn image is an abstraction that must be interpreted. We assume, not always correctly, that the viewer can supply the three-dimensional translation.

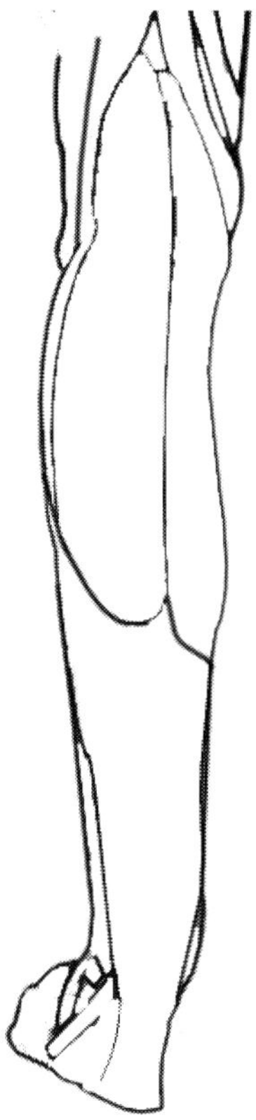

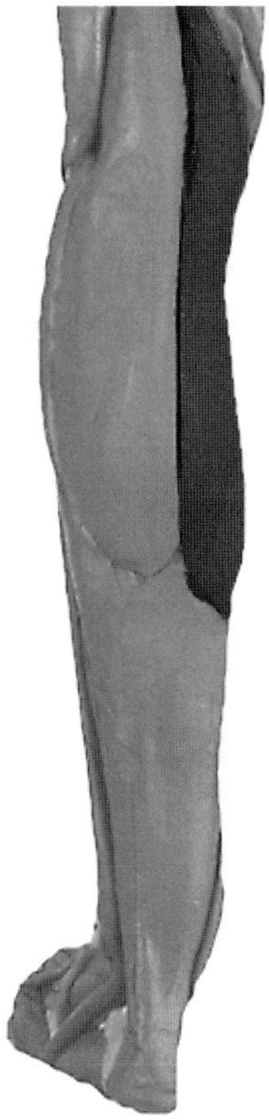

The following illustrations present an example of the kind of problems everyone shares in translating illustrations of anatomical form.

If we need an understanding of the width of the medial head of Gastrocnemius, the typical point of view presented is a posterior view. Clearly, the medial head is visible and the visual evidence shows it to be a narrow muscle.

If we turn the leg to a more medial view, we see a fuller representation of the proportion of the muscle. The muscle is wider than the earlier visual evidence indicated. it all depends on the view chosen by the illustrator.

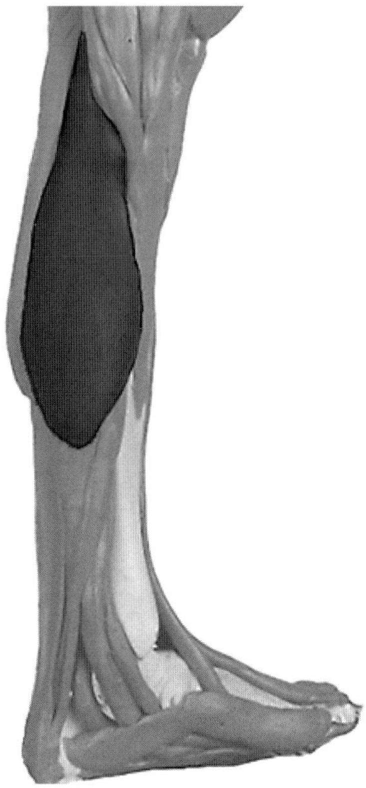

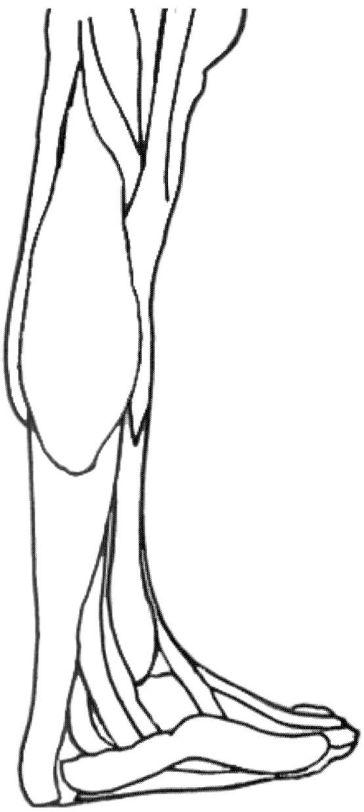

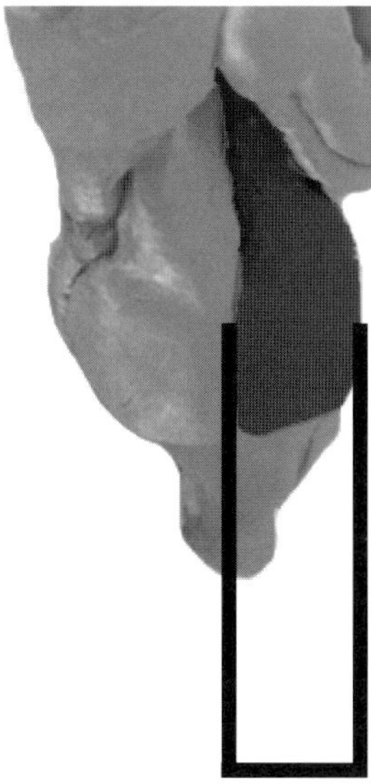

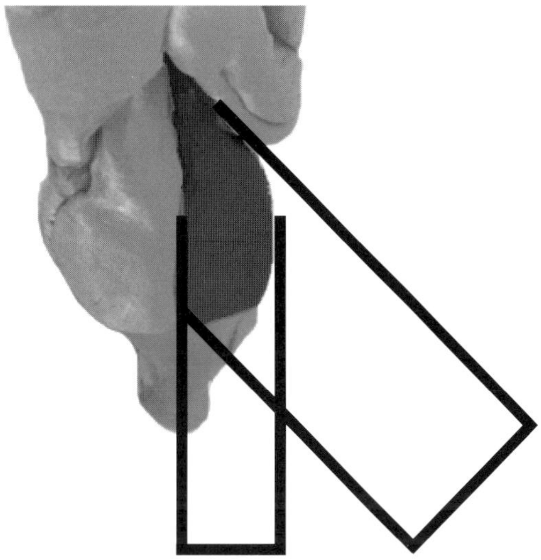

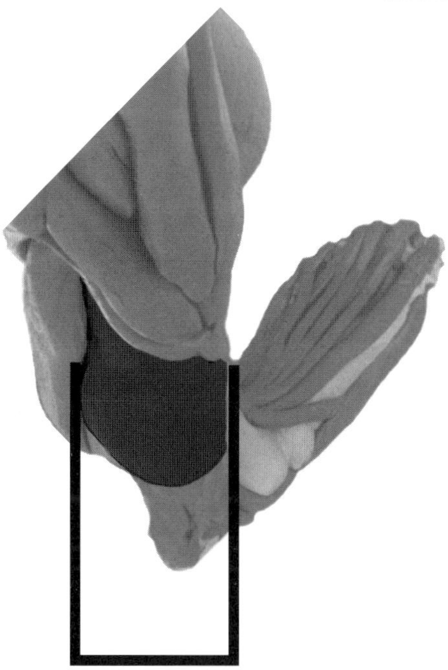

Photographic images provide neutral data, but are still only two-dimensional. It is as if you are stood absolutely still before an object, then opened *one* eye. Opening both eyes, providing a second view, helps us triangulate an object in space. To gather three-dimensional information about an object, either the viewpoint must relocate or the object must be moved.

When you have the opportunity to study anatomy in a fully three-dimensional way—through models or actual bodies—it's vital to develop an array of viewpoints of the same structure. Multiple data-sets, or ideas about a form can actually construct a "model" in your understanding.

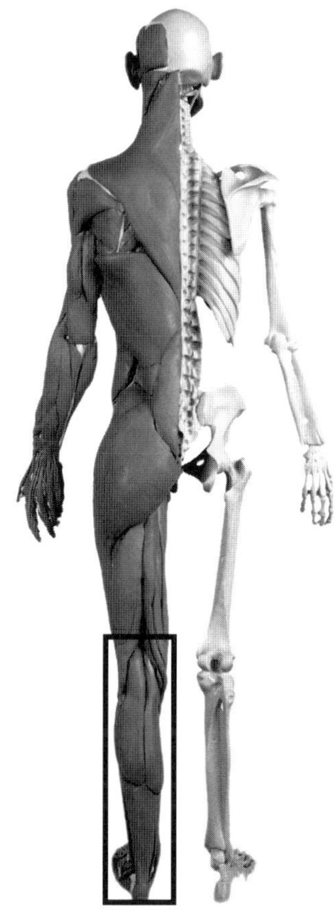

Illustration of this Atlas

This atlas presents real, three-dimensional forms seen from multiple views. In most of the illustrations created for this series of atlases, the muscles are simply formed, in clay onto Maniken™I Professional and Scholastic models of the human skeletal system.

Like the body, these Myologik™ atlas are organized into sub-systems. Each volume of the atlas presents one subsystem. This volume is devoted to the muscles of the Visceral or Branchial subsystem, providing the forms of jaws, mouth, throat and vocal apparatus. Other volumes cover Axial, Pectoral, Pelvic and Extrinsic (in which attachments are shared between subsystems) musculo-skeletal subsystems.

Muscles are illustrated one-at-a-time, starting with the deep muscles, working outwards to the more superficial muscles. By building the muscles individually on both the left and right sides of Maniken, and then turning the model about in space, we present each muscle in what I hope to be the simplest, most direct and informational manner. Atlases are basically "picture" books. This one is crafted to show you clearly the shape and attachments of each muscle. Use it alone and in support of any other anatomy textbooks, atlases, charts, videotapes, computer programs, lectures or dissection opportunities.

Information and ideas in this atlas are approximations. Anatomy, in actual development, is individual expression of the genetic blueprint, and it is extremely variable from one person to the next. There is so much variation, that anatomy is more of an art than a science. The actions and muscle attachments described and shown here are subject to many individual variations—some people even lack some of these muscles. Even the pronunciation is a matter of choice. No one knows the correct way to sound Greek and Latin terms, but language is an important part of the way we know things, and communication should be easy regardless of accent or pronunciation. The pronunciations presented in this atlas are all from one source, and are only offered as a suggestion.

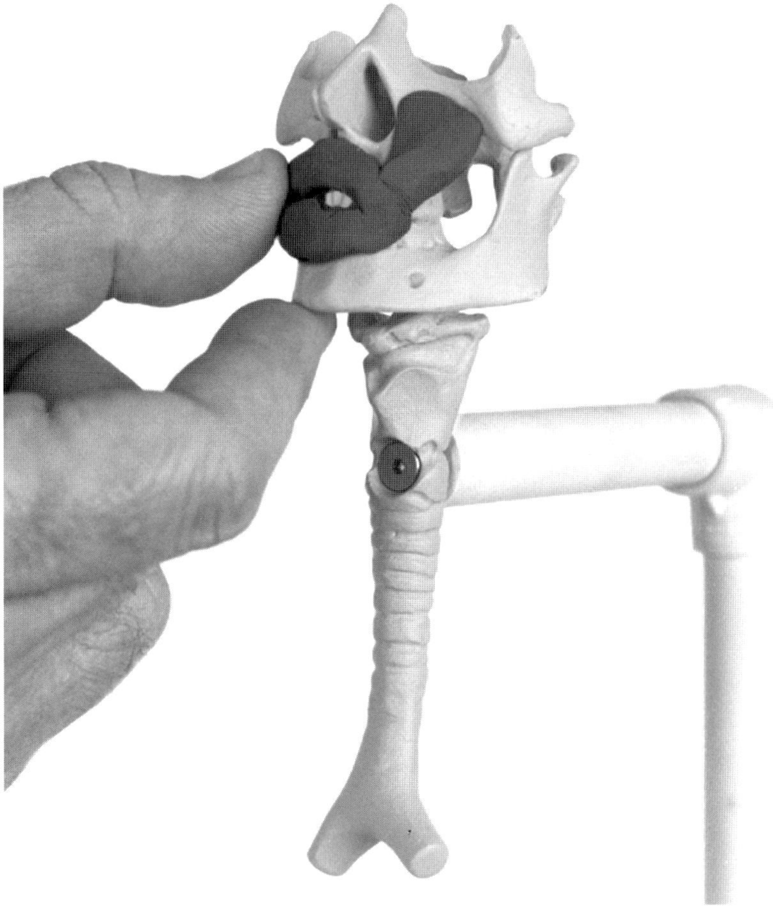

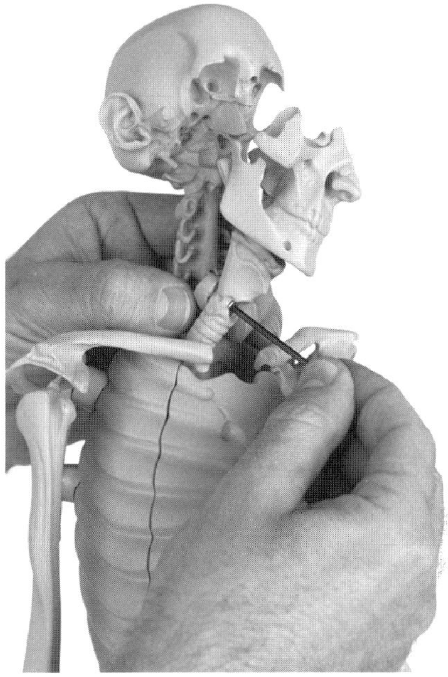

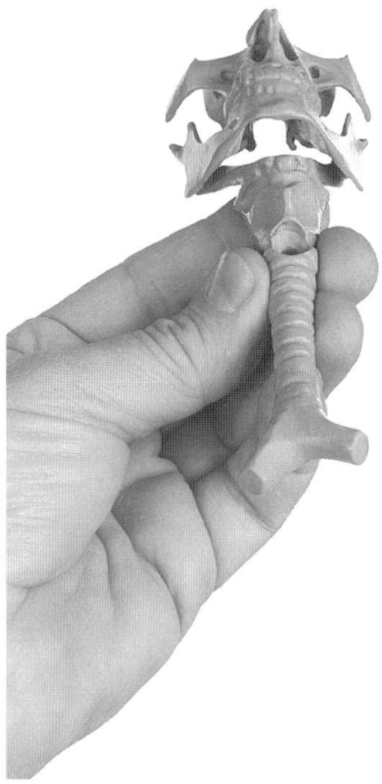

If you want to build anatomy on a MANIKEN™, this atlas can serve as a guide. By far the easiest way to build the visceral musculature is to detach the visceral subsystem from the axial and mount it separately on a stand (as shown above). Or you can embed the bronchii (the distral trachea) in a lump of clay.

The visceral muscles are mostly small and delicate, and are quick and easy to build. You will gain a much better understanding if you build all the muscles on both sides rather than just one side, and be sure to include the fibrous and cartilage structures shown in this atlas.

The goal of this atlas is to help understand the ideas of anatomy. It is not about how beautifully a muscle might be crafted in clay. Accordingly, the clay muscles have been formed quickly and simply with an emphasis on big idea rather than detail. In addition, I was more concerned with spatial form and with skeletal attachments than with the muscle architecture. Often a lighter color of clay shows non-osseus structures that are so much a part of visceral anatomy.

MANIKEN®

A Conceptual Skeleton

The classic classroom skeleton is only a partial representation of the human skeletal system. Some vital parts of skeletal anatomy almost always disappear in the processes of skeletonization, and only selected parts are restored in the rearticulation.

Traditionally, the rib-cage is augmented with some kind of resin to replace the costal cartilages without which there is no place to attach the sternum. Without the sternum, there is no place to attach the pectoral girdle. In other words, no costal cartilages; no arms. No form equals no function. Even real bones must be assembled into a conceptual model of zoological ideas.

The standard skeleton—real, replica or miniature—emphasizes the elegance of the limbs and the nobility of the human brow, but misses the point of vertebrate existence. Axial and appendicular (**somatic**) anatomy evolved to support the anatomy of metabolism and respiration: (**visceral** anatomy.) In my ZOOLOGIK™ System, it is the visceral model that most needed reinterpretation.

In order to find the most understandable form, I looked at function and worked backwards to discover the relevant form.

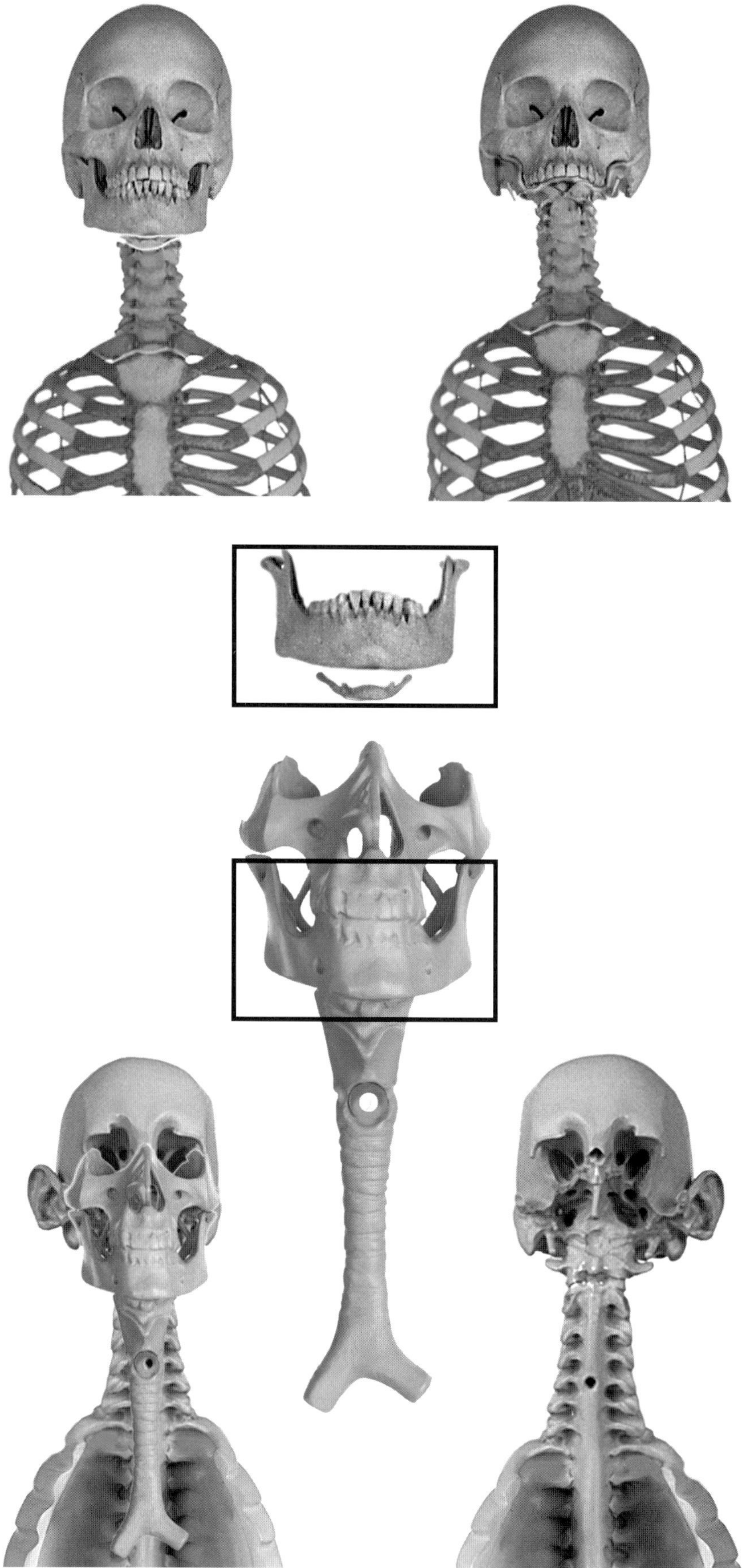

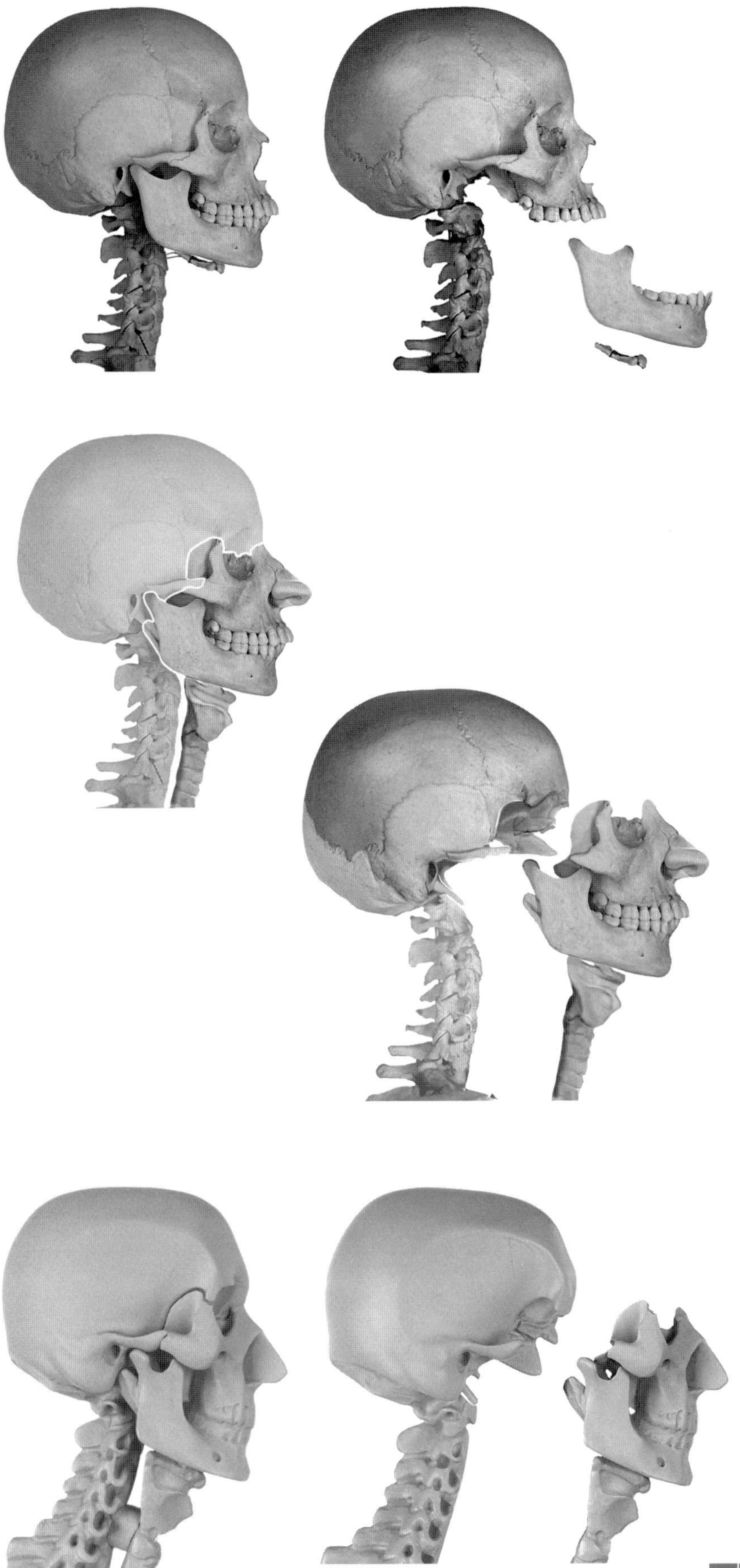

We are Chordates. There are a handful of ideas that humans share with other animals that make us so. We probably evolved from a archetype called a ***tunicate,*** *or sea squirt. The adult tunicate is a muscular bag, glued to a rock. It lives to eat and reproduce. Its gut is a tube with both ends open: an incurrent mouth, and an excurrent anus.*

The adult tunicate's only rigid, skeletal structure is an internal sleeve of slitted cartilage that develops inside the mouth. Beyond this, the gut is a simple stomach/ intestine. The chamber at the mouth is called the ***pharynx****.*

Visceral anatomy belongs to the gut. It is the passenger of our somatic body. But it is Nature's first idea in all chordates. With our prodigious capacity for words both as abstract symbols and as kinesthetic realities, we humans experience a unique integration of the simple neurology dedicated to the sensitivity of the primordial mouth with the awesome expansion of that neurology needed for kinesthetic analysis.

Much of the richness of vertebrate evolution is seen in the folding and unfolding of the muscle and cartilage of the pharyngeal tube into gill arches that, in turn, become arched jaws, mouth, throat and even middle ear. Other sensory mechanisms joined the sentinel near the mouth, creating a head. The cranial segments of axial anatomy became an interface between brain and pharynx.

Visceral anatomy as modeled in the MANIKEN™ PROFESSIONAL

Long recognized as a "keystone" element of the vertebrate skull, the **sphenoid** bone is unusually complex. It seems to me to have two basic forms, dorsal and ventral, each functionally divergent. The dorsal sphenoid is the key of the neurocranium; clearly axial.

The ventral sphenoid is the key of the facial bones, and to the jaws. It buttresses both the palatine bones and the maxillae with its ventral projections, the ***pterygoid processes.*** The sphenoid is a composite bone made of axial replacement, dermal and branchially derived bone.

In MANIKEN™PRO *models, I divided the sphenoid into dorsal and ventral laminae. The dorsal portion is part of the base of the skull of the axial subsystem; the ventral portion is given to the visceral subsystem. This includes the orbital part of the greater wing of the sphenoid and the pterygoid plates dorsal to the palate. Note the pommel-like hamulus at the distal end of the medial pterygoid plate shown at the right.*

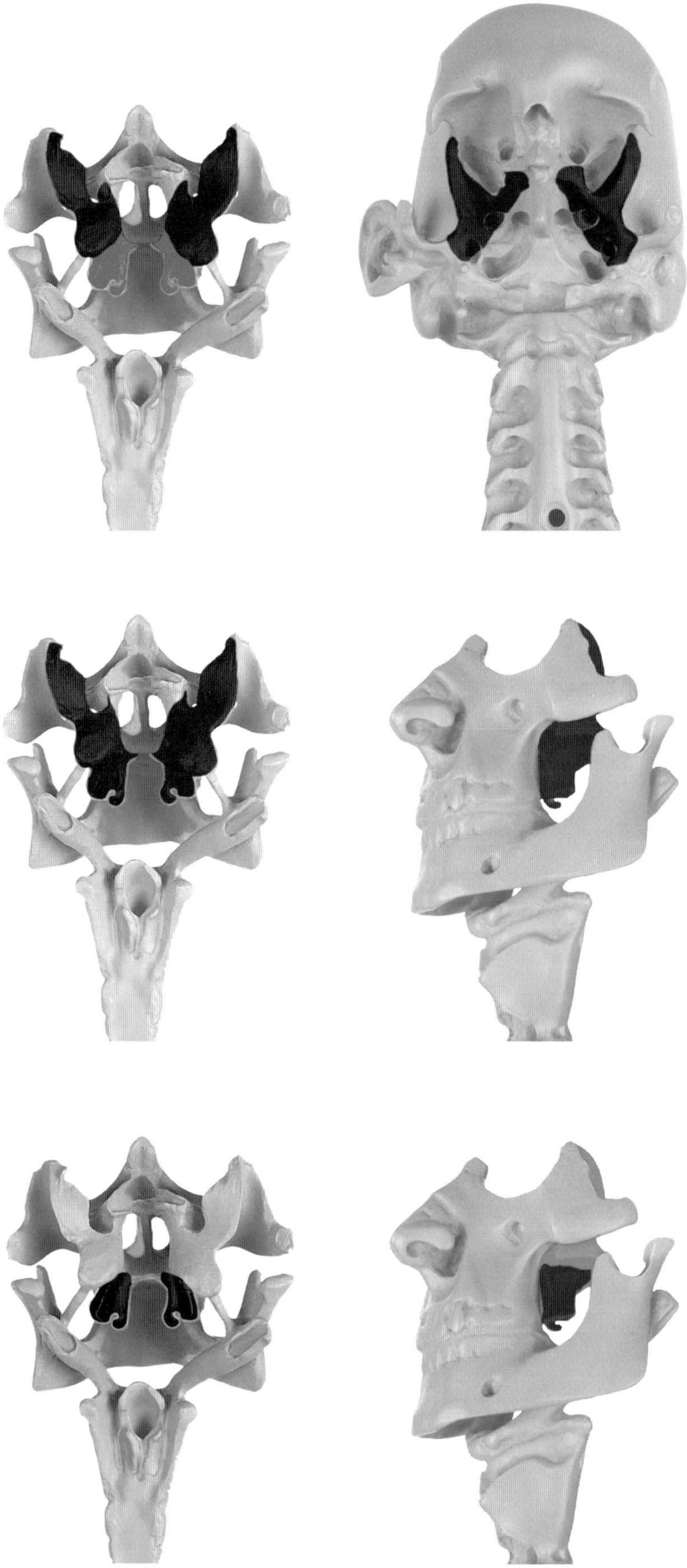

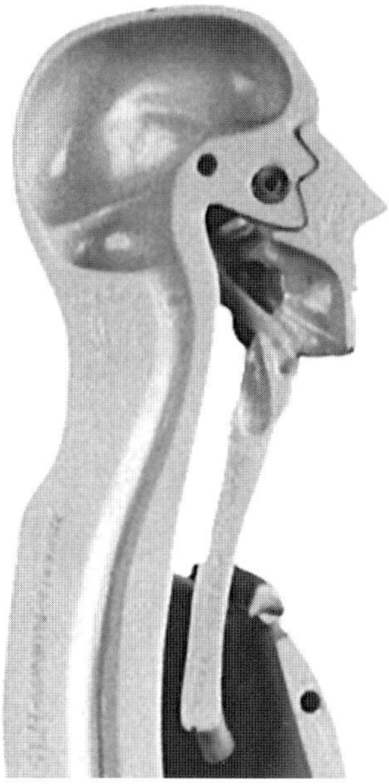

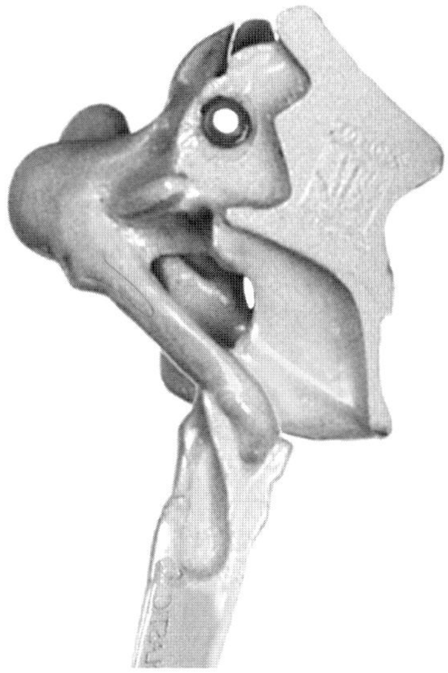

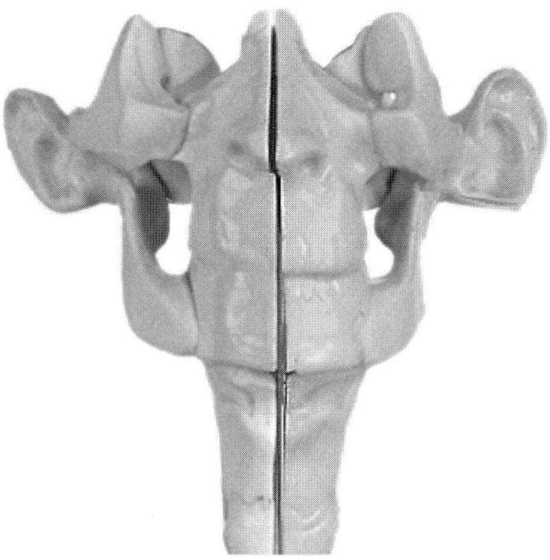

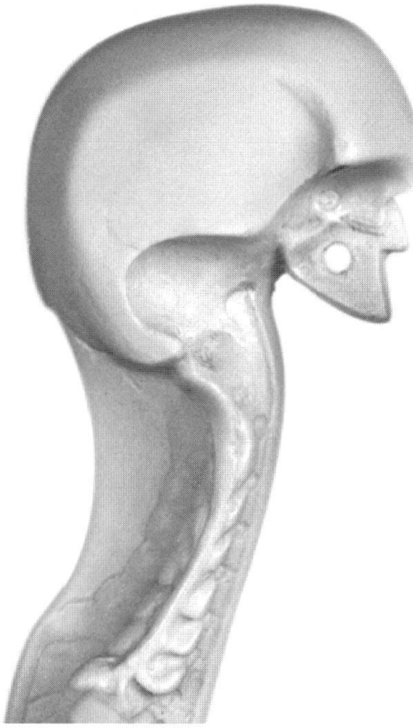

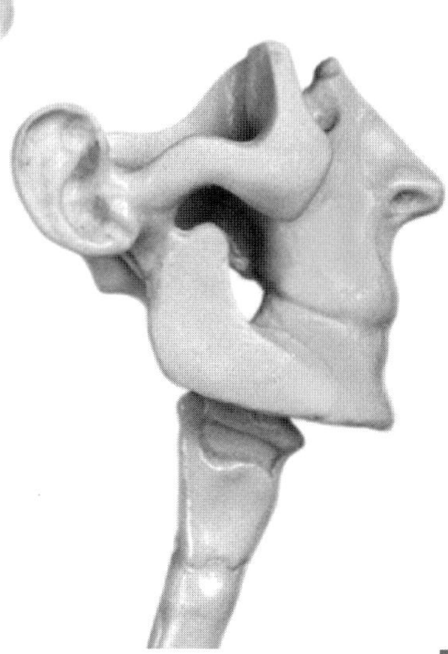

Visceral anatomy as modeled in the MANIKEN™ SCHOLASTIC

In the SCHOLASTIC *model which I have used extensively in this atlas, I show both the sphenoid and the temporal bones as dually derived interface structures. The squamous part of the temporal bone, and the petrous capsule of the ear function as parts of the neurocranium and are part of the Scholastic axial subsystem. The middle ear capsule, auditory tube cartilage, mastoid and styloid processes, mandibular fossa and zygomatic process of the temporal bone are all included in the visceral subsystem.*

In this interpretation, the entire zygomatic arch is visceral, as are the spines and ligaments of the sphenoid bone and styloid process. The external ear is part of the visceral subsystem in this model, although I suspect that it belongs to the axial subsystem, as shown in the PRO *model.*

Note that the SCHOLASTIC *model divides into left and right sides. For illustration purposes, this atlas also includes several photos of the* PRO *visceral subsystem sawn in half to provide medial views.*

Post-mandibular visceral anatomy

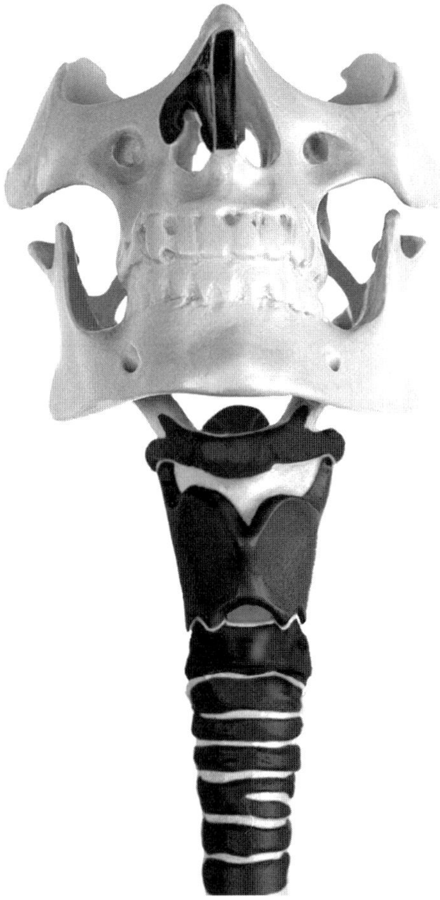

This section illustrates my interpretations in MANIKEN™ models of the post-mandibular cartilages of pharynx, larynx and trachea. Given the absence of visceral cartilages in the traditional classroom skeleton, this section also provides a brief review of their forms and functions.

*Except for the hyoid bone, the skeleton of the **visceral** sub-system, also called **branchial**, is missing from the traditional classroom skeleton. The skeleta of the **larynx** and **pharynx** demand our awareness.*

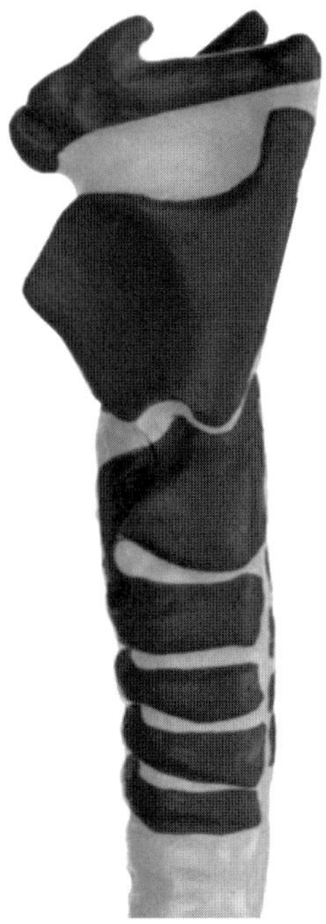

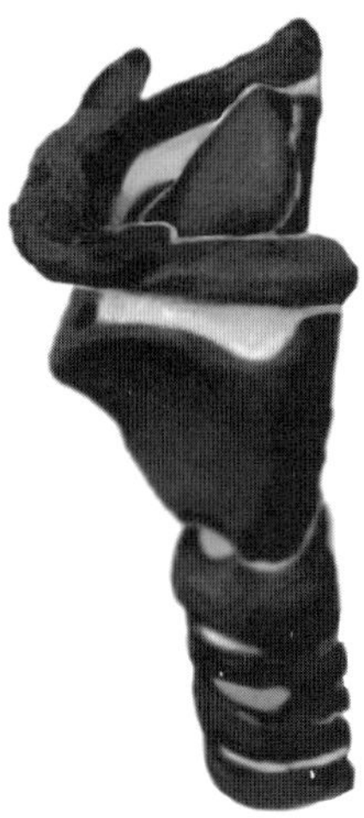

Hyoid bone

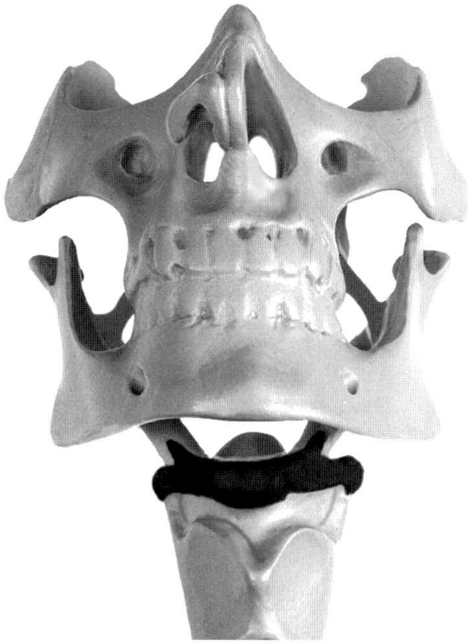

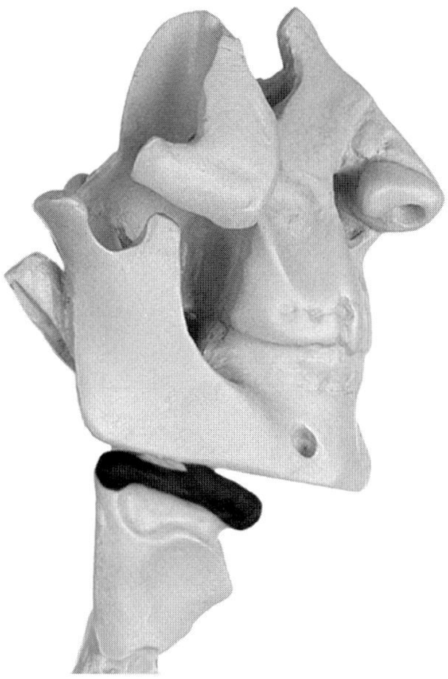

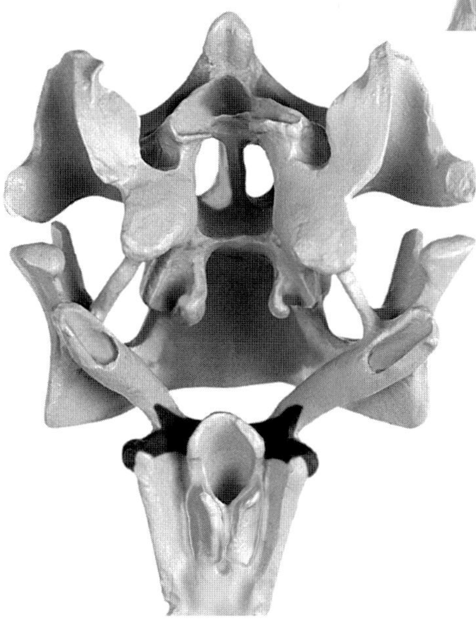

The hyoid is the only bone in post mandibular visceral anatomy. Apparently, human speech requires so much mobility for the laryngeal skeleton, that all the other parts do not ossify, but remain cartilage life-long.

If handled carefully, the hyoid can survive masceration (the skeletonizing process) intact or in pieces: body and two sets of cornua, or horns: the greater and lesser.

In superior traditional skeletal preparations, the hyoid is often mounted by wire, floating ventral to the cervical skeleton.

Thyroid cartilage

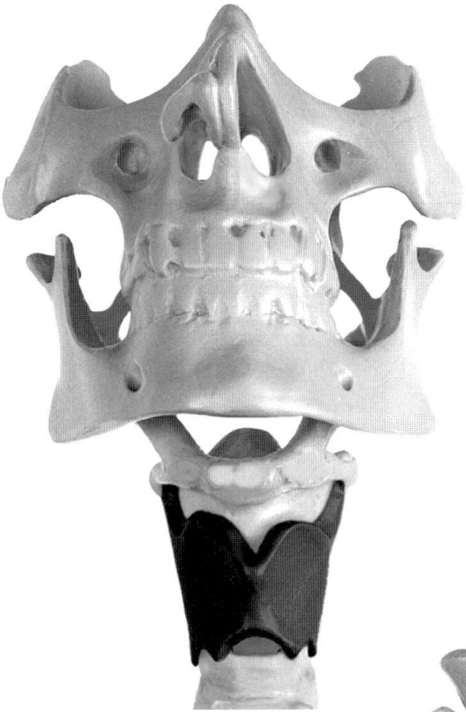

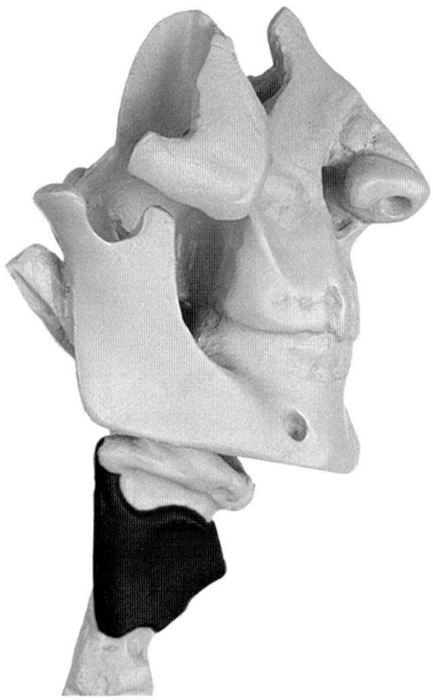

Like the hyoid bone, this broad cartilage also has two sets of cornua. The superior horns are suspended from the dorsal tips of the greater cornua of the hyoid bone by lateral thyro-hyoid ligaments. The inferior horns overlap the proximal cricoid cartilage and articulate on a pair of facets.

The two U-shaped elements are also linked by a sheet of ligament called the thyrohyoid membrane.

No thyroid cartilage
=no tongue
no windpipe
no vocal cords

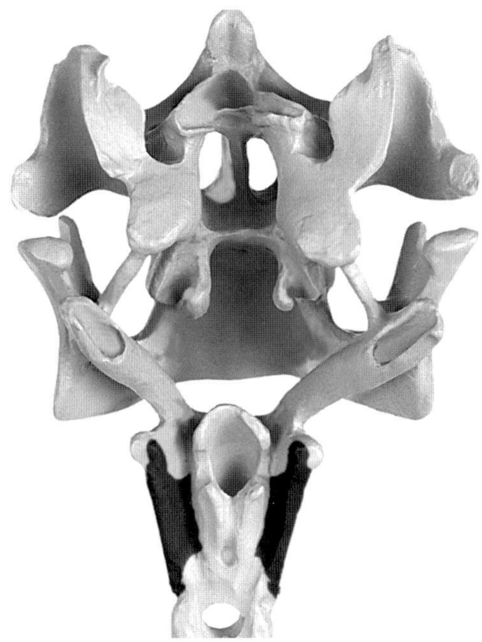

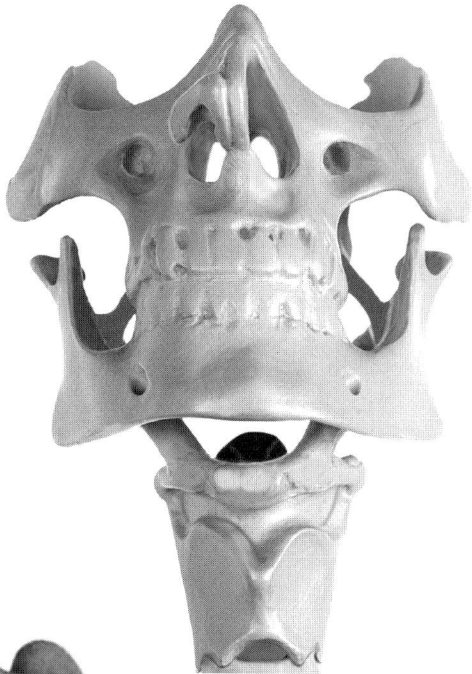

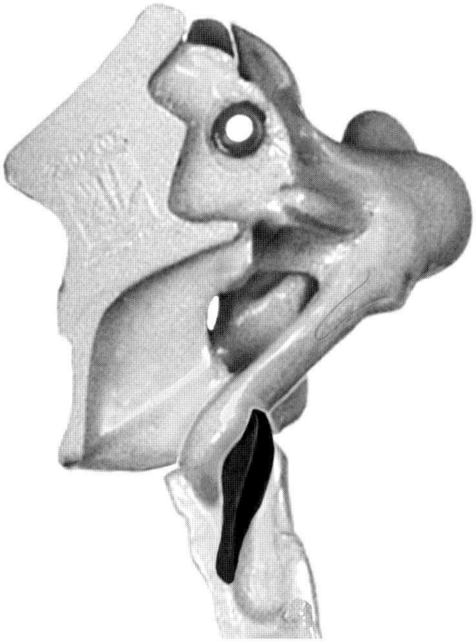

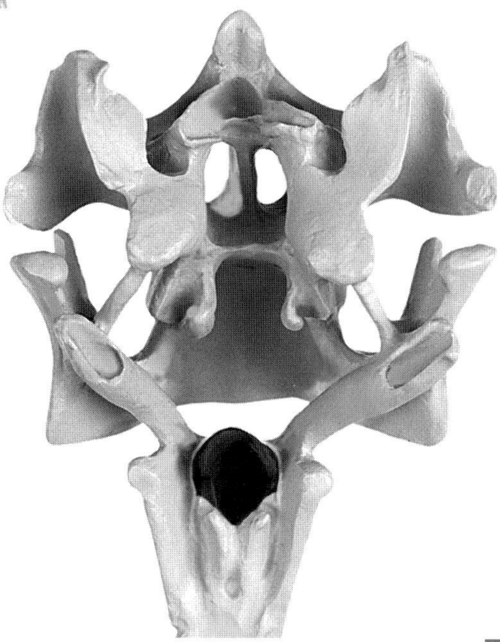

Epiglottis cartilage

This cartilage is leaf-shaped, and articulates with the thyroid cartilage at the dorsal surface of its angle.The distal surface of this cartilage closes off the inlet of the larynx.

The epiglottis is embedded in sheets of mucous membrane that sweep from both sides dorsally, to meet at the midline. The membranes also embed the arytenoid cartilages. When the epiglottis closes, these aryepiglottic sheets fold at the sides and are therefore called the ***aryepiglottic*** *folds.*

No epiglottis cartilage
=no tongue
no windpipe

Cricoid cartilage

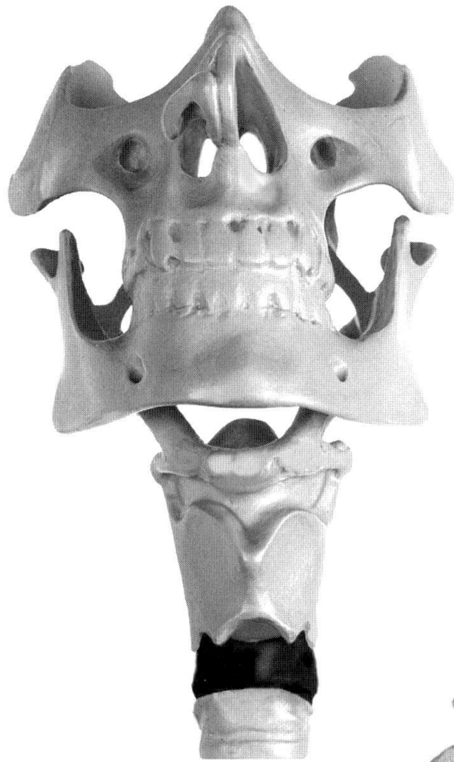

This ring of cartilage is named for a signet ring which is a narrow band that expands to a broad decoration for use as a signatory seal. The narrow part of the ring faces ventrally. The broad expansion forms the sides and dorsal laminae of the larynx.

The paired ***arytenoid*** *cartilages surmount its dorsal proximal border at articular facets. The cricoid and thyroid cartilages rock at their connecting articular facets.*

Attached to each arytenoid cartilage is one side of a slit in a membrane that helps keep things like food out of the lungs. The edges of the slit thicken into vocal "cords."

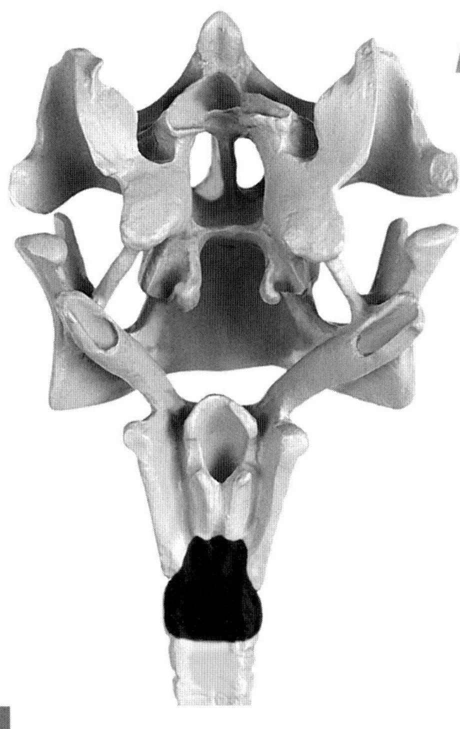

The windpipe or trachea is attached to its distal margin.

No cricoid cartilage
=no vocal cords
no windpipe

Arytenoid cartilages

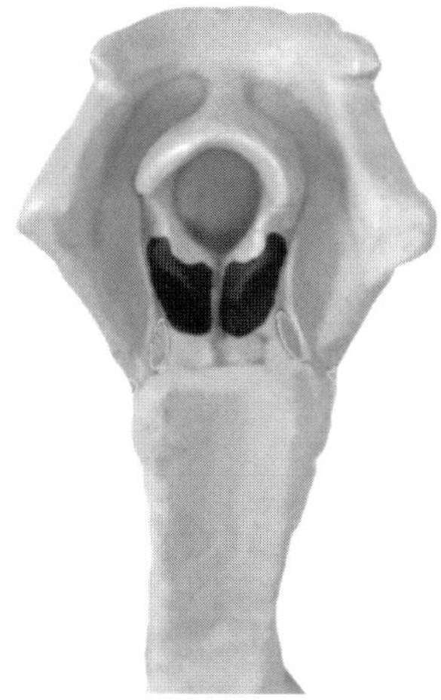

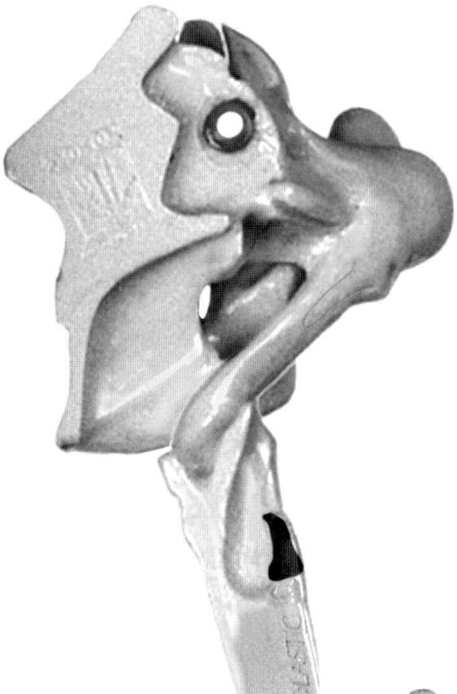

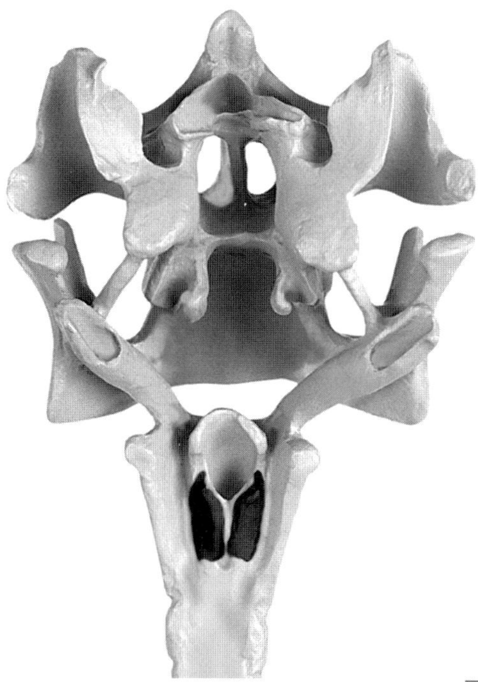

These cartilages pivot singly or simultaneously on the promontory facets on the dorsal expansion of the cricoid cartilage. The vocal cords attach to them, and their pivoting movements tauten and relax the cords. "Racking" one or both away from the vertical position modulates the vocal apperture in many combinations.

These cartilages are surmounted by very small cuneiform and corniculate cartilages which seem to provide transitional support and flexibility on into the aryepiglottic folds of mucous membrane in which all the associated arytenoid cartilages are embedded.

The flexibility of the cricoid and thyroid cartilages from which the arytenoid muscles arise allow for extremely subtle variations on all possible movements.

No arytenoid cartilages
=no vocal cords
no windpipe

Soft palate

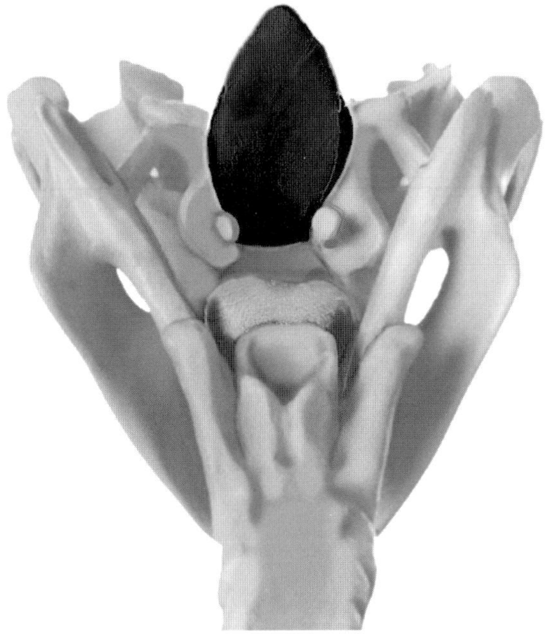

Characteristic of all vertebrates are internal nares. Originally, the olfactory receptors were mostly in the lining of a pair of pits, or ***nares****, located between the nasal and maxillary dermal bones. One of the critical changes that made air breathing possible was for the external nares to penetrate into the pharynx. This allows for breathing when the jaws are closed. This development of* ***internal nares*** *contributed greatly to adaptation for life on land.*

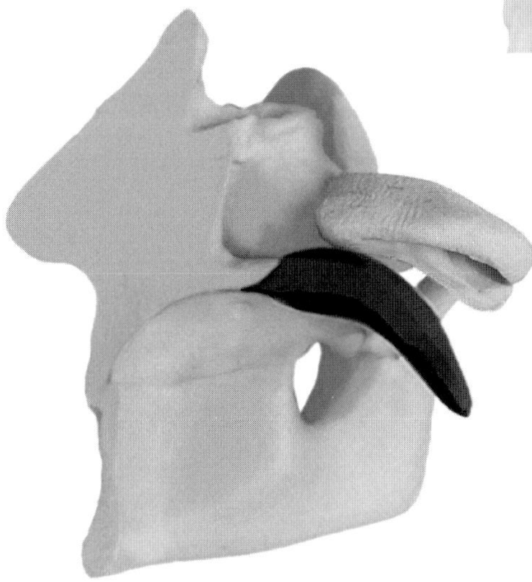

The hard palate is a key mammalian adaptation that creates a separation between the nasal airway and mouth. This allows sustained simultaneous chewing and breathing, interrupted only by swallowing.

The soft palate (also called ***veli palatini)*** *is a fibrocartilaginous continuation of the hard palate. Its flexibility allows it to be moved like a flap valve so that its free end moves cranially to close off the pharyngeal entrance of the airway, or conversely, to prevent food from exiting the pharynx through the nose instead of down the esophagus.*

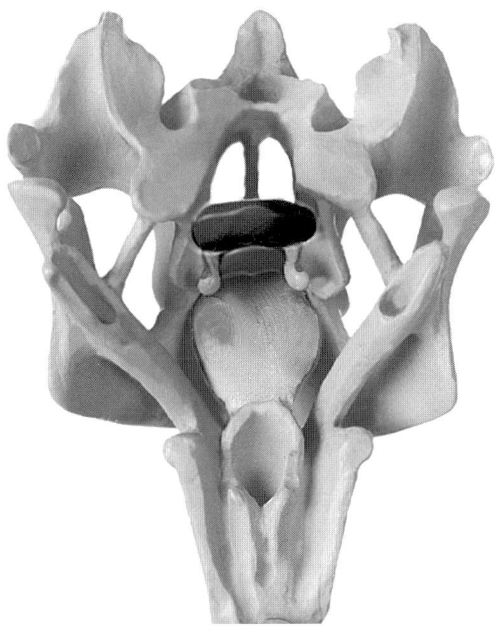

No soft palate
=no breathing while chewing

Sphenomandibular ligament

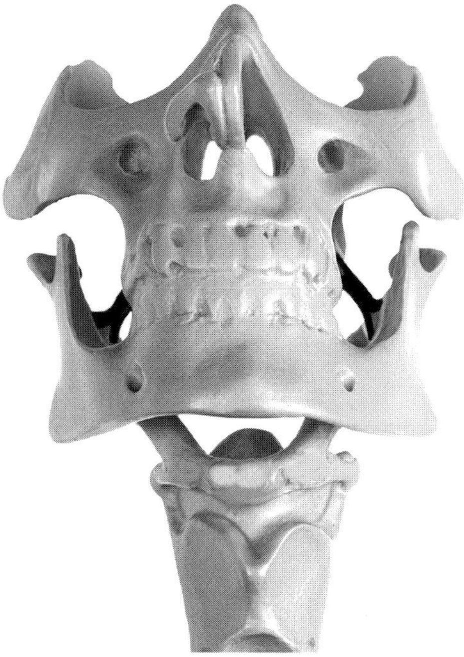

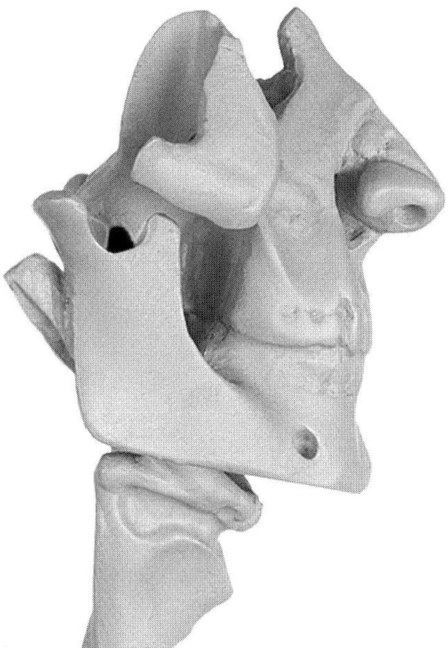

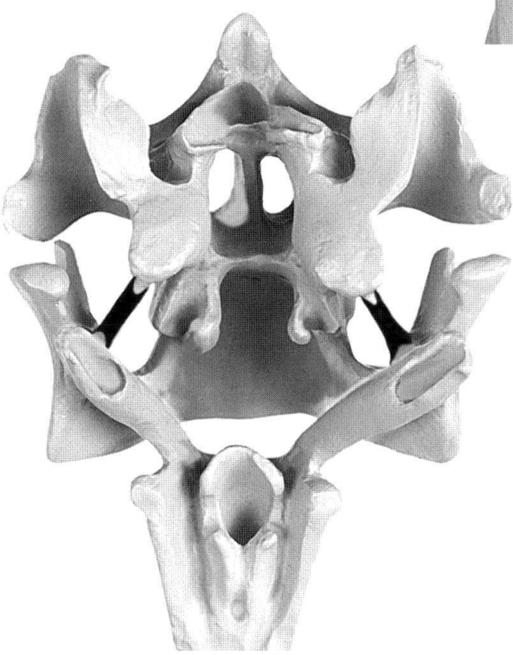

The ligaments that hold the visceral skeletal elements together have been emphasized. The sphenomandibular ligament runs fom the spine of the sphenoid to a barb on the medial mandible next to the mandibular foramen. The foramen admits the mandibular branch of the trigeminal nerve into the ramus of the mandible.

Stylohyoid ligament

The spectacular spine at the base of the head is the styloid process of the temporal bone. It points caudoventrally toward the series of U-shaped elements of the visceral subsystem.

The stylohyoid ligament runs from the styolid process to the lesser horn of the hyoid. In some animals, this ligament is a stylohyoid bone.

In all the Maniken™ models, this ligament is much thicker than in reality in order to hold the oral and laryngeal elements of the pharynx together.

In the Maniken™ Pro, the styloid processes are steel pins embedded in the base of the head. The dorsomedial ends of the ligament are notched to fit the styloid processes.

In Maniken™ Scholastic, used for many of the medial views, the temporal bone is divided so that all its visceral elements are part of the actual model of the visceral subsystem. Therefore, the syloid process and stylohyoid ligament or bone are included in the visceral component.

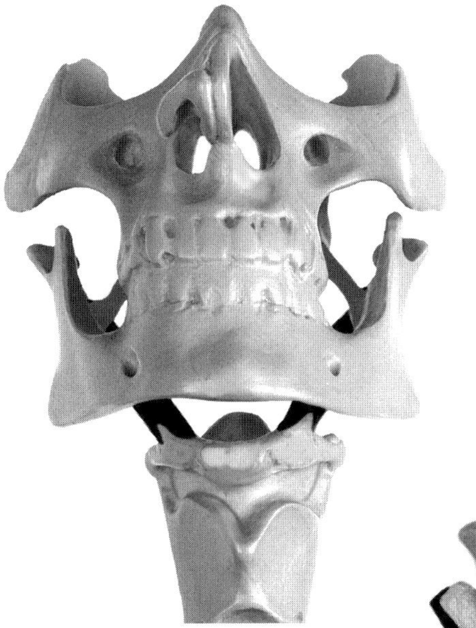

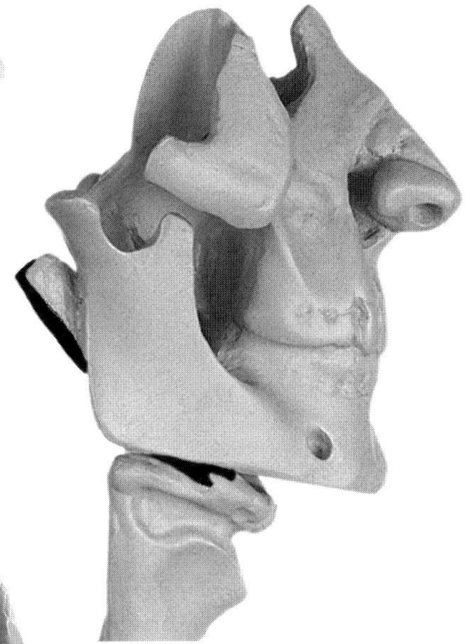

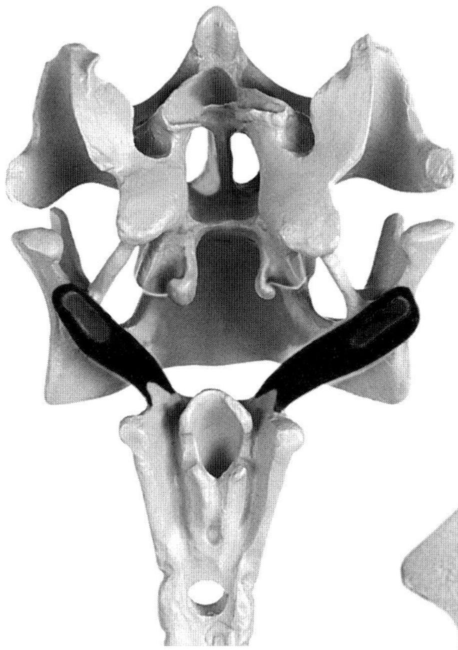

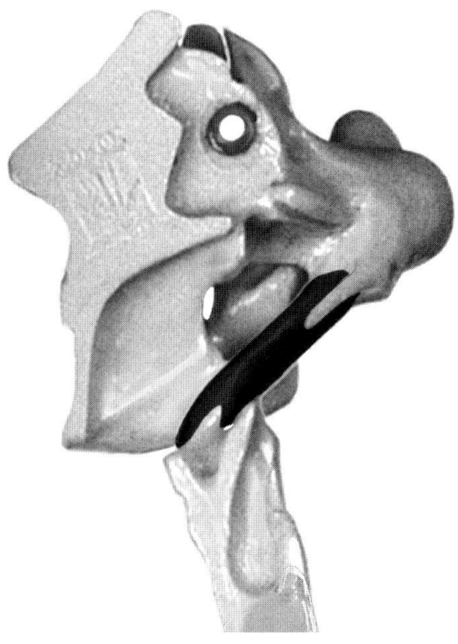

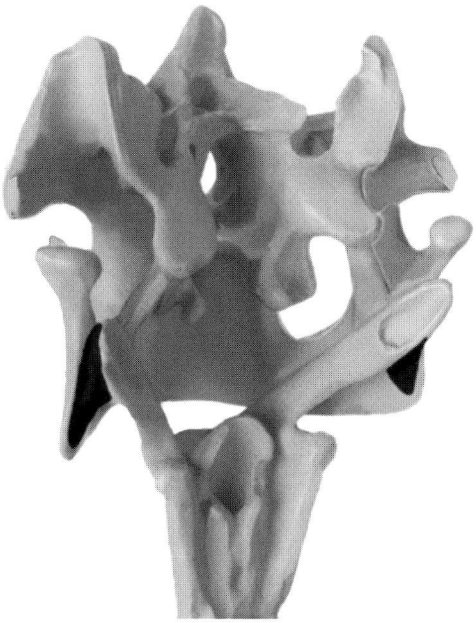

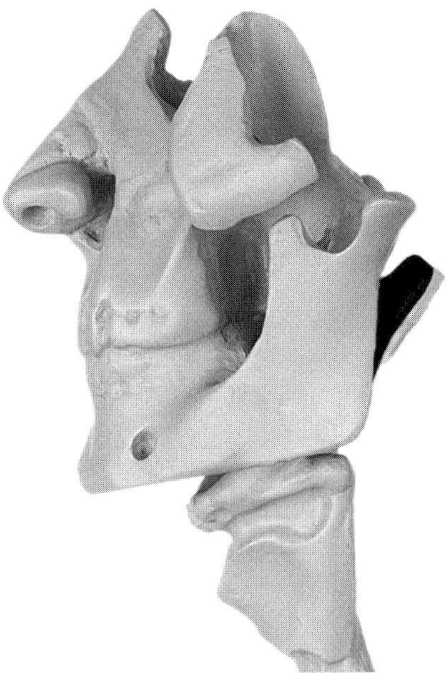

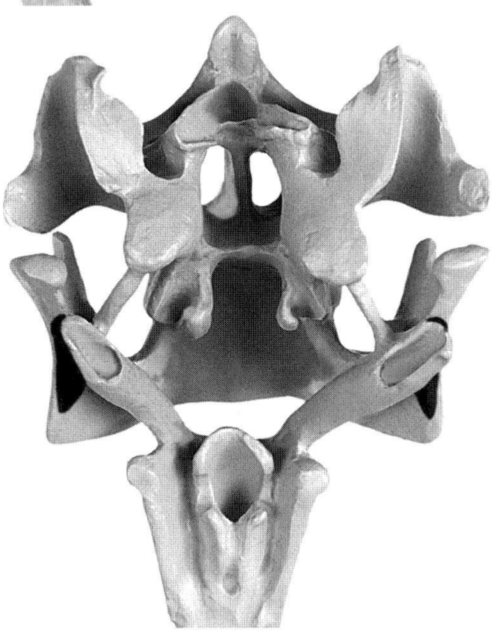

Stylomandibular ligament

The stylomandibular ligament is a much smaller verison of the stylohyoid ligament. It arises from the styloid process of the temporal bone and fans out over the dorsomedial edge of the ramus of the mandible.

Auditory tube cartilage

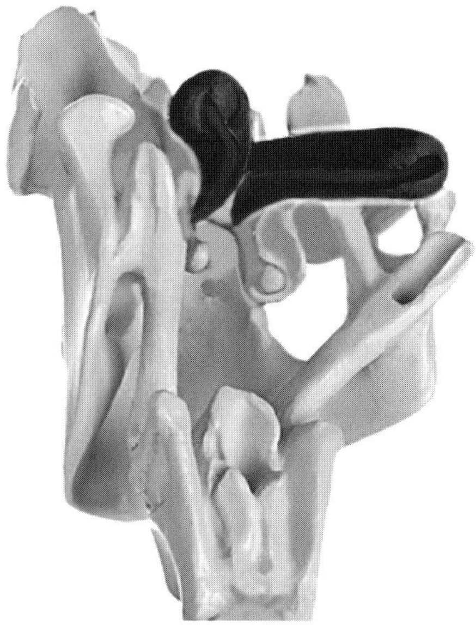

*The ear drum of reptiles covers a foramen on the side of the head. In mammals, the brain and skull expand laterally, leaving the ear drum behind, inside an auditory tunnel or **meatus**.*

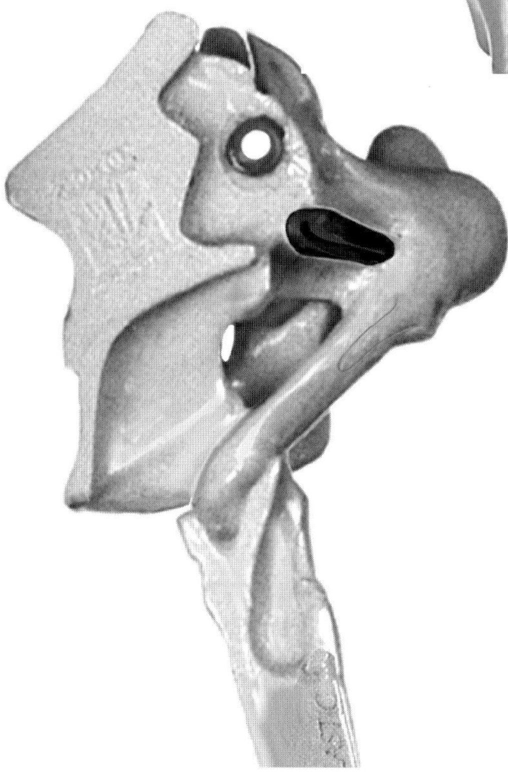

The reptile maxilla and mandible are each comprised of two bones. The dorsal pair are the articulation between maxilla and mandible. The mammalian middle ear is adapted to amplify the movement of the eardrum and translate it to the inner ear. The skeleton of the middle ear is created from the that pair, which fold into a capsule in the base of the head just medial to the original jaw joint.

The ability to close and practically seal the oral and pharyngeal vestibule is fundamental to swallowing. With the eardrum now embedded in the head, it is necessary to direct a relief valve into the pharynx so that pressure can be equalized on both sides of the eardrum.

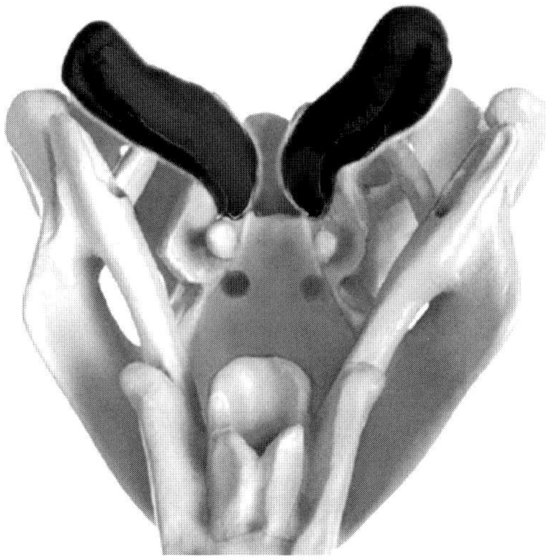

This relief valve is a fibrocartilage tube variously called the auditory tube, nasoauditory tube or eustrachian tube. Its opening is just dorsal to the internal nares. Muscles arise from the tube that connect to other, more mobile parts of the pharynx. Their motion can help pull the tube open to facilitate the pressure equalization.

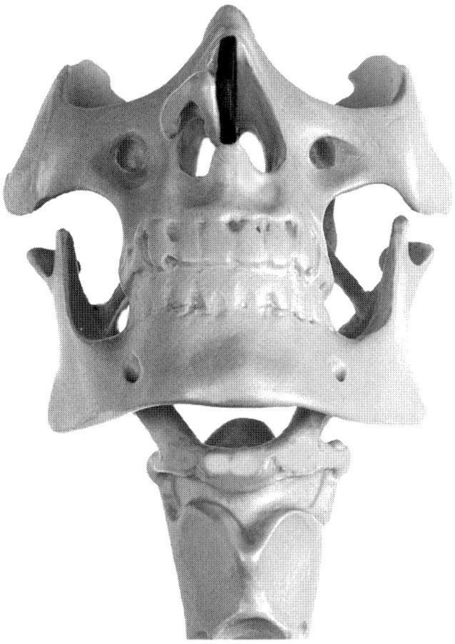

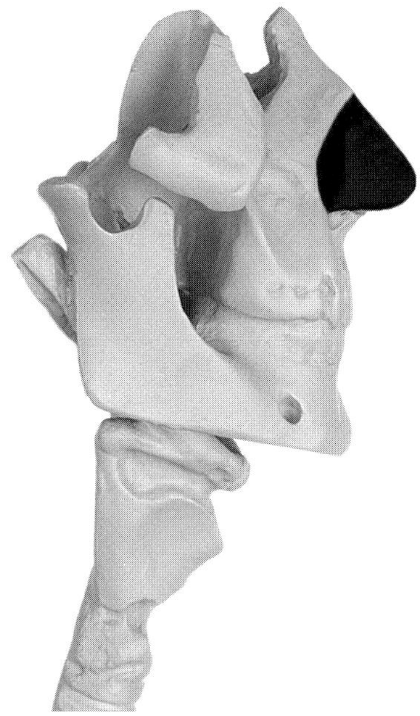

Nasal cartilages

Our infraorder of anthropoid primates is catarrhine (rather than platyrrhine). Descendant from Old World monkeys, our external nares or nostrils are narrow, and open downward or caudally. The term catarrhine means a downward flowing, hooked form. Our nostrils are partially masked by a vestibule created by cartilages of the external nose so that the ventrally opening nares "hook" downward or caudally.

Whether the nostrils are narrow or widely flared depends on the length of the nasal septum, the cartilage that continues the bony divider of the nasal passage (provided by the vomer and perpendicular plate of the ethmoid bone).

Attached to the tip and caudal edge of the septum, and to the nasal spine, are two comma-shaped cartilages called the alar, or wing-shaped cartilages. The cranial and dorsal edges of the alar cartilages attach to the cranial nasal apperture and septum by a fibrous sheet which also contains small, parietal cartilages. The nostrils are completed dorsally, by fibrous fat and skin.

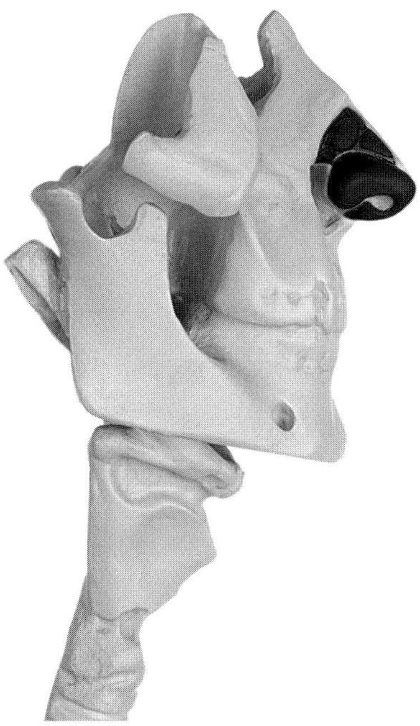

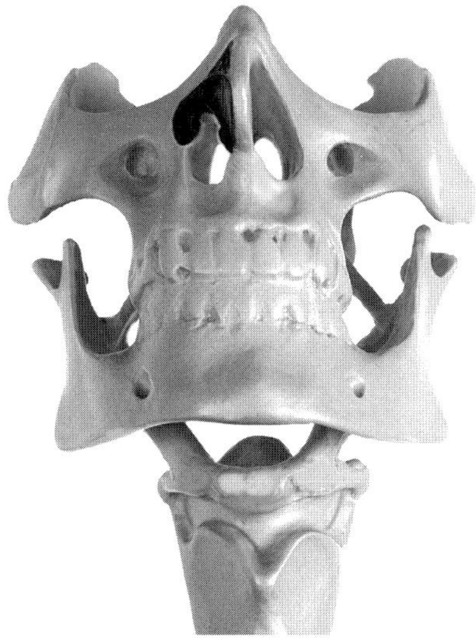

Anatomical Organization

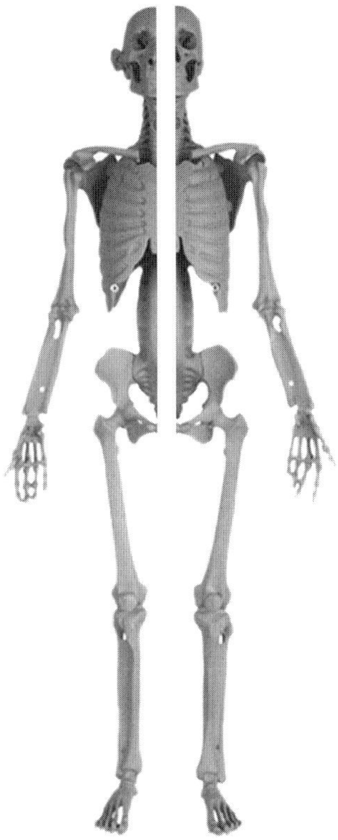

"Visceral" refers to viscera: lungs, intestines, livers and other gooey things. Visceral anatomy is usually considered as part of splanchnology, the organs of respiratory and metabolic function, and of reproduction. The anatomy of the face, mouth and throat are entrance to those systems. But anatomical tradition usually obscures the simple, clarity of anatomical organization. While the deep forms have been described as splanchnic, the musculature on the *outside* of the face and throat is included with the somatic anatomy of movement. It is time to take into account the functional *interactivity* and developmental relationships of face and neck with those deeper forms of the ear, nose and throat. At that point, *all* the muscles intrinsic to the visceral system can be considered interrelated, as in this atlas.

The visceral system belongs to the gut, the passenger of our axial and appendicular (*somatic)* bodies created for motion. The typically somatic chordate anatomy is bilaterally symmetrical, which is the most logical way to build a body for reciprocal, sinusoidal motion in water.

Note that the gut is *not* about locomotion. The primordial plan of viscera has no symmetry. It is at the mouth end of the tube that the visceral system connects most interactively with the axial system. At this point, there are even composite bones like the sphenoid and temporal that are platforms of function for both. The gill arches *are also* bilaterally symmetrical in order to interface between the big ideas of metabolism/movement. Because most of the visceral skeleton is adapted from the branchii of the gill arches, the visceral subsystem is also called the ***branchial*** skeleton.

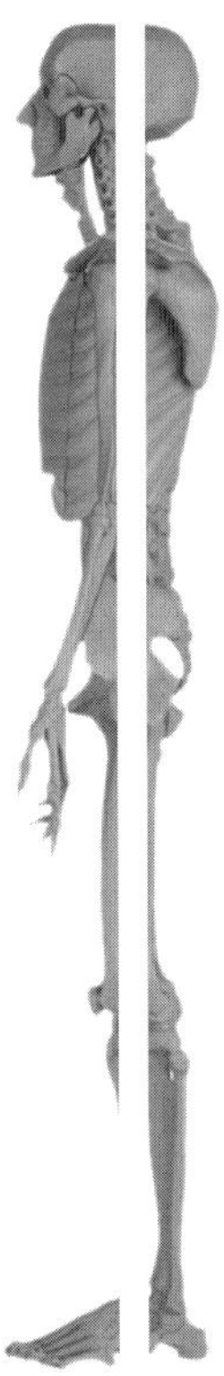

Bilateral symmetry allows us to describe each side from medial to lateral. In the horizontal, fish attitude, each half also divides into upper and lower zones: *ventral,* for the lower compartment occupied by the guts, and an upper, *dorsal* zone of musculoskeletal array—quite different ideas. Dividing the body into dorsal/ventral parts yields very different, unmirrored zones. The vertebral column is arched in mammals, and in bipedal human behavior, the heritage of the mamalian arch is in the dorsal curve of the thorax. To compensate for this curve, other segments have to recurve, as in our lumbar lordosis.

The top illustration on the next page shows a human skeletal system in a reptilian position that approximates our terrestrial standardbearer. Since, with this model, we easily can separate the visceral from axial subsystems, it becomes clear that this visceral hitchhiker must account for its own anatomical orientation. Even though the entire visceral subsystem depends on the ventral division of the central nervous system, and is, therefore, neurologically entirely ventral, its anatomical patterns have their own, separate branchial identity, and consequently their own dorsal/ventral/medial/lateral orientation.

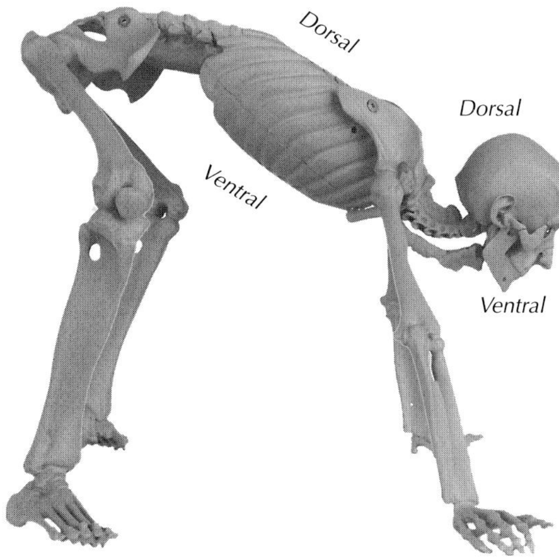
Dorsal
Dorsal
Ventral
Ventral

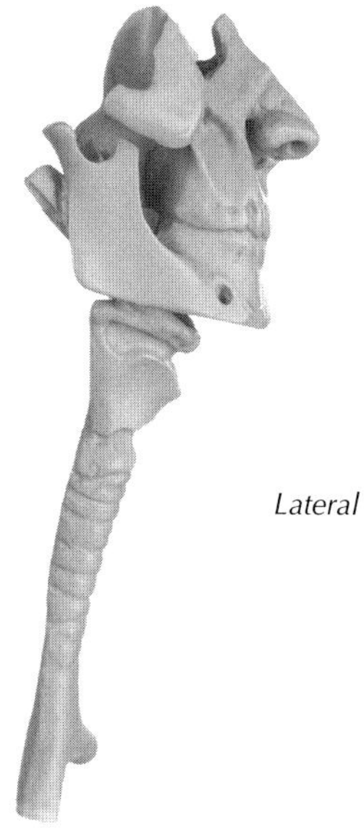
Lateral

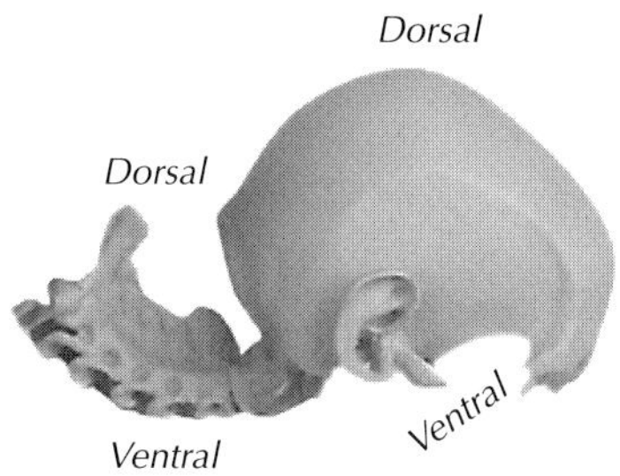
Dorsal
Dorsal
Ventral
Ventral

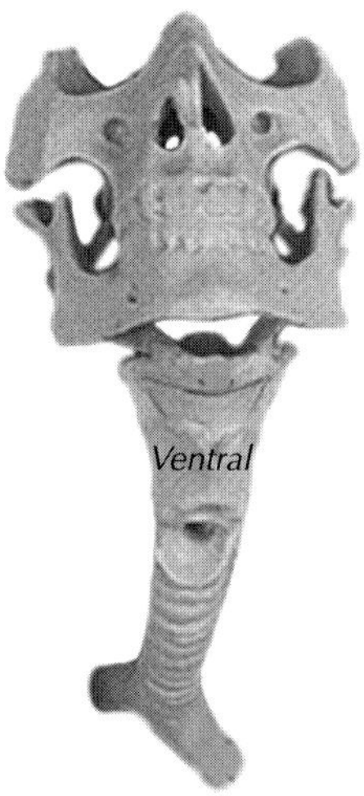
Ventral

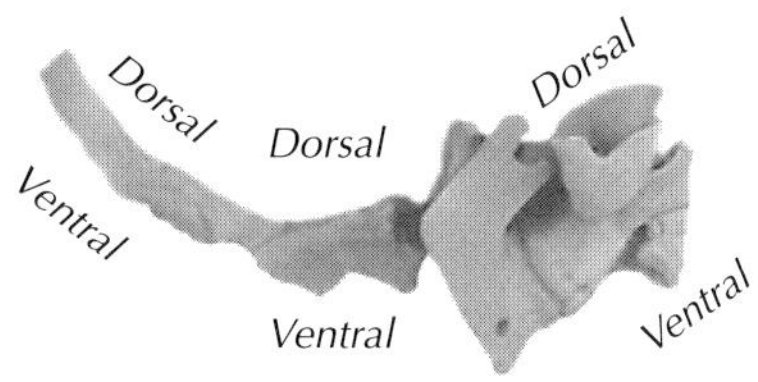
Dorsal
Dorsal
Dorsal
Ventral
Ventral
Ventral

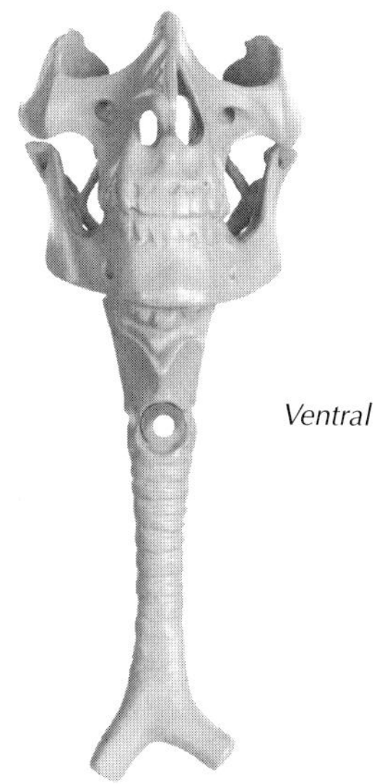
Ventral

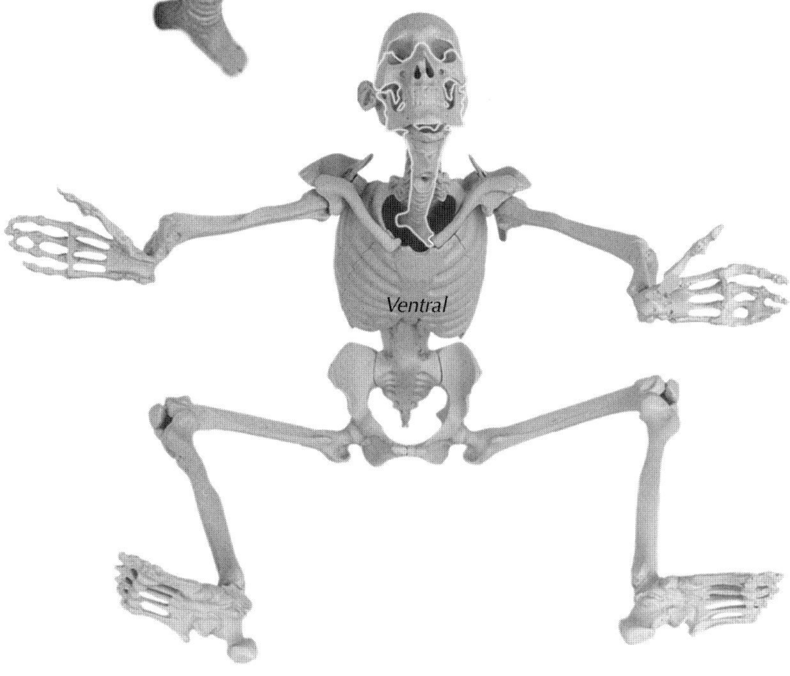
Ventral

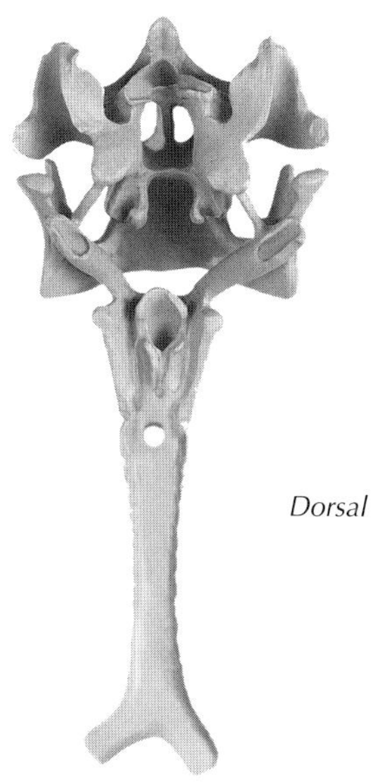
Dorsal

Orientation

Flexion of the limb is any movement that brings the ventral surfaces toward one another. *Extension* approximates dorsal surfaces. That simplicity of flexion vs extension tends to slip away when we become bipedal. See how rotation of the hip caudally reverses dorsal and ventral. The dark zones shown on these limbs are ventral branches of ventral neurology, and the light zones represent the dorsal branch of ventral neurology.

In branchial or visceral musculoskeletal anatomy, the logic of flexion and extension is much more difficult, and is often avoided. Instead, language that describes visceral function dominates. Usually functional descriptions are non-technical terms, such as "opens jaw" or "closes epiglottis." Many references are made to "narrowing" or "constricting" unopposed by any antagonists except for relaxing the same muscle to return to a neutral state. When terms like "lift" or "elevate" occur, they refer to the standard Anatomical Position.

I prefer to use stance-neutral terms that are relative to anatomy rather than how the animal occupies space. No matter what the attitude of the body, left is still left, and right is right; ventral belly and dorsal back do not change.

The head, *cranial* end is unchanged as is the *caudal* tail, so this atlas states ***cranial*** rather than "up" or "superior." "Lift" will also point cranially. Terms like ventral, dorsal, medial, lateral, cranial, caudal all refer to the axial origins of the Chordate body.

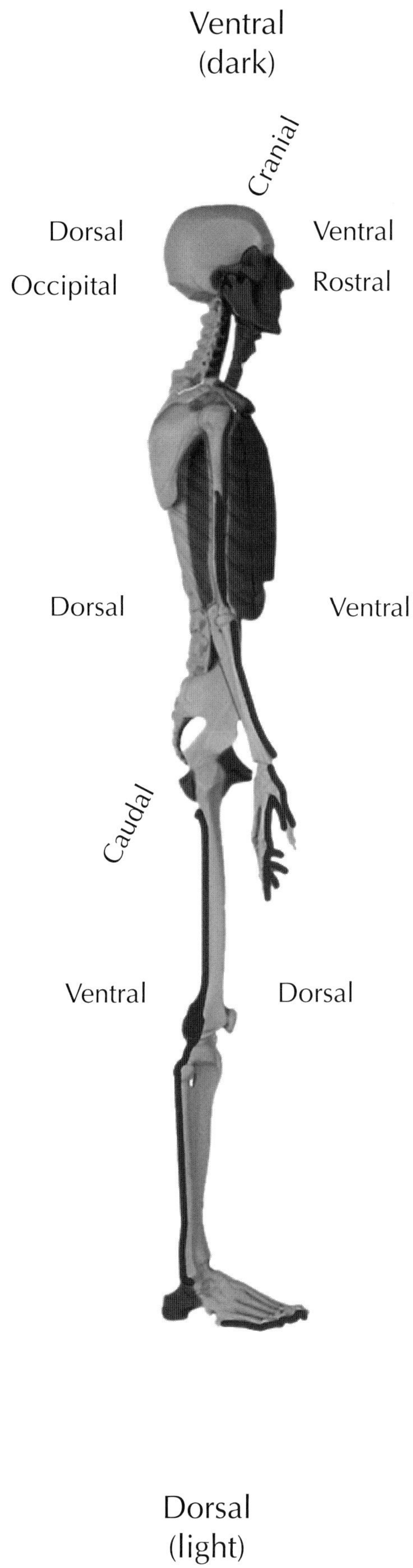

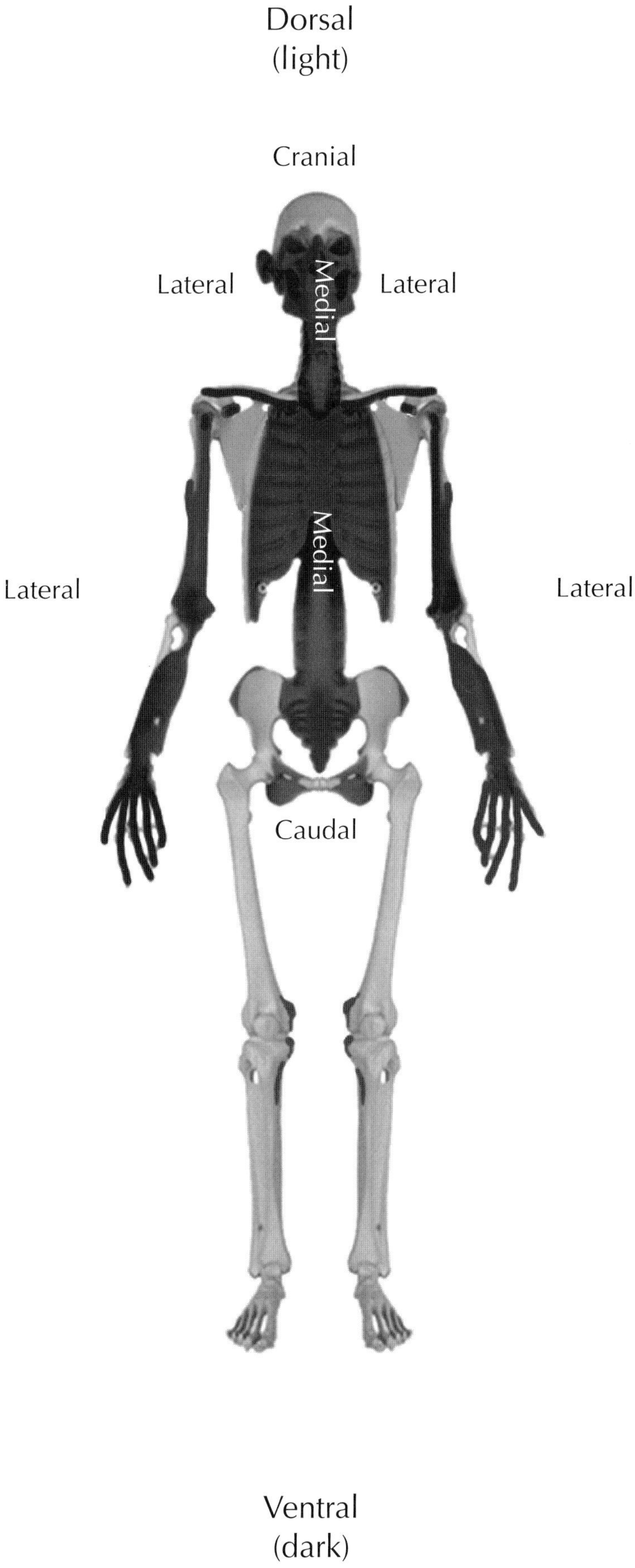

A view from the side looking toward the median is called the *lateral* aspect, while the *medial* aspect looks at either the left or right side of the body from the middle or median. We'll also use the standard language of *proximal* and *distal*. Sequence of position based on nearness to the axial foundation and its closely associated 1st gill arch.

The term "arch" in visceral anatomy often refers to an isthmus in the pharynx, where mucous membrane covered muscles narrow the throat. They are often called the first, second and third pharyngeal arches. These arches are not to be confused with the embryologically observed series of branchial gill arches that serve as the source of mesenchyme for the anatomy of the visceral subsystem.

Because of the subtleties of organization of the head and face, "dorsal/ventral" are interchangeable with a more local set of terms. Structures approaching the "nose-tip" ventrally are called *rostral* (referring to the prow of a ship. Those moving dorsally, toward the back of the head are *occipital* (referring to the occipital bone.)

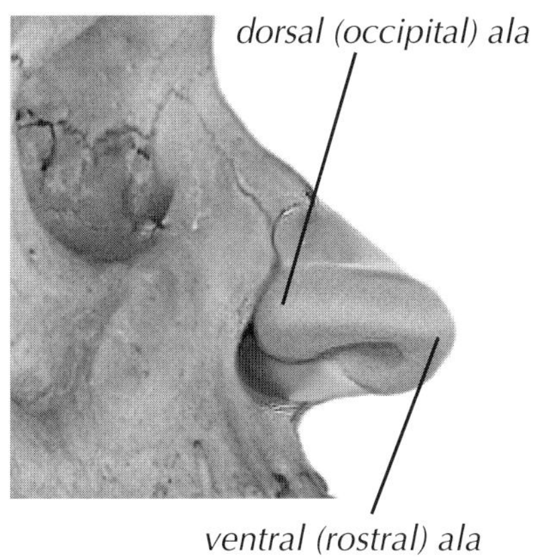

Exercises in Orientation

When the body is viewed from a compound direction, such as a 3/4-view of the front, it's labeled as "ventrolateral" or "anterolateral." Sometimes *three* view compounds are labeled, such as the rather cumbersome but informational term, "craniodorsolateral."

This exercise consists of a set of illustrations that shows the visceral system in various compound views. Each part of the orienting terminology is addressed, step-by-step in order of actions taken to achieve the desired viewpoint.

Every orientation shown in this atlas may be worked out this way, one increment at a time.

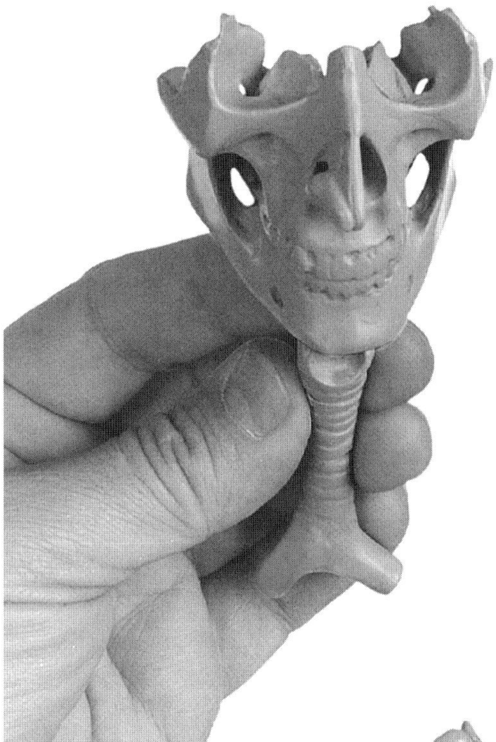

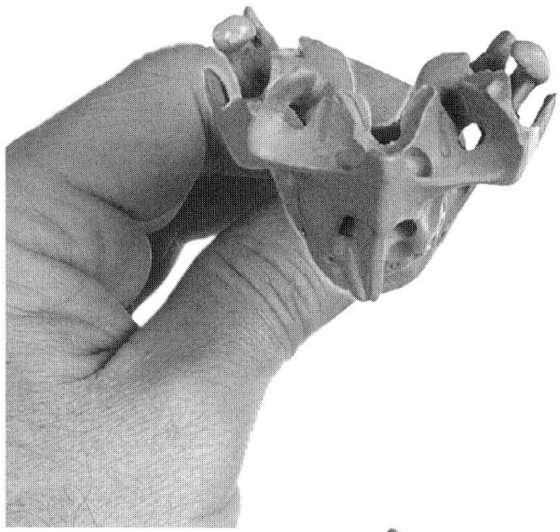

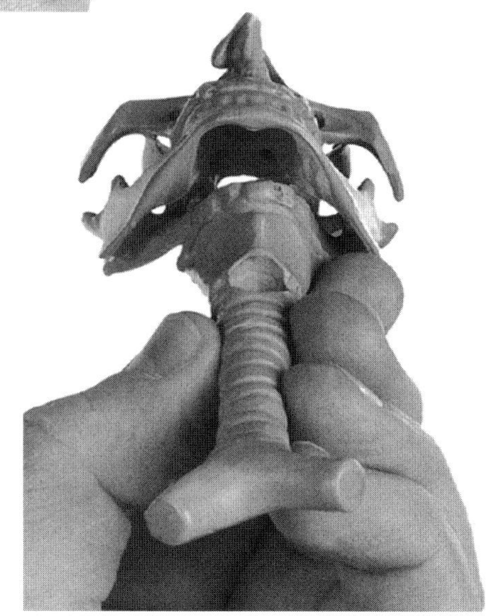

A Step 1: *this front , or anterior, or ventral view is turned toward the reader for cranial+ventral view called "cranioventral."*

A Step 2: *Tipping the model even more toward the reader produces a "cranial" view.*

A Step 3: *tipping the model away from the reader delivers a nearly "caudal" view that we will call "caudoventral."*

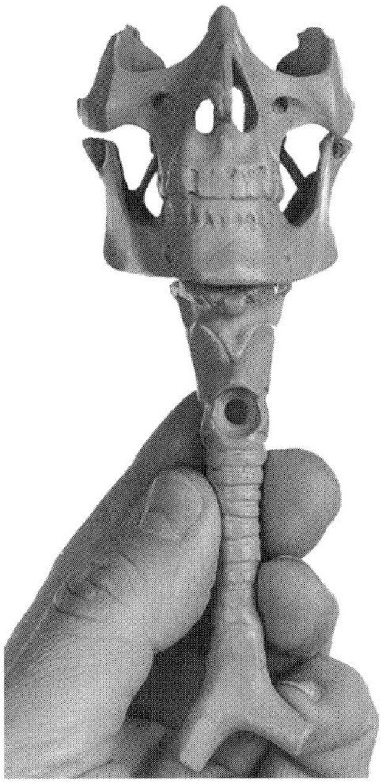

B **Step1**: *"ventro…", first establish a front view toward the reader.*

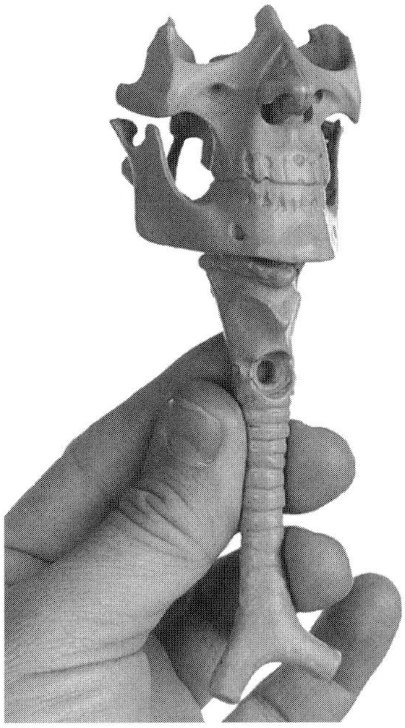

B **Step 2**: *"…latero…", or side view is added to the front view becoming "ventrolatero…", producing a 3/4 frontal view for the reader.*

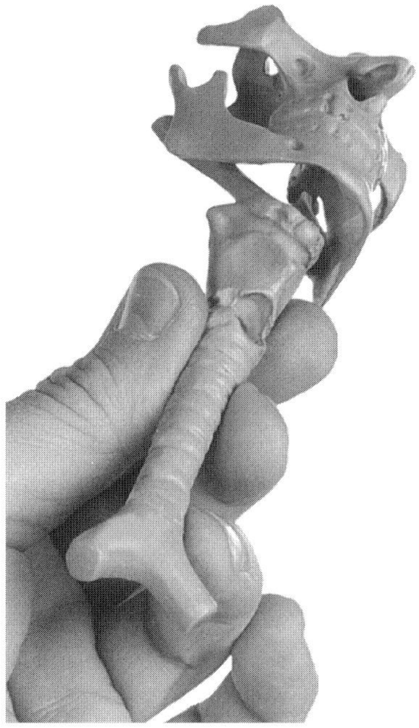

B **Step 3**: *"…caudal", or tail-end view is added to the 3/4 frontal so that it is seen by the reader from below, becoming "ventrolaterocaudal". The term used rearranges the parts to become "caudoventrolateral."*

C **Step 1**: *"dorso…", first establish a rear view for the reader.*

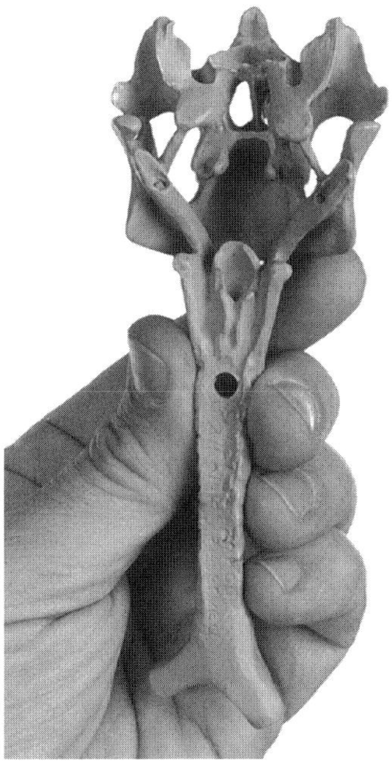

C **Step 2**: *"…latero…", or side view is added to the front view becoming "dorsolatero…", producing for the reader a 3/4 rear view.*

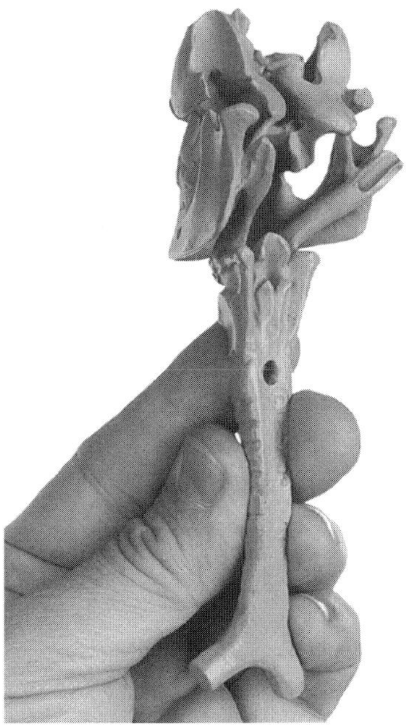

C **Step 3**: *"…caudal", or tail-end view is added to the 3/4 rear so that it is seen from below: now it's the reader's "dorsolaterocaudal", or alternately, "caudodorsolateral."*

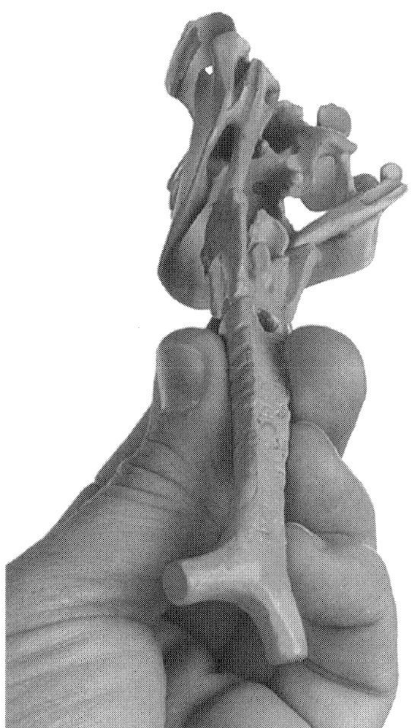

Organization Of this Atlas

This ZOOLOGIK™ atlas is organized as the body, into subsystems. Each volume of the atlas presents one subsystem; this volume is devoted to the "extrinsic" muscles that connect the subsystems. Muscles are illustrated one-at-a-time, starting with the deeper muscles, and working outward to the more superficial muscles.

If you are building clay muscles onto the MANIKEN™, you can follow the muscle order from the first to the last muscle. But there are also many other sequences that will allow you to successfully build the anatomy.

1/Muscle Name

Current usage as called for in *Nomina Anatomica.*

2/Attachments Illustrated

Muscle attachments are shown as dark areas painted on the skeletal surface of only one side of the MANIKEN™. Left alternates with right, one spread showing the left side, the next spread showing the right side.

3/ Muscles illustrated

Photographs of each muscle built upon the MANIKEN™ model are taken from different angles, as if you were turning the model in your hands in order to gain a better three-dimensional understanding. Occasionally additional information is drawn onto the clay muscle. Some photos show the muscles on both sides of the body. When only one side is shown, the choice of side coincides with the illustrated attachments shown in the left column. Muscles shown on the left side alternate, 2-page spread-by-spread, with muscles shown on the right side.

4 /Orientation

The view, or aspect, presented in each illustration is labeled. If there is confusion about the view, turn anytime to the chapter on orientation and carry out the action in question shown there.

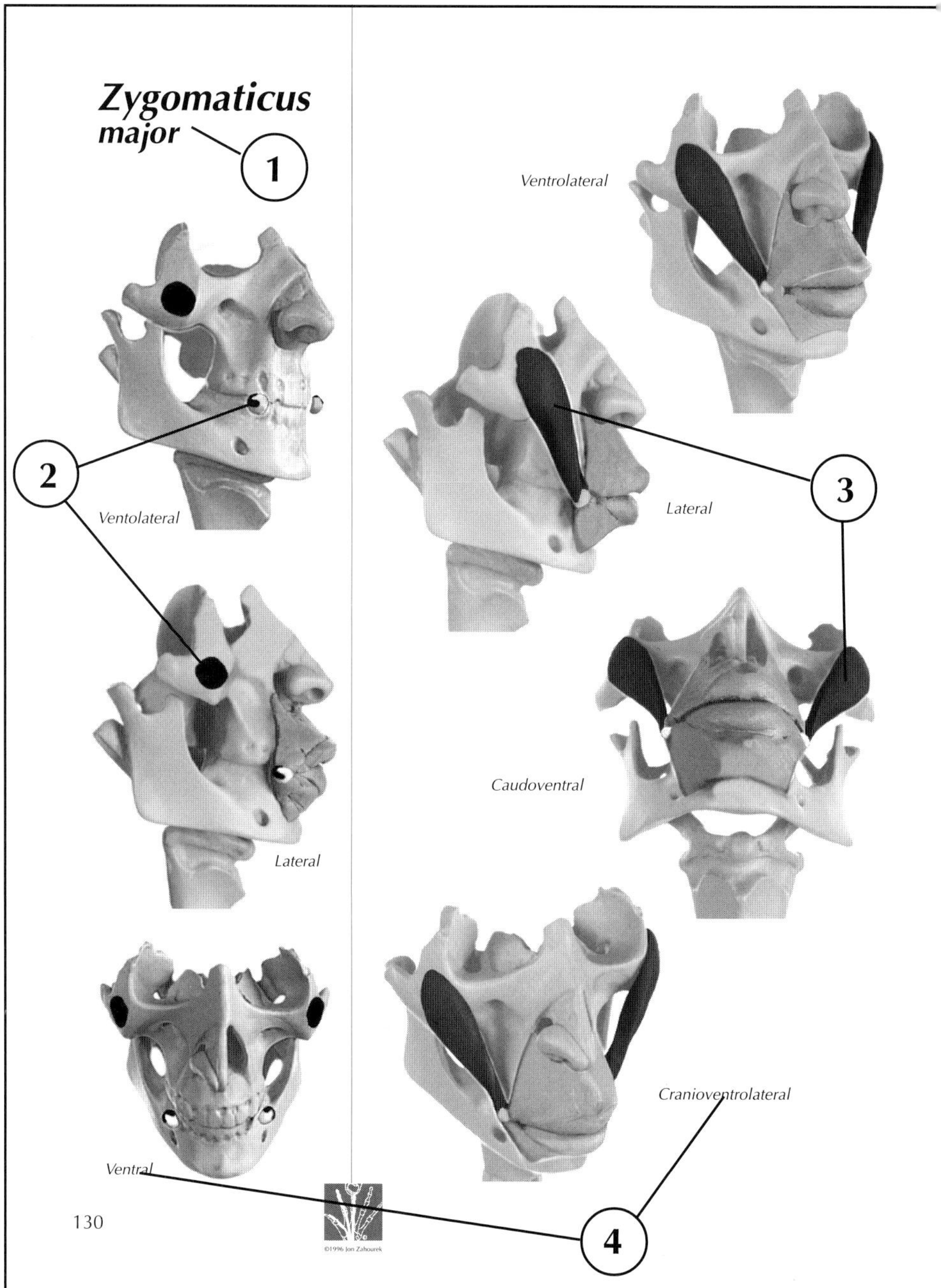

Though this atlas, is primarily a "picture-book", I urge you to read the chapters that precede the Lessons. You should find in them an explanation of the premises that underlie our thinking and language.

The explanation of the aspect of each illustration can, as you have seen, involve several steps. Any time you get *dis*oriented about the view, refer to the Orientation chapter, pages 18–19.

The key to getting the most possible from this or any other atlas of anatomy, if to do your best to *internalize* the information. Place the ideas in your own body to understand them, and to know your body.

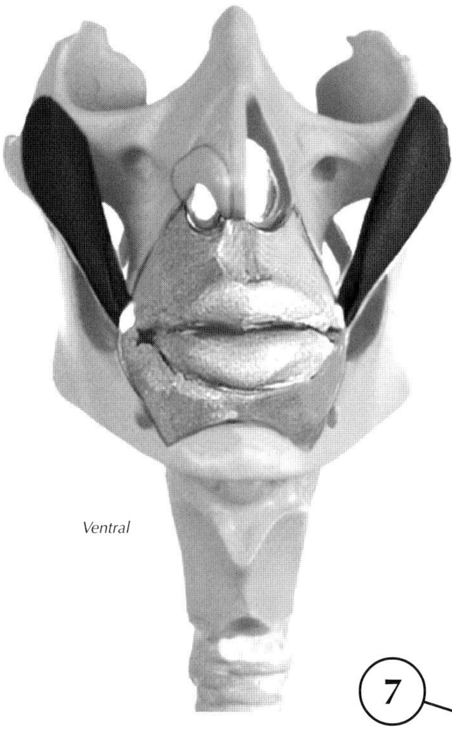

Ventral

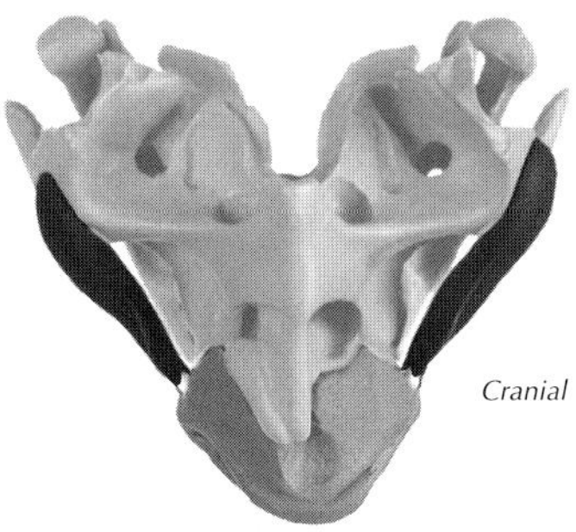

Cranial

Zygomaticus major

5

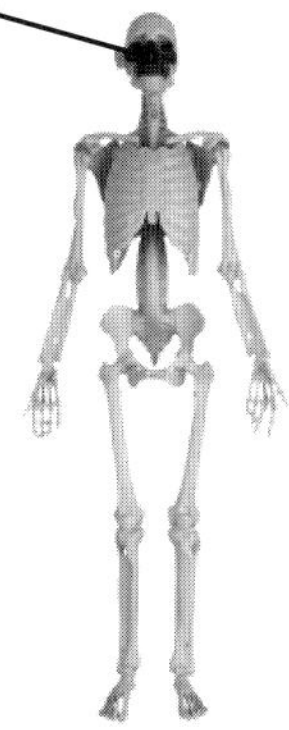

6

Actions
Pulls upward and laterally on the modiolus at the corner of the mouth.

7

Attachments
Cranial
Anterior (zygomatic) part of the zygomatic arch.
Caudal
Modiolus at the corner of the mouth.

8

9

*zy•go•**mad**•i•kus*
From the Greek, "zygon" + "ikos", meaning related to the yoke or connector.

***may**•jur*
Latin, meaning larger.

10

131

5/Skeleton Involved
The surface areas of attachment to the muscle are darkened.

6/Actions Described
Commonly agreed upon functions of the muscle.

7/Attachments Described
"*Proximal*" (meaning that attachment closer to the trunk), and "*Distal*" (farther from the trunk). Fascia attachments, while of immense importance, are not treated in detail in this atlas. Where necessary for significant muscle attachments, ligaments are built, and then treated as any other skeletal attachment.

8/Notes
Occasionally, a point of interest about the muscle is included.

9/Phonetic Pronunciation
The phonetic pronunciation given here is our everyday alphabetic translation of the standard pronunciation system from one source: *Webster's Third New International Edition (G.C. Merriam, 1976)*, an excellent three-volume dictionary. Other reasonable pronunciations are equally correct.

10/Derivation of Terms
The meaning of each term in the muscle name is given by the derivation of each word from Greek or Latin. These translations are from the same source as the Pronunciations.

VISCERAL Muscles

Thyroepiglotticus

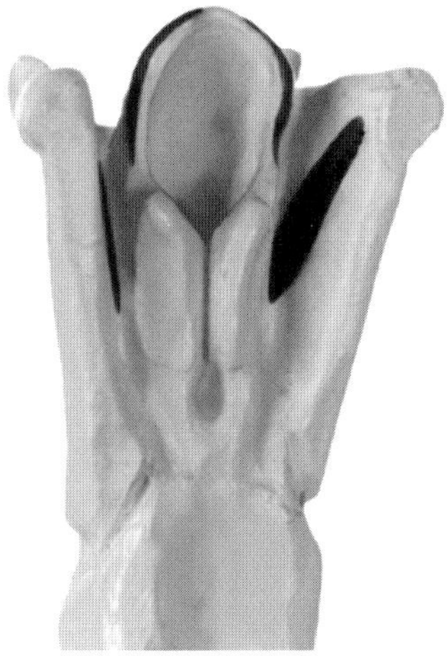

Dorsolateral

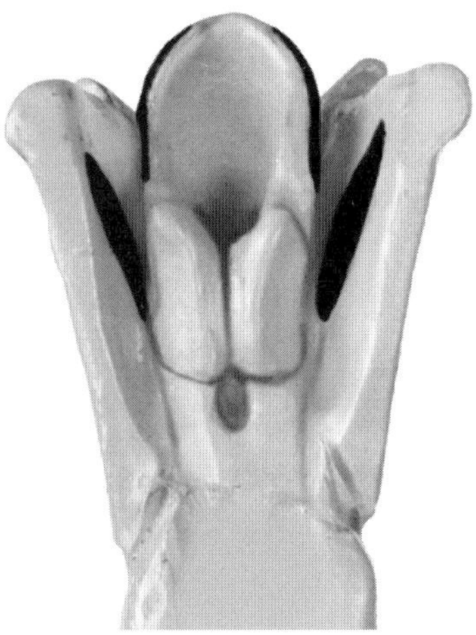

Dorsal

Cranial

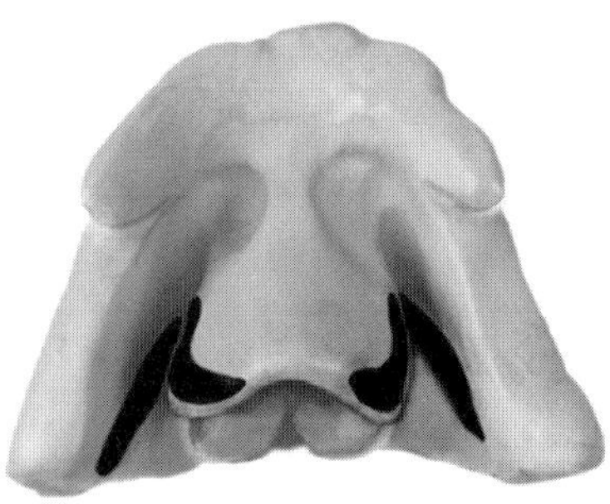

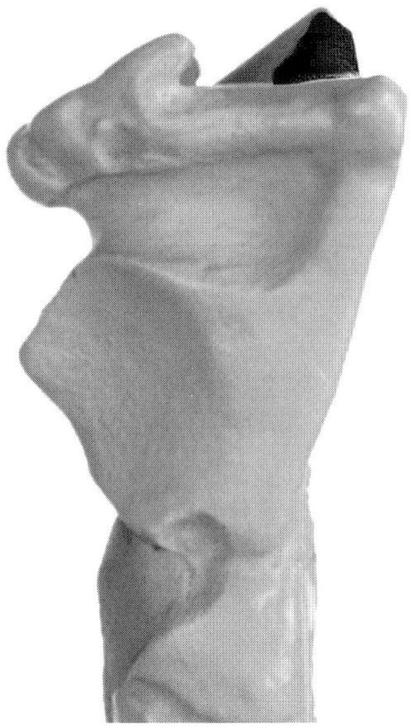

Lateral

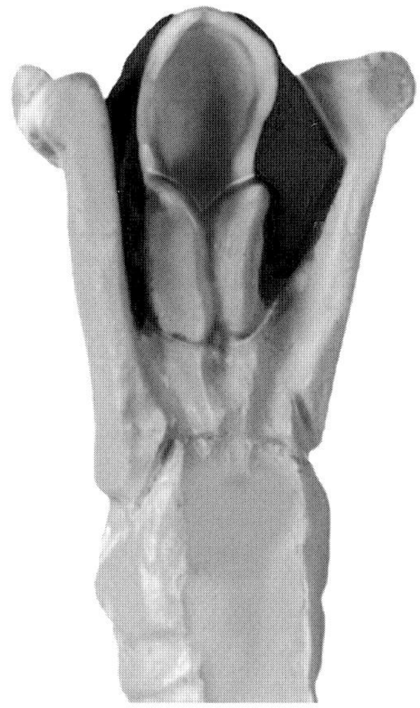

Dorsolateral

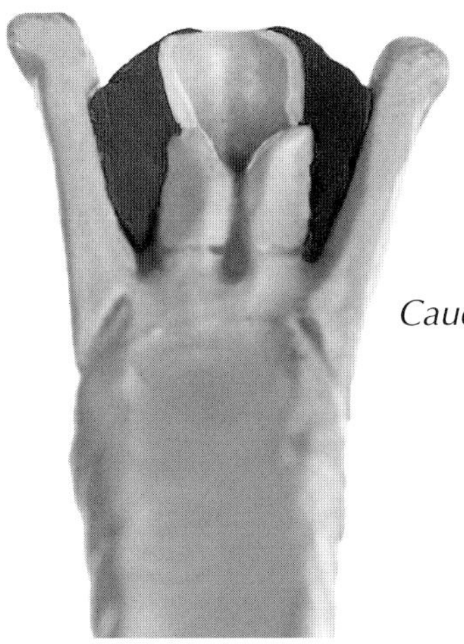

Caudodorsal

Craniolateral

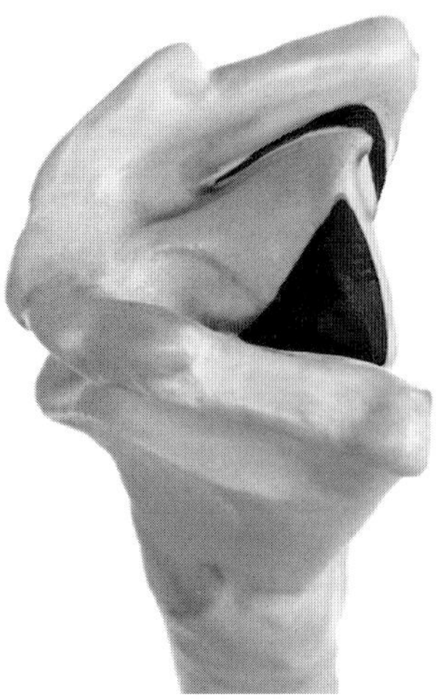

Craniodorsolateral

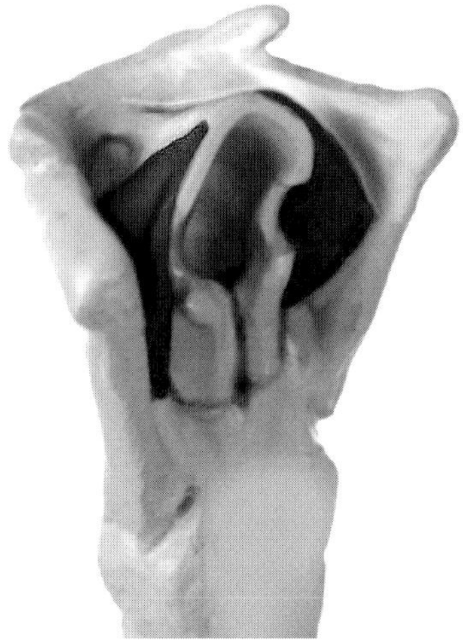

Thyroepiglotticus

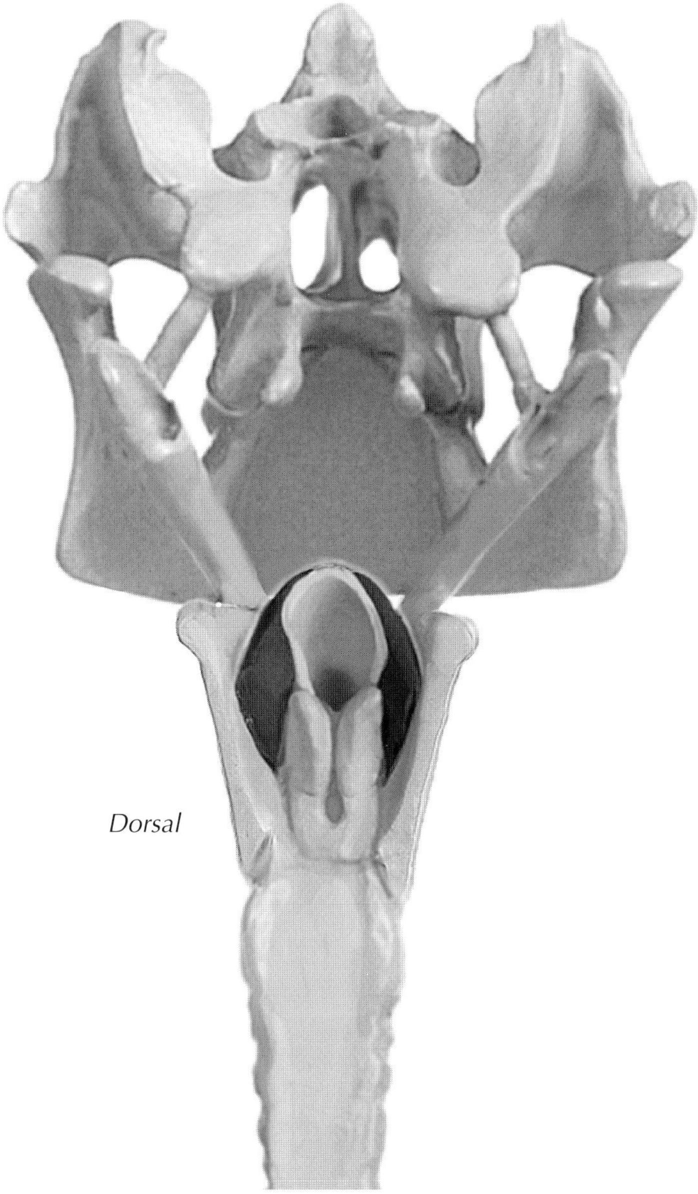

Dorsal

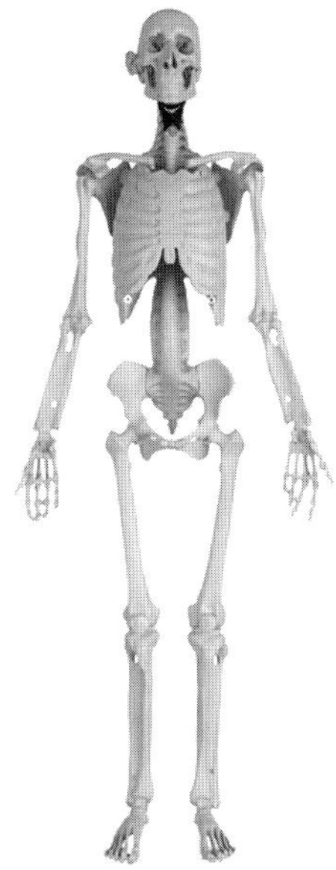

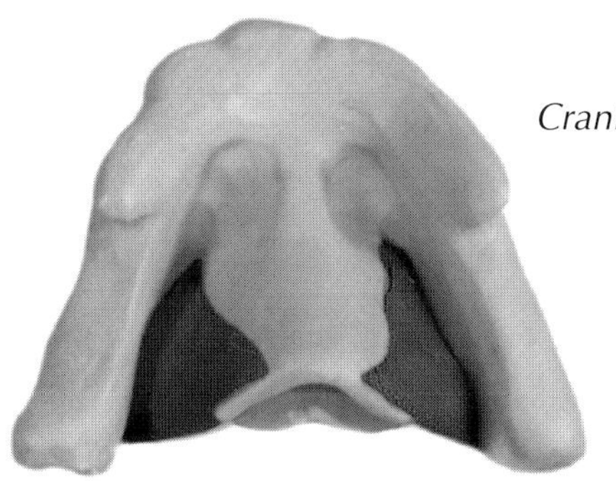

Cranial

Actions

Widens laryngeal opening.

Attachments

Caudal

Inner surface of angle of thyroid cartilage.

Cranial

Runs diagonally dorsocranially to margin of epiglottis.

Note

This muscle forms the cranial edge of the thyroarytenoideus.

*thy•ro•ep•i•**glot**•uh•cus*

From the Greek, "thyreos," meaning a shield, and "epi" + "glotta" meaning near or against the tongue.

Thyroarytenoideus

Dorsal

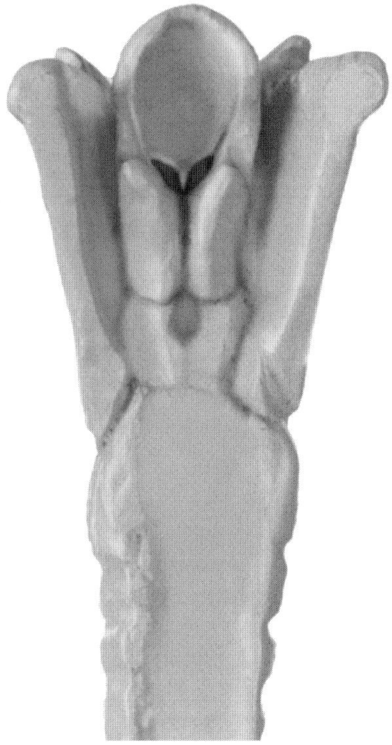

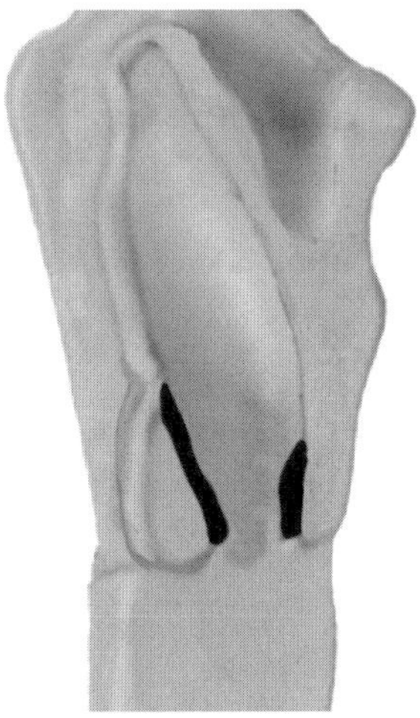

Medial

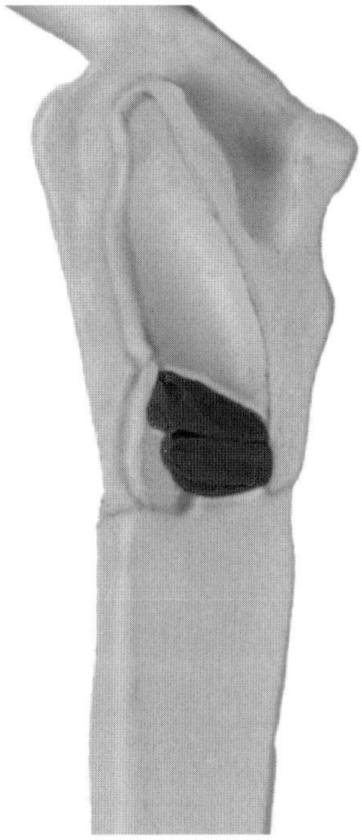

Medial

Craniodorsal

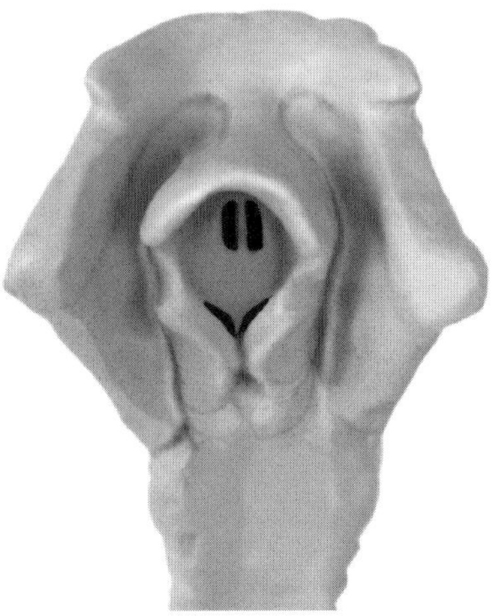

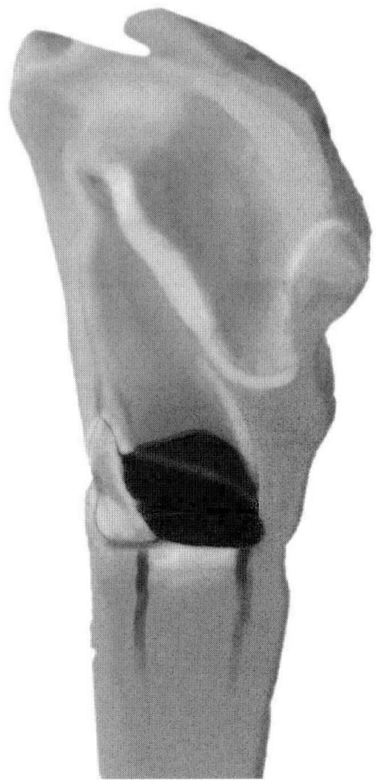

Craniomedial

Anterocaudal

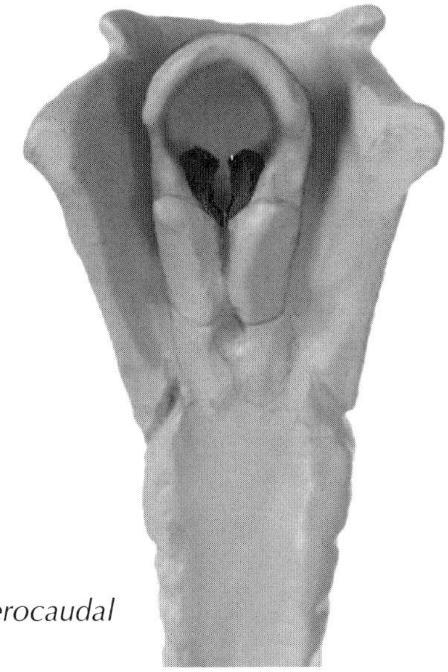

Thyroarytenoideus

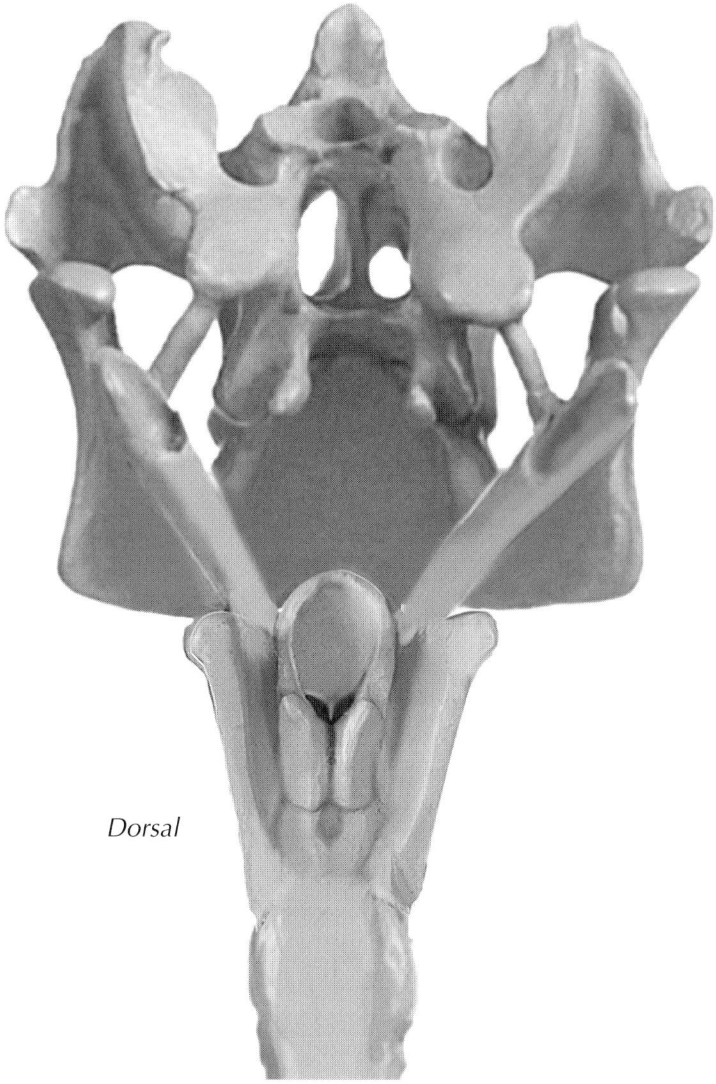

Dorsal

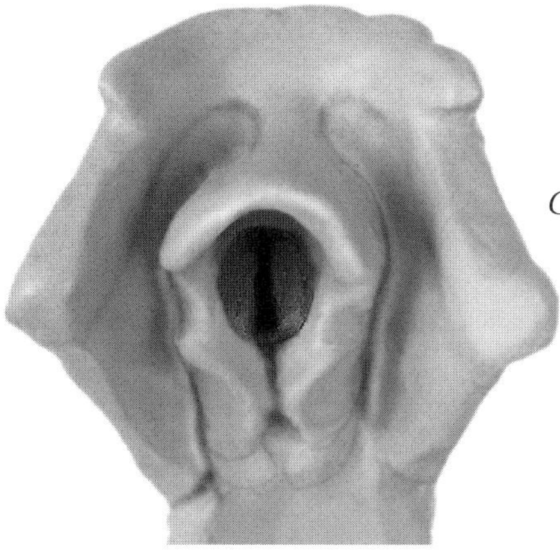

Craniodorsal

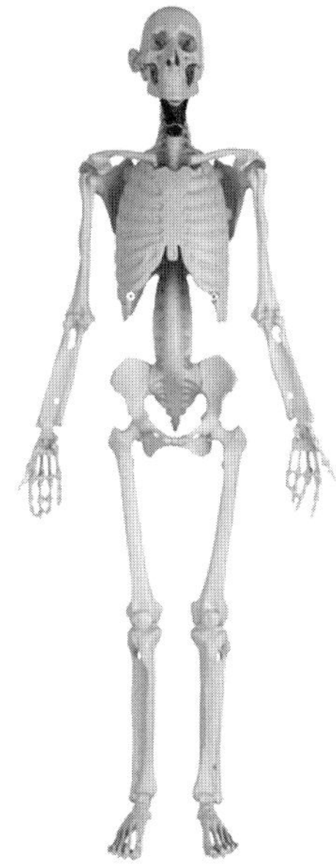

Actions

Brings arytenoid cartilages forward and towards the midline, approximating and shortening the vocal folds.

Attachments

Caudal

Inner surface of angle of thyroid cartilage.

Cranial

Ventrolateral surface of arytenoid cartilage.

*thy•ro•a•ret•uh•**noid**•ee•us*

From the Greek, "thyreos" meaning a shield, and "arytaina"+"eidos," meaning shaped like a ladle.

Arytenoideus transversus

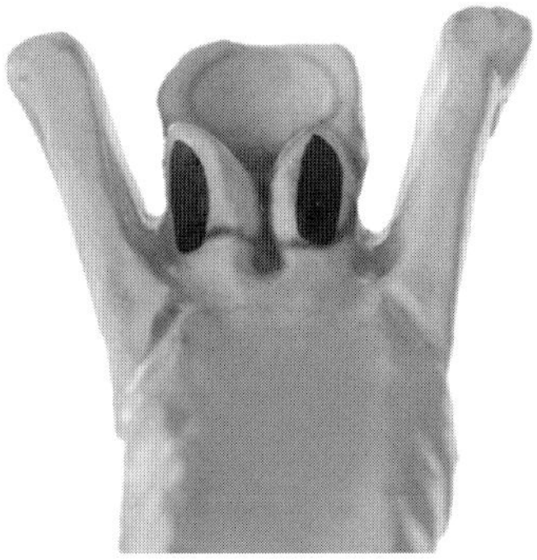

Caudodorsal

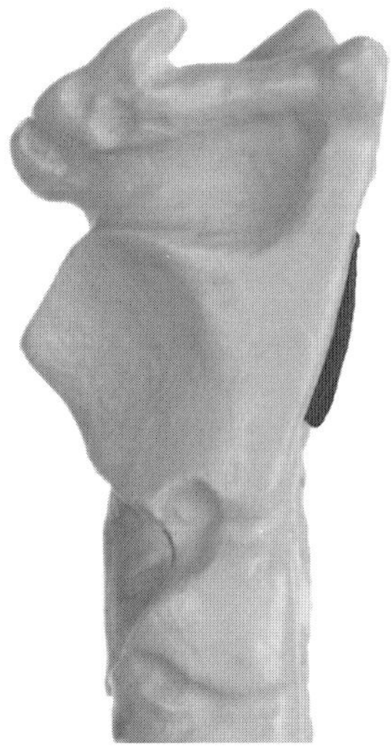

Lateral

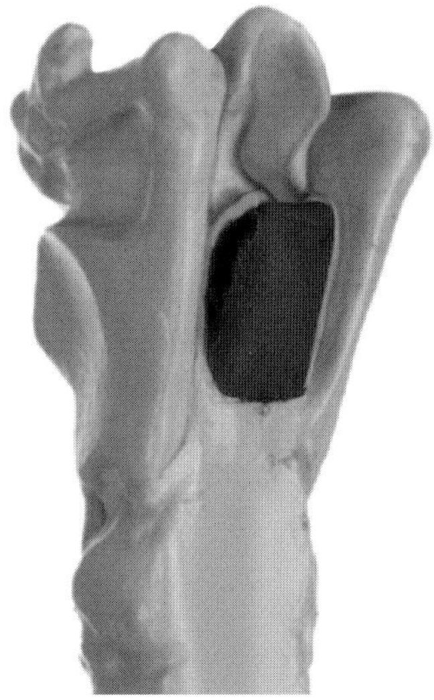

Dorsolateral

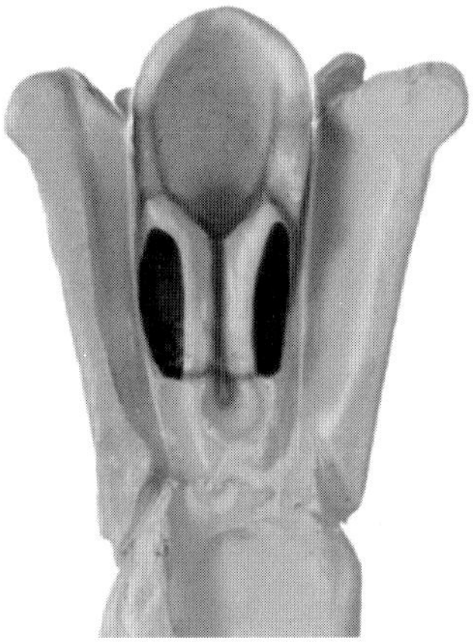

Dorsal

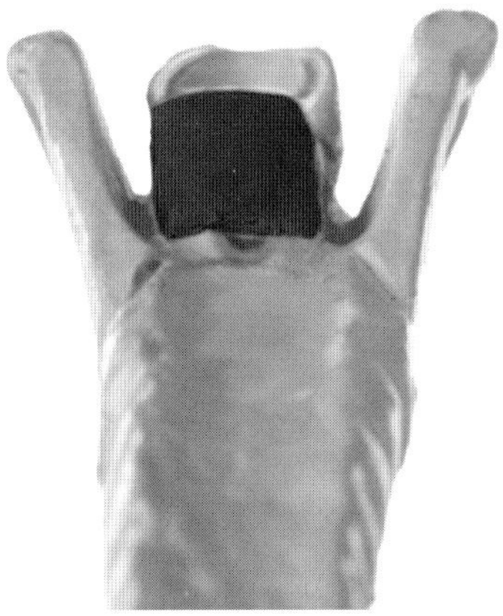

Caudodorsal

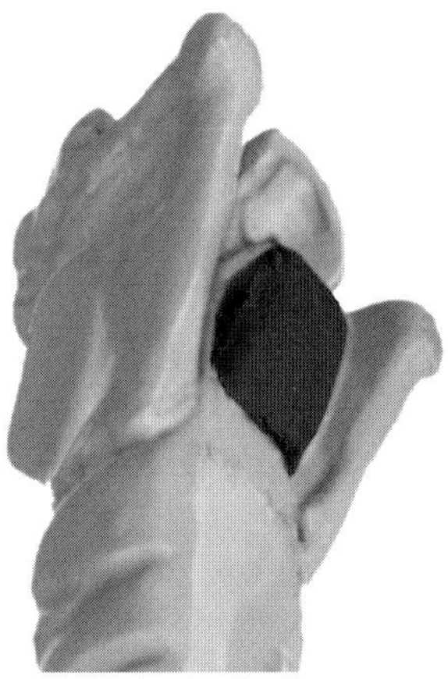

Caudodorsolateral

Dorsolateral

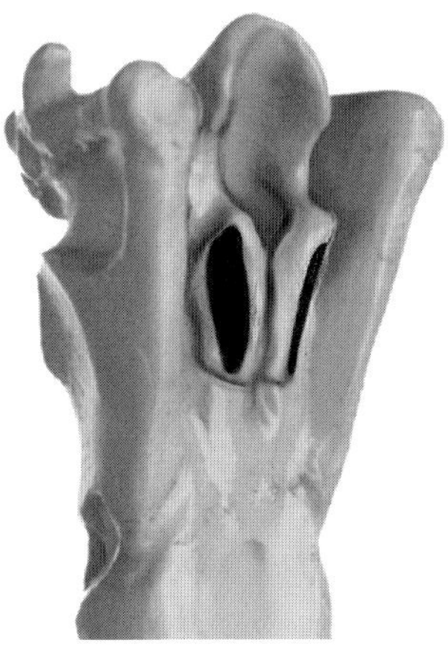

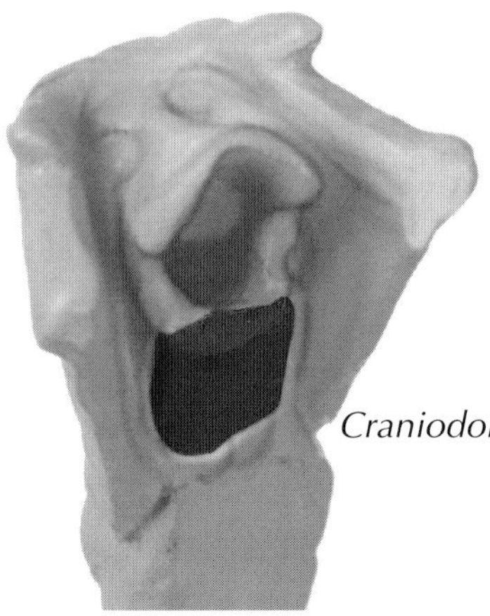

Craniodorsolateral

Arytenoideus transversus

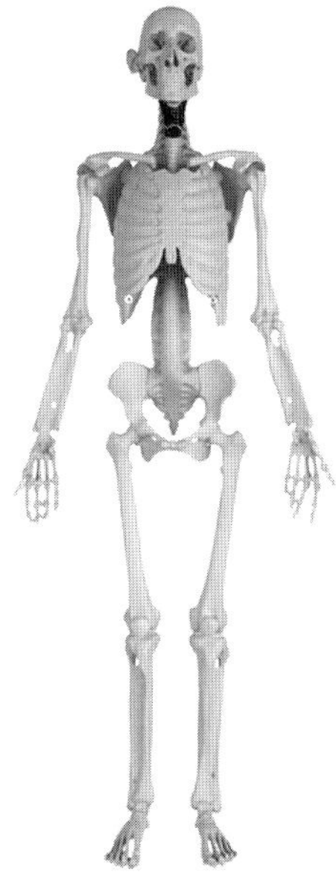

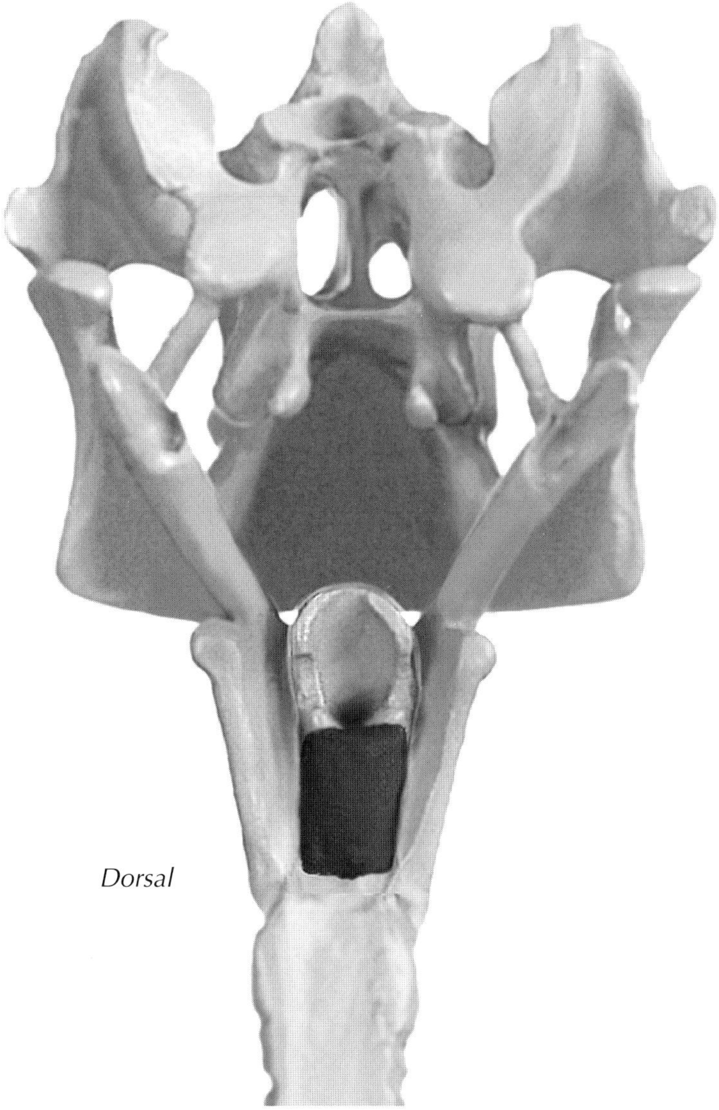

Dorsal

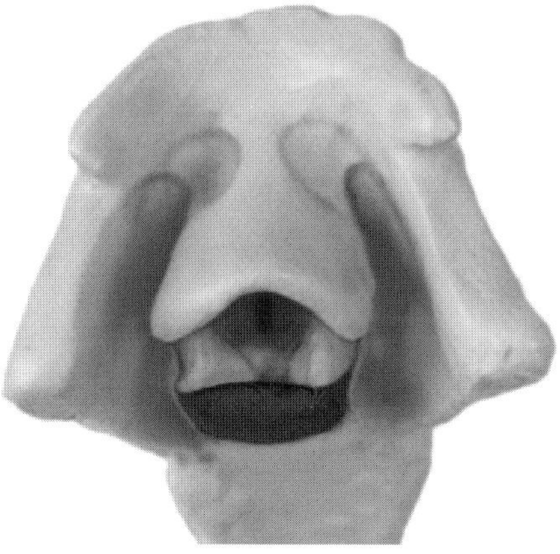

Cranial

Actions

Approximates arytenoid cartilages, closing the glottis.

Attachments

Lateral border and posterior surface of muscular process of both arytenoid cartilages.

Note

This is a single quadralateral muscle that crosses the midline, a rarity.

*a•ret•uh•**noid**•ee•us*

From the Greek, "arytaina"+"eidos," meaning shaped like a ladle.

*trans•**ver**•sus*

Latin, meaning crossing.

Arytenoideus obliquus

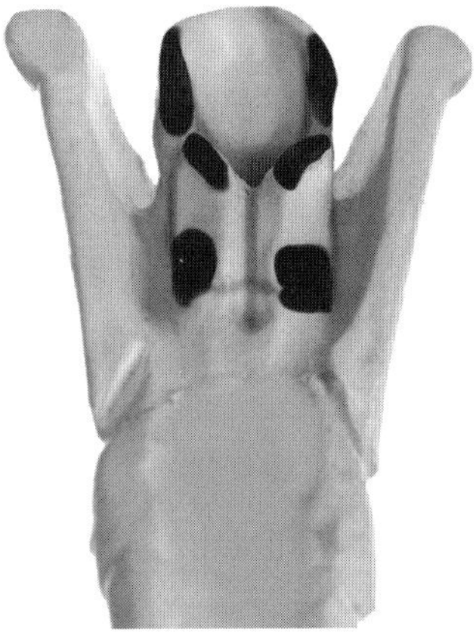

Caudodorsal

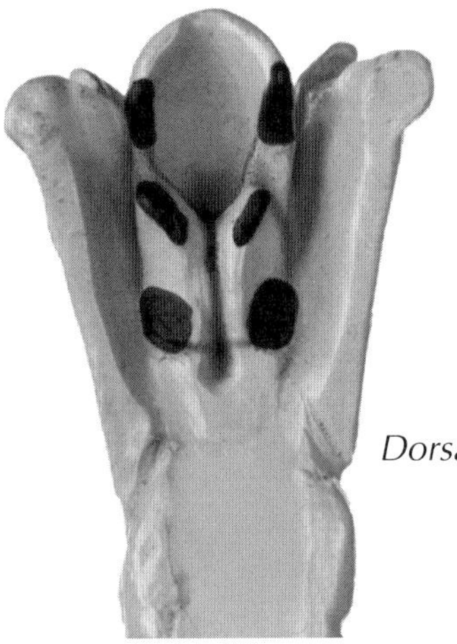

Dorsal

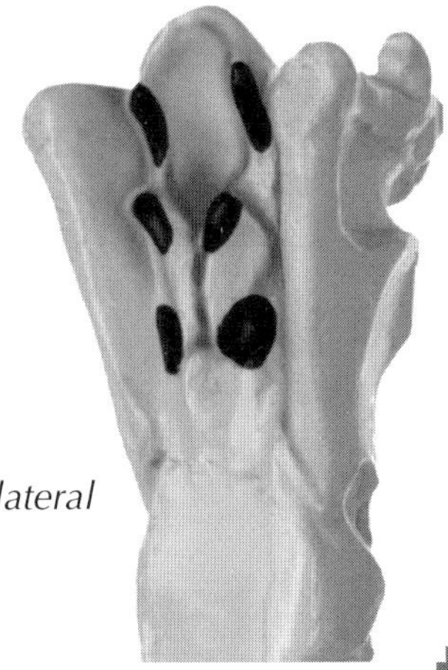

Dorsolateral

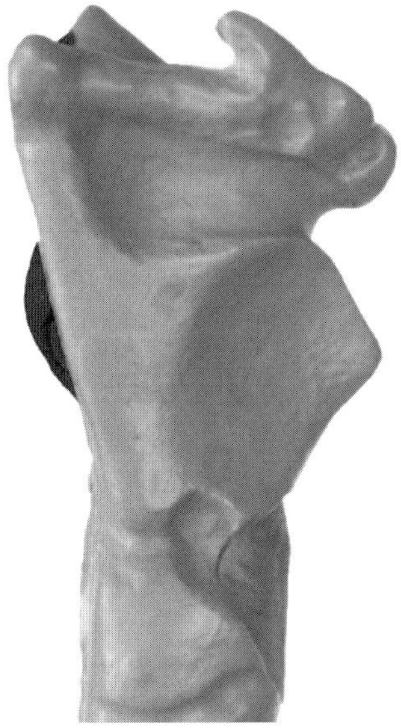

Lateral

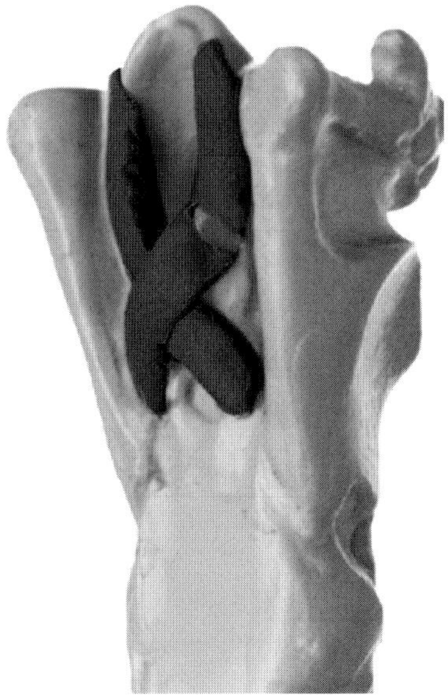

Dorsolateral

Caudodorsal

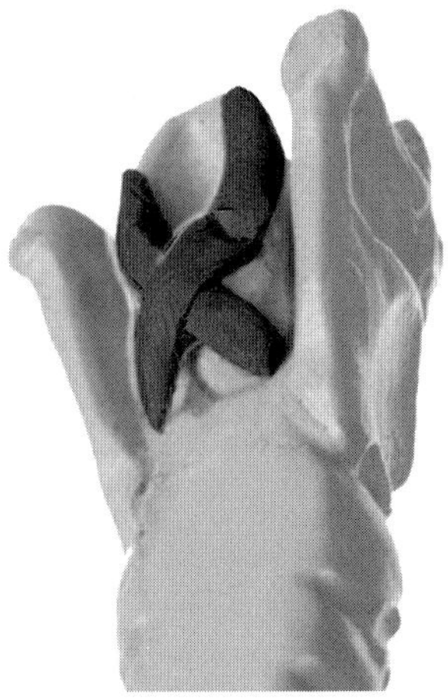

Caudodorsolateral

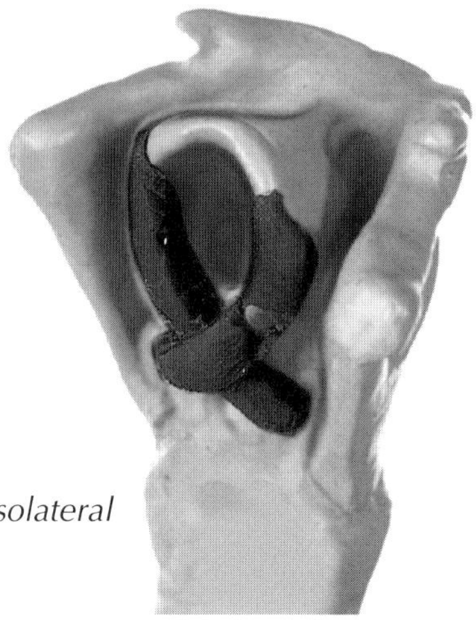

Craniodorsolateral

Arytenoideus obliquus

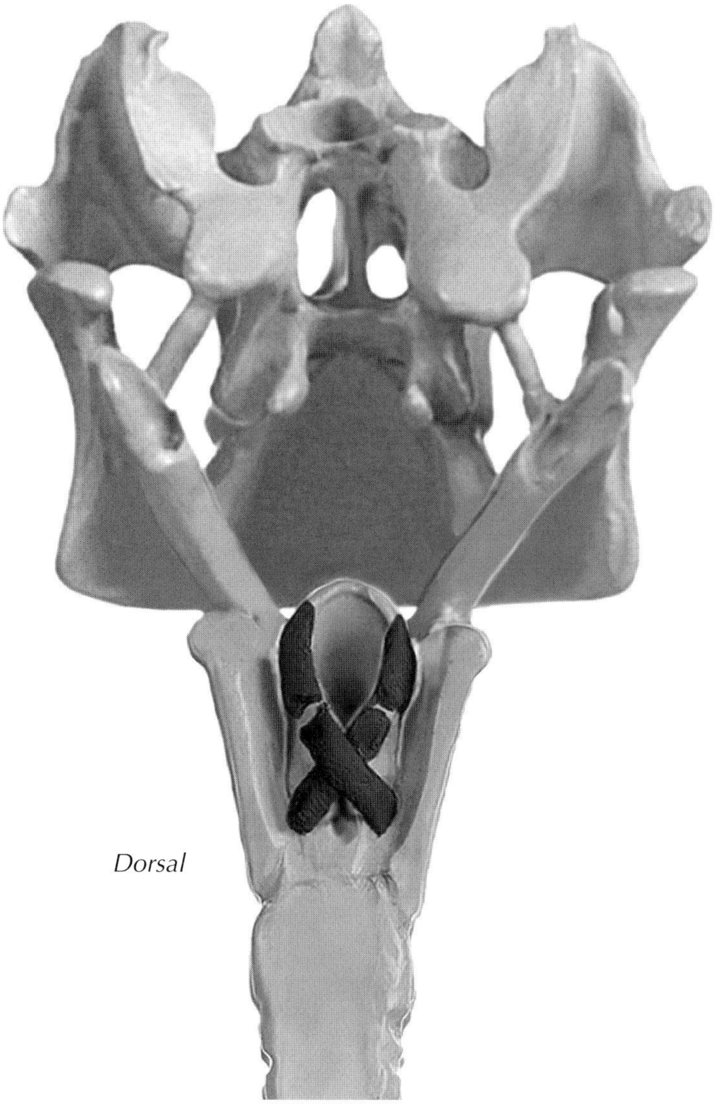

Dorsal

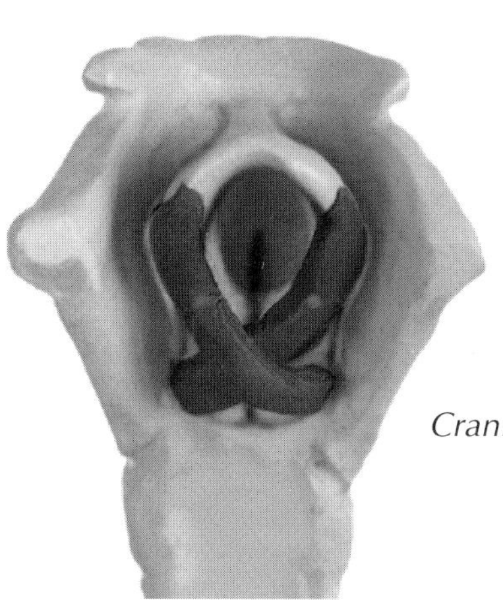

Craniodorsal

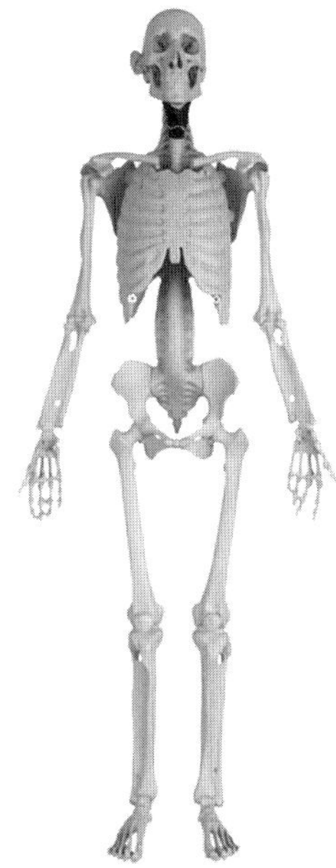

Actions

Approximate the arytenoid cartliges, closing the glottis.

Attachments

Caudal

Posterior muscular process of arytenoid cartilage.

Cranial

Apex of contralateral arytenoid cartilage.

Note

These muscles form an "X" across the midline superficial to the arytenoideus transversus. The most cranial slips, from the arytenoid cartilages to the margin of the eipglottis is sometimes separated as the "aryepiglotticus."

*a•ret•uh•**noid**•ee•us*

From the Greek, "arytaina"+"eidos," meaning shaped like a ladle.

*oh•**bleek**•us*

Latin, "obliquus," meaning slanting.

Cricoarytenoideus lateralis

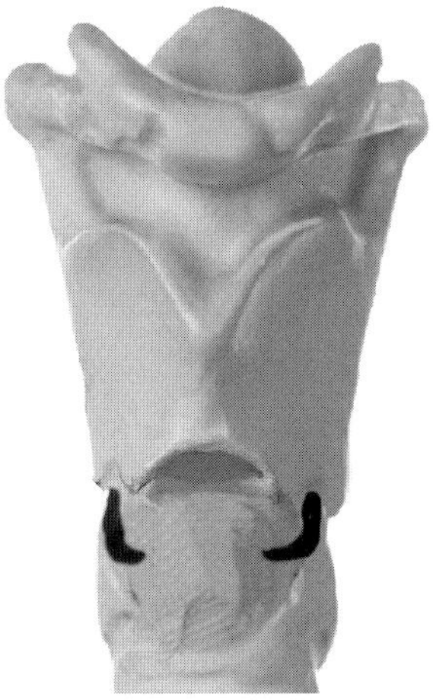

Ventral

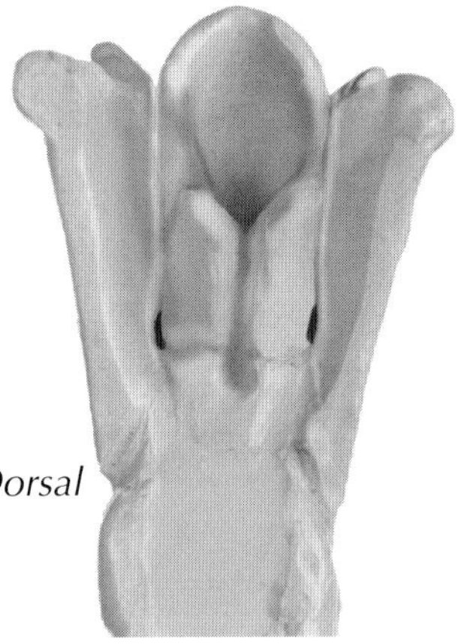

Dorsal

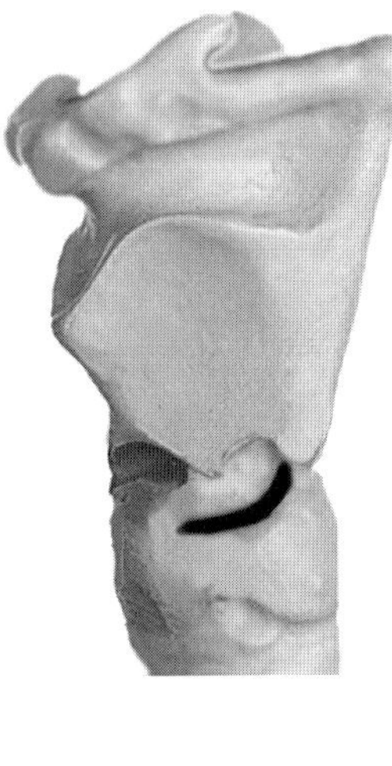

Lateral

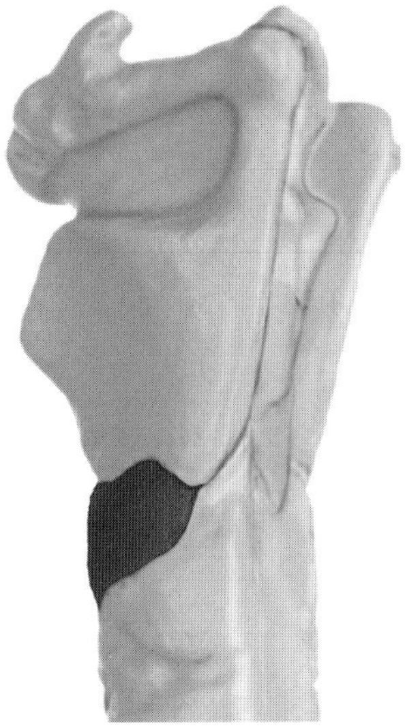

Dorsolateral

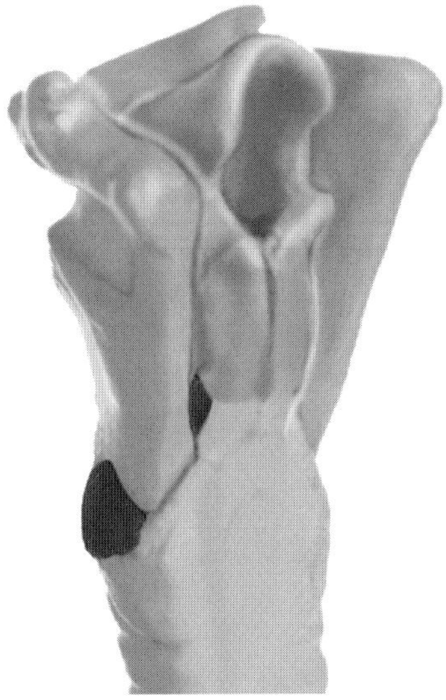

Craniodorsolateral

Dorsal

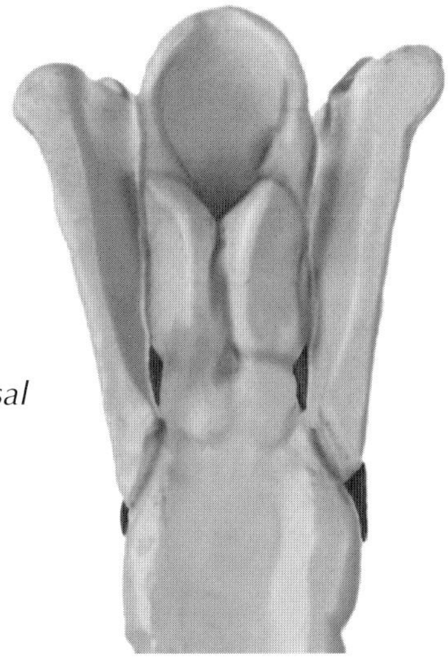

Ventral

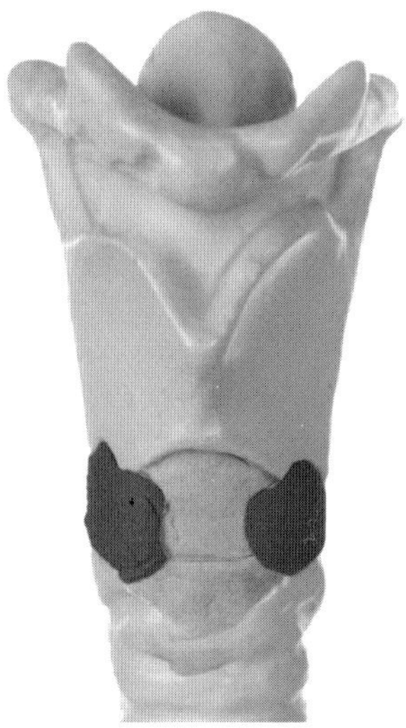

Ventrolateral

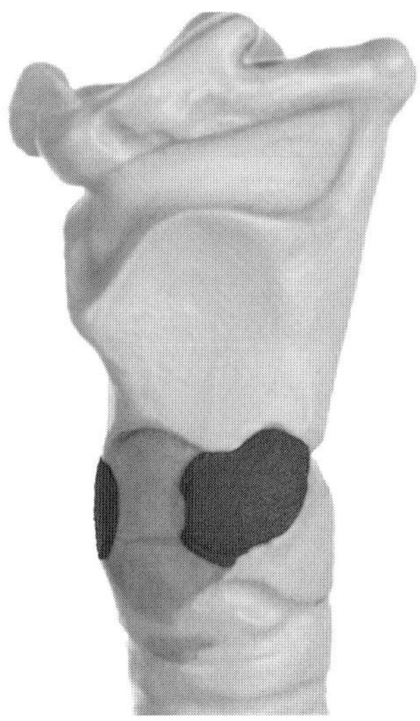

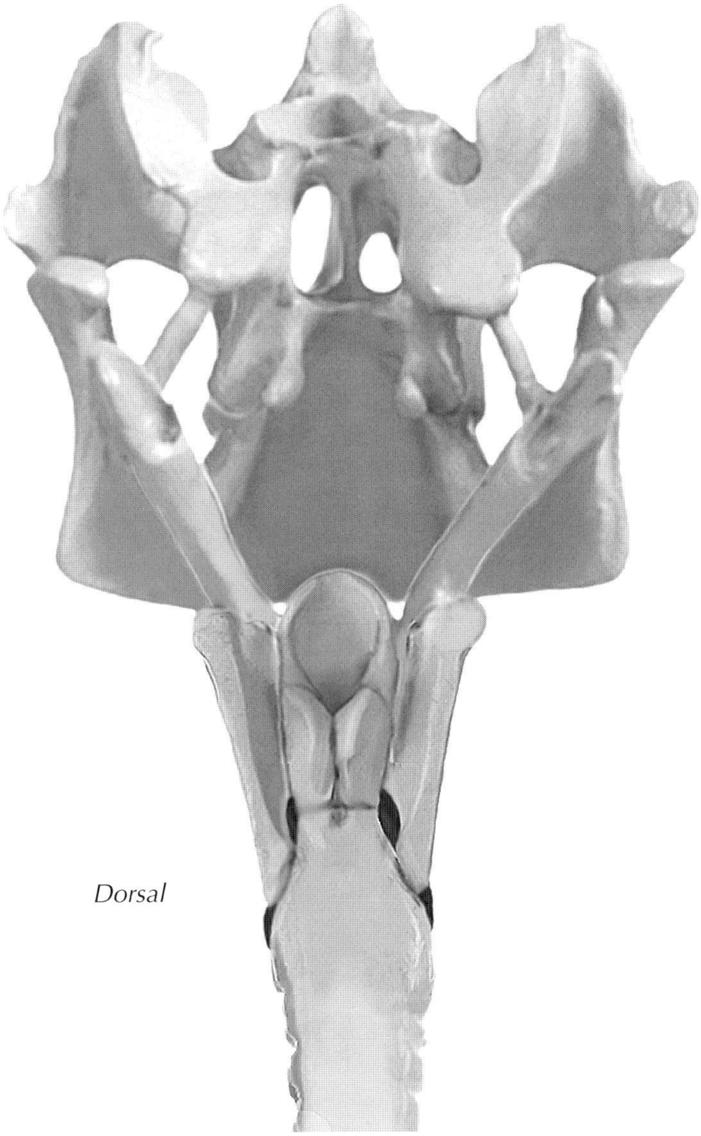

Dorsal

Cricoarytenoideus lateralis

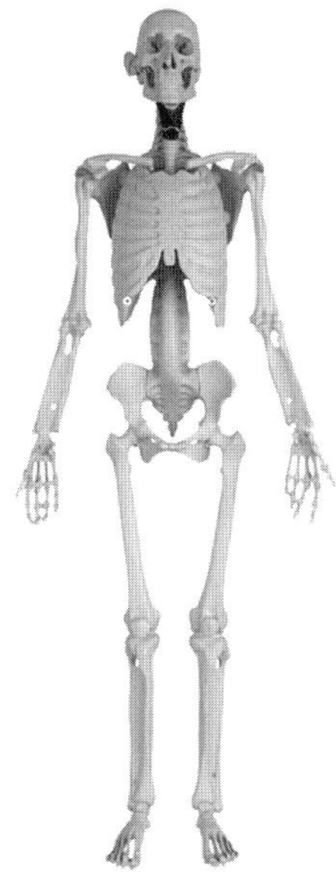

Actions

Rotate arytenoid cartilages medially, approximating the vocal folds and closing the glottis.

Attachments

Ventrocaudal

Cranial margin of lateral part of the cricoid cartilage.

Dorsocranial

Anterior surface of the muscular process of the arytenoid cartilage.

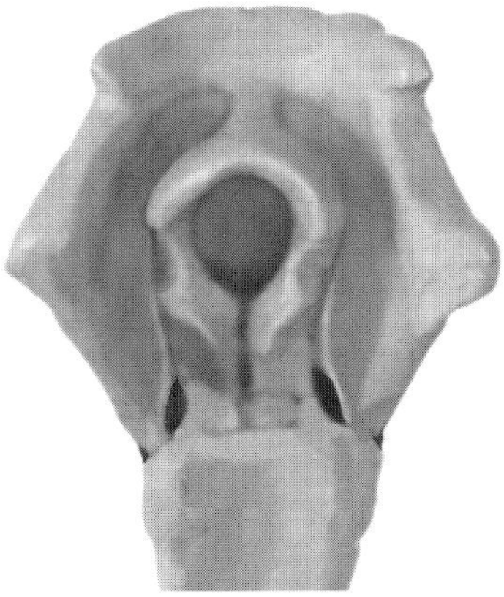

Craniodorsal

*cry•ko•a•ret•uh•**noid**•ee•us*

From the Greek, "krikos," meaning ring, and From the Greek, "arytaina"+"eidos,"meaning shaped like a ladle.

*lat•er•**al**•us*

From the Latin, "latus"+"alis," meaning related to the side.

Cricoarytenoideus posterior

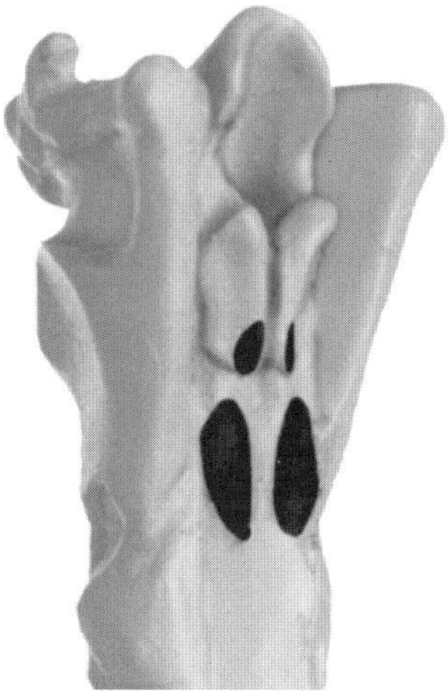

Dorsolateral

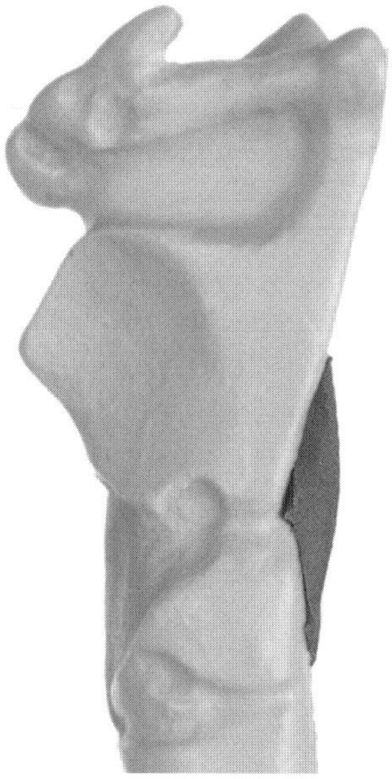

Lateral

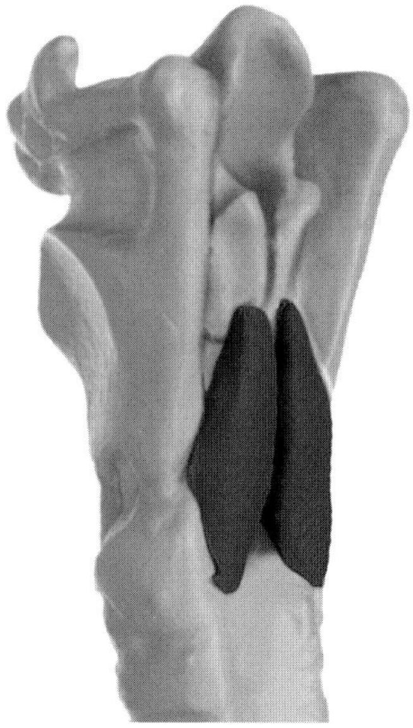

Dorsolateral

Dorsal

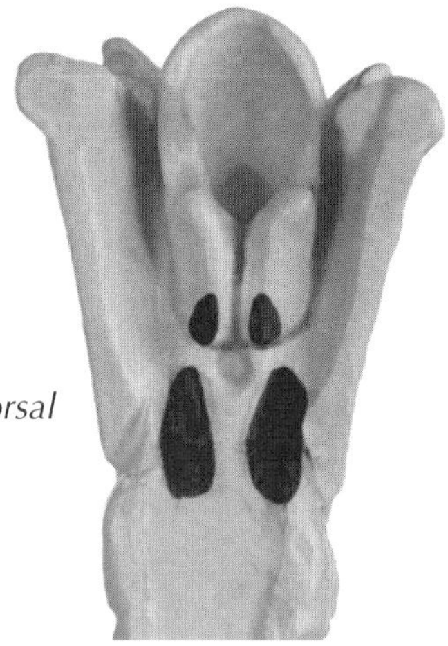

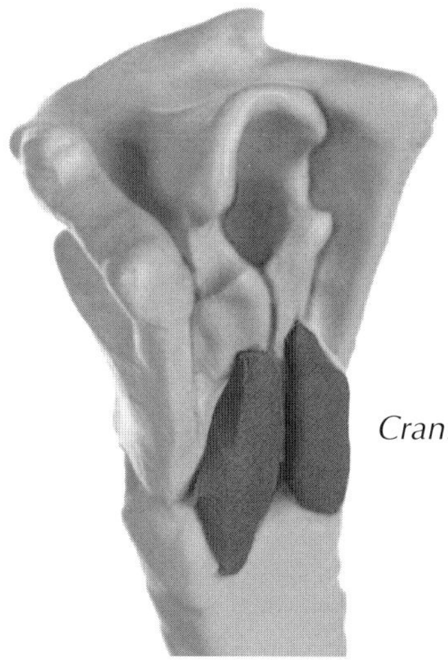

Craniodorsolateral

Caudodorsal

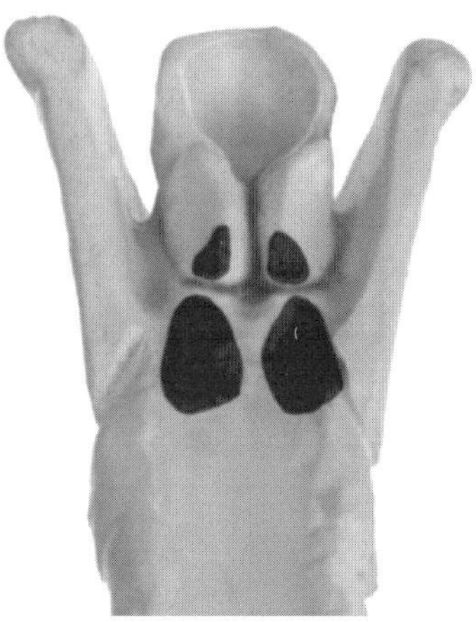

Craniodorsal

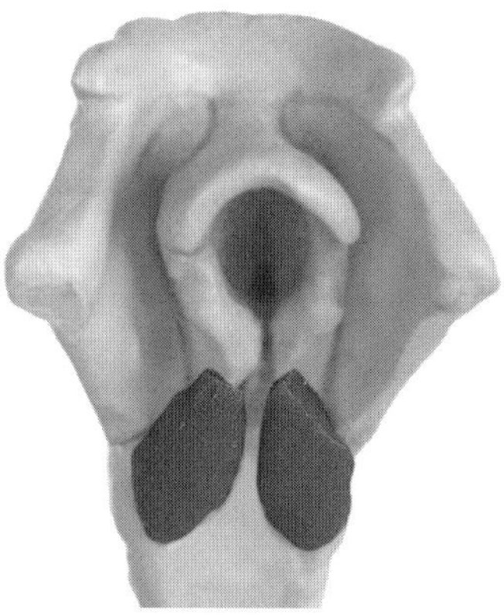

Caudodorsal

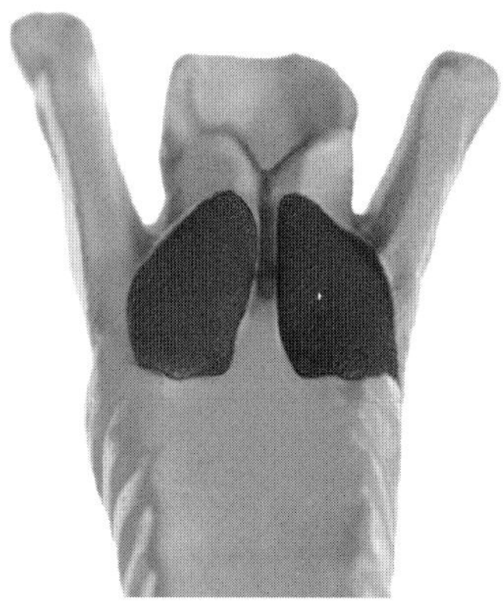

Cricoarytenoideus posterior

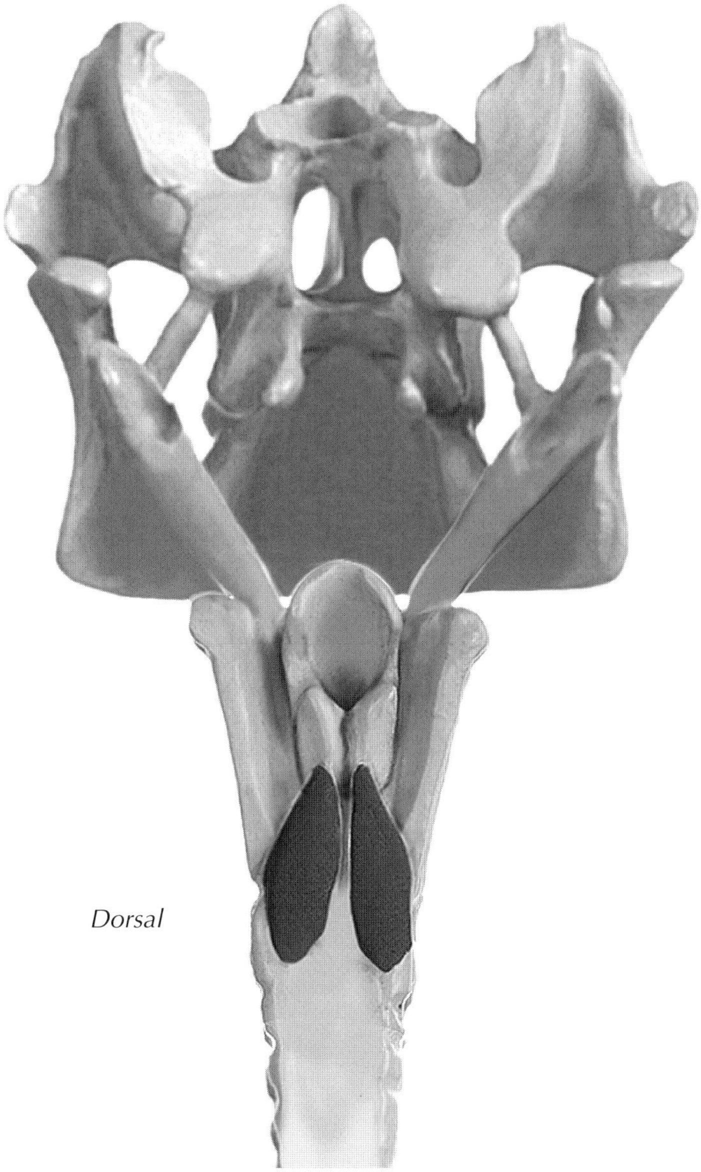

Dorsal

Caudodorsal

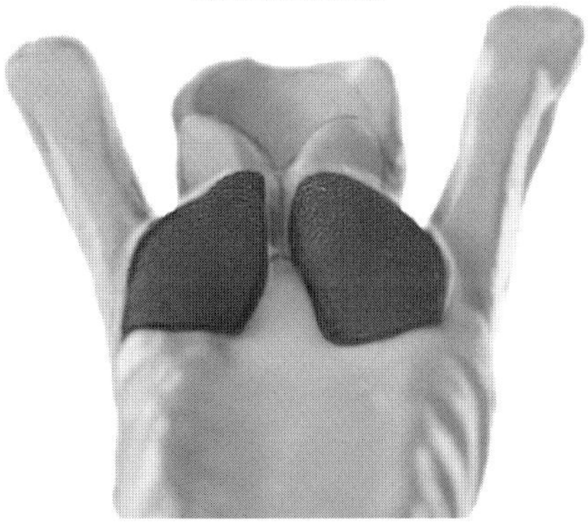

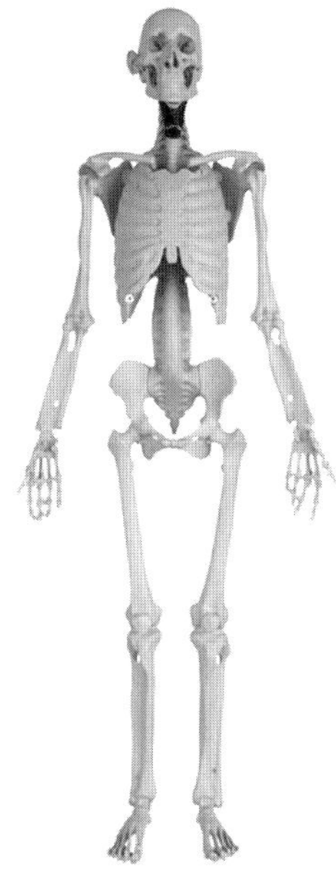

Actions

Rotate arytenoid cartilages laterally, opening the glottis and stretching the vocal folds.

Attachments

Proximal

Posterior surface of cricoid cartilage.

Distal

Muscular process of the arytenoid cartilage.

*cry•ko•a•ret•uh•**noid**•ee•us*

From the Greek, "krikos," meaning ring, and "arytaina"+"eidos," meaning shaped like a ladle.

*pos•**teer**•ee•ur*

Latin, meaning behind or following.

Cricothryoideus

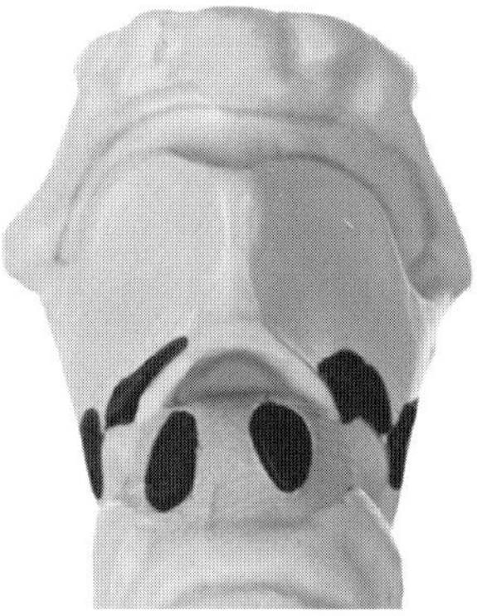

Caudoventral

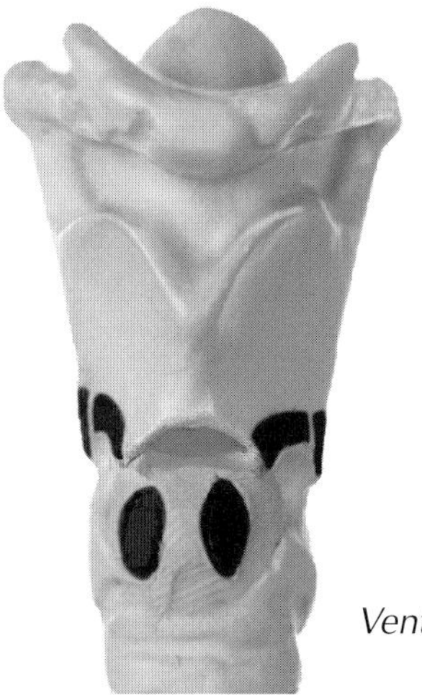

Ventral

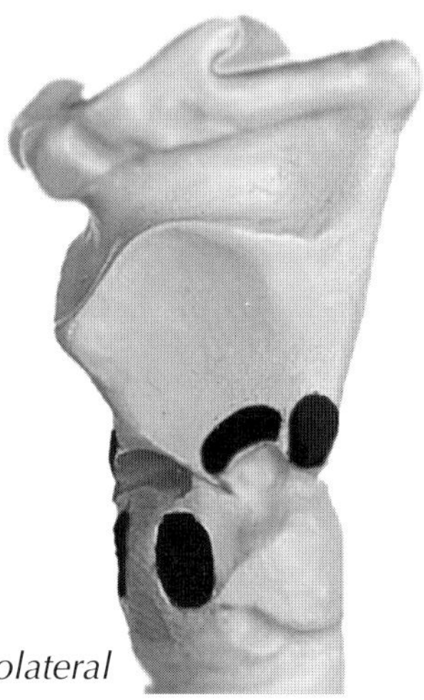

Ventrolateral

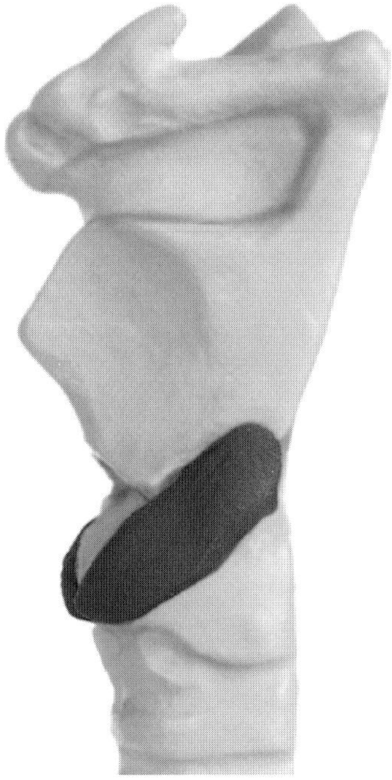

Lateral

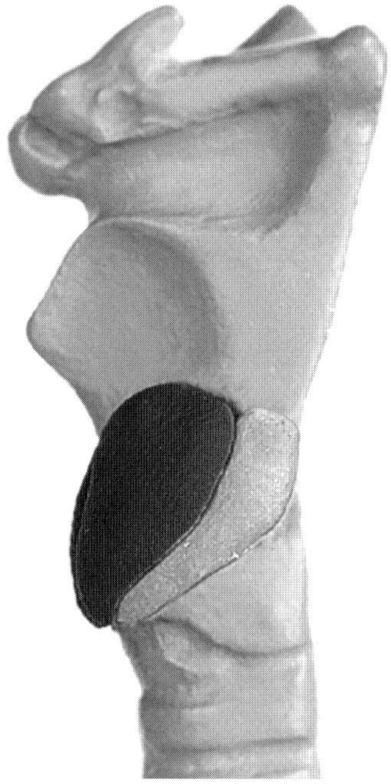

Lateral

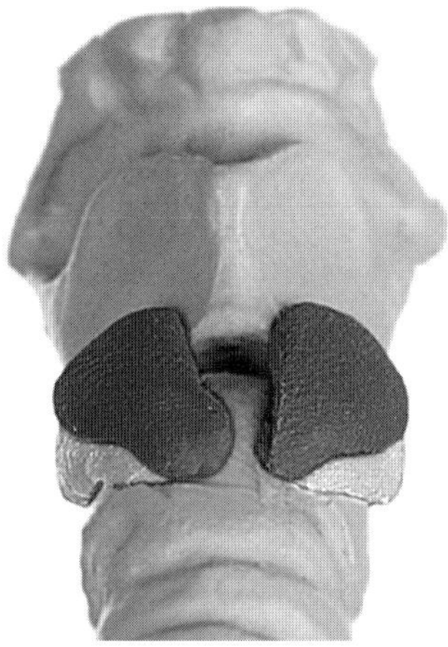

Caudoventral

Ventrolateral

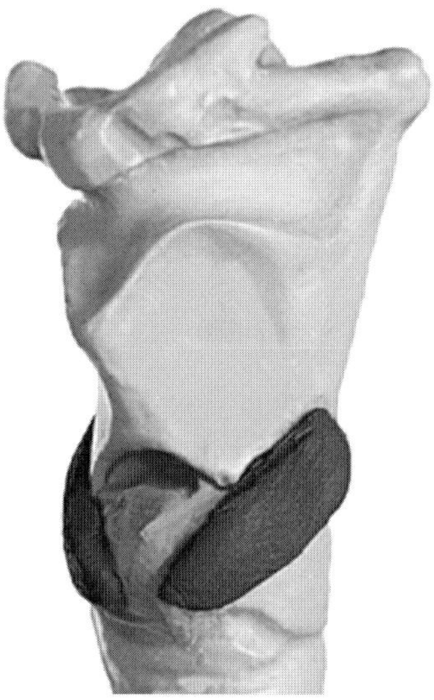

Ventrolateral

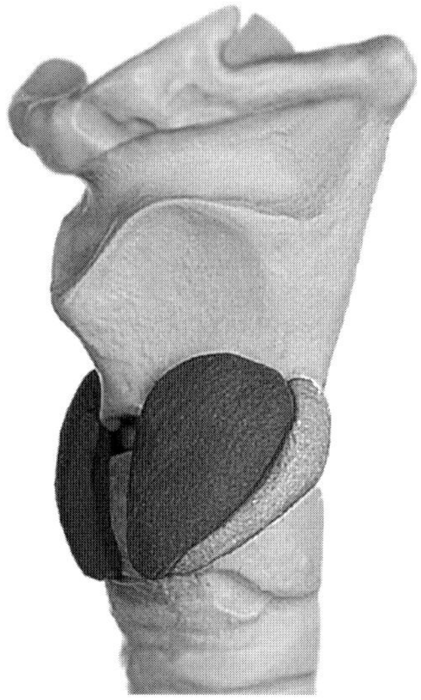

Cricothryoideus

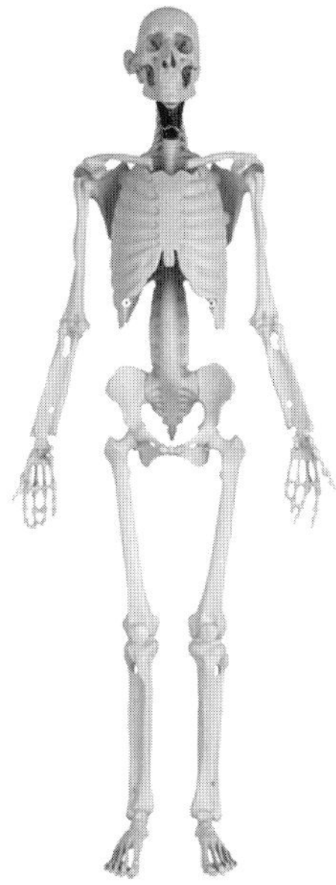

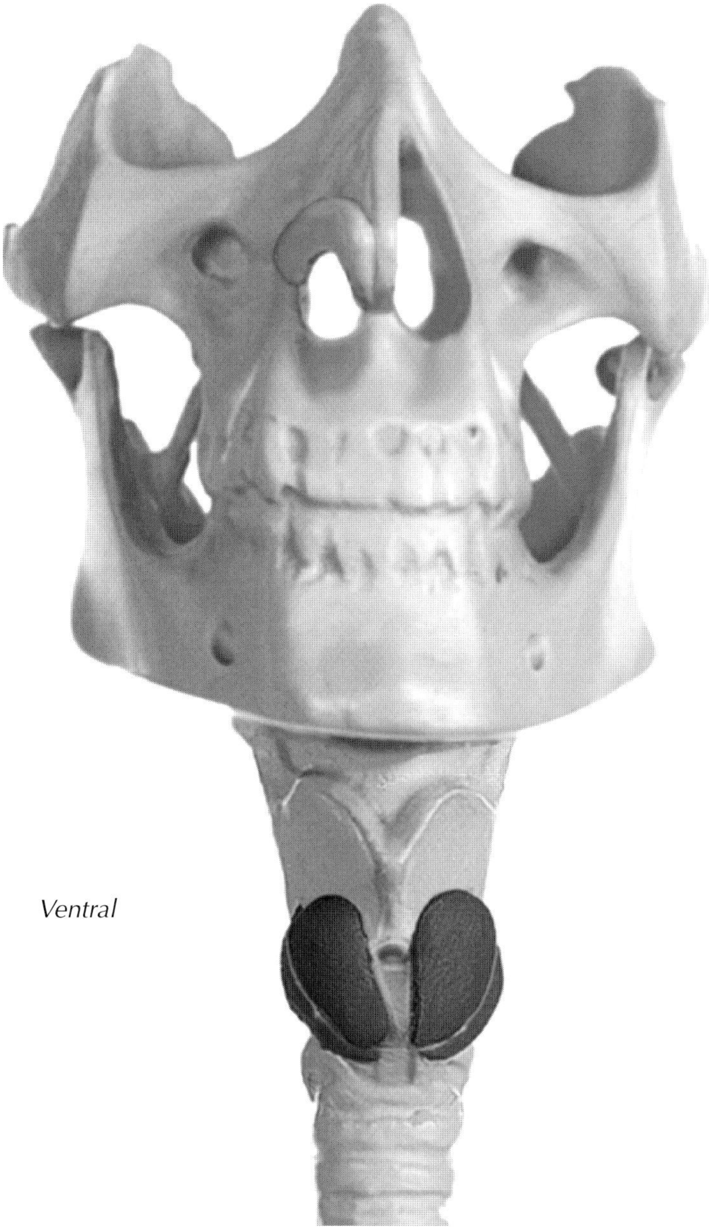

Ventral

Actions

Elevates ventral part of cricoid cartilage, tilting it up and stretching the vocal folds.

Attachments

Caudal

Ventrolateral surface of cricoid cartilage.

Cranial

<u>pars obliquus</u>: Anterior border of the inferior cornu of the thyroid cartilage.
<u>pars rectus</u>: Caudal border of lamina of thyroid cartilage.

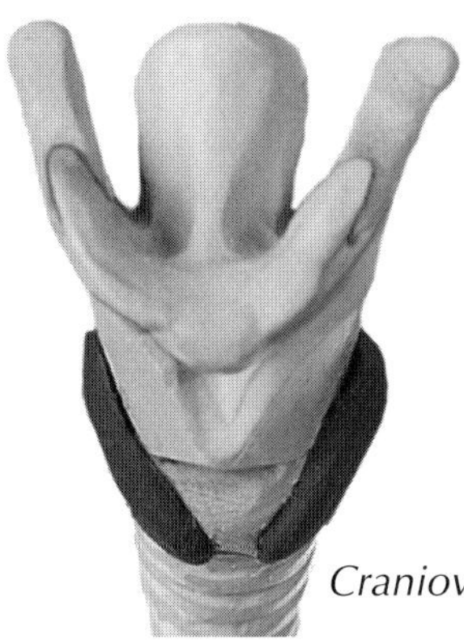

Cranioventral

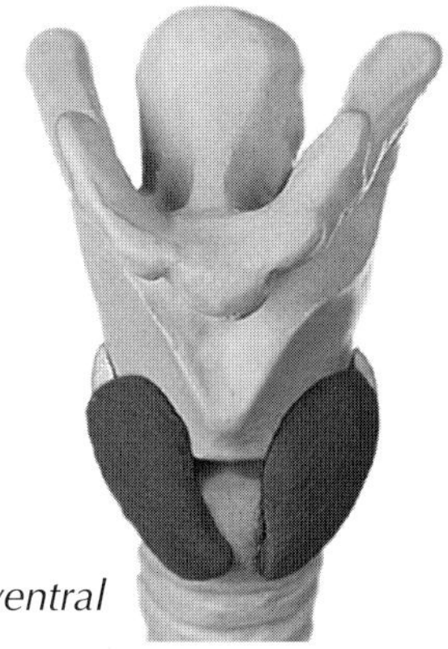

Cranioventral

*cry•ko•thy•***roid***•ee•us*

From the Greek, "krikos," meaning ring, and "thyreos"+"eidos," meaning shaped like a shield.

Thyrohyoideus

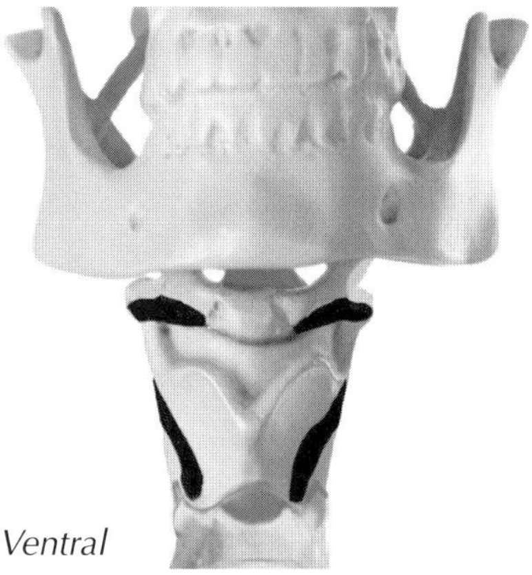

Ventral

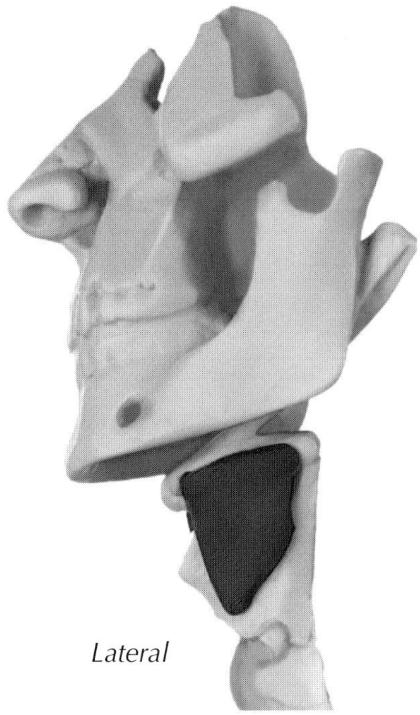

Lateral

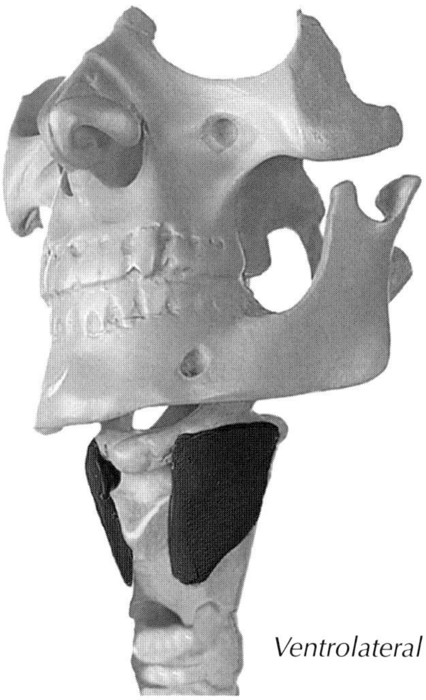

Ventrolateral

Lateral

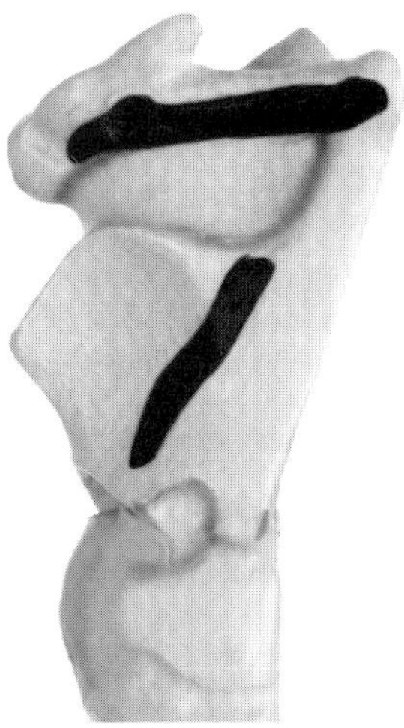

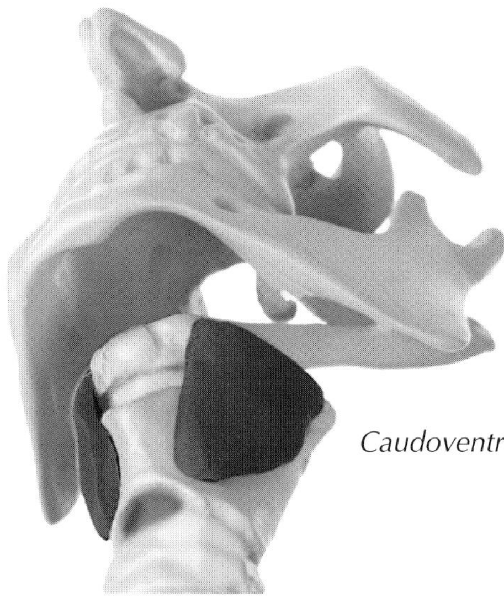

Caudoventrolateral

Caudoventral

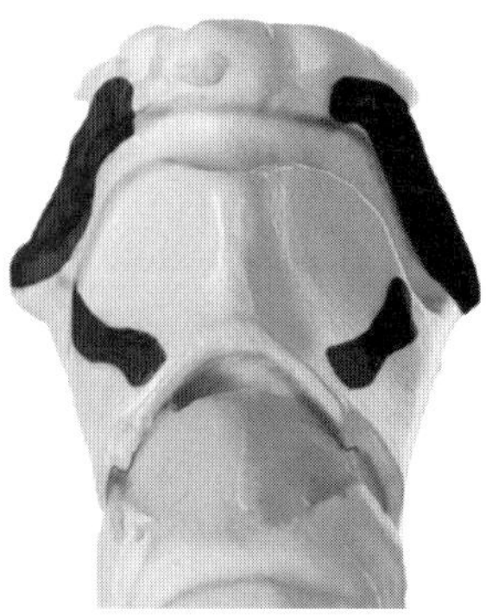

Cranioventral

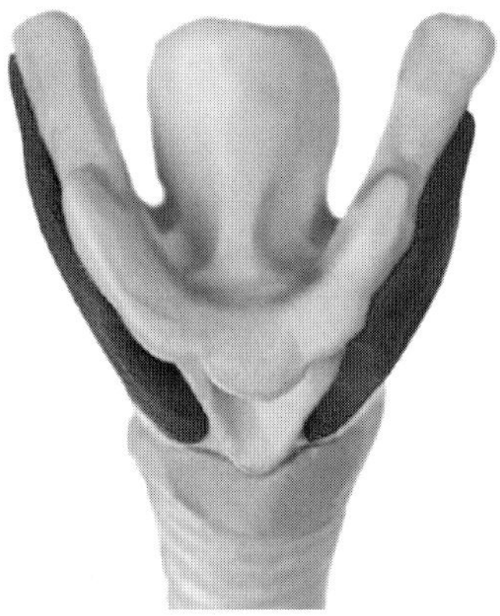

Cranioventrolateral

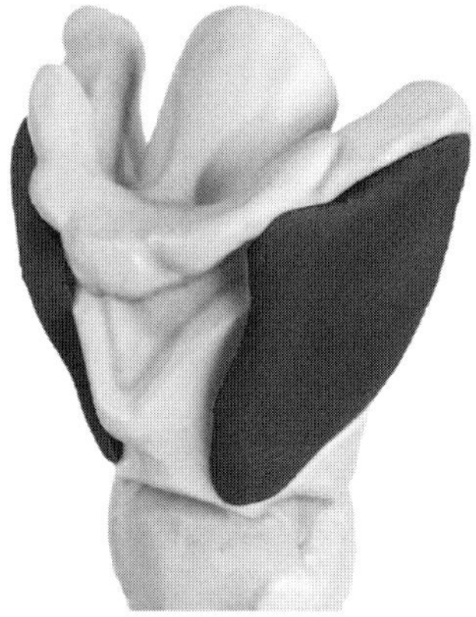

Thyrohyoideus

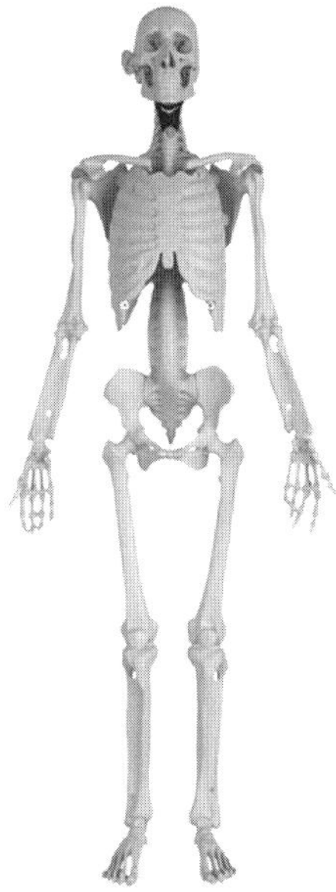

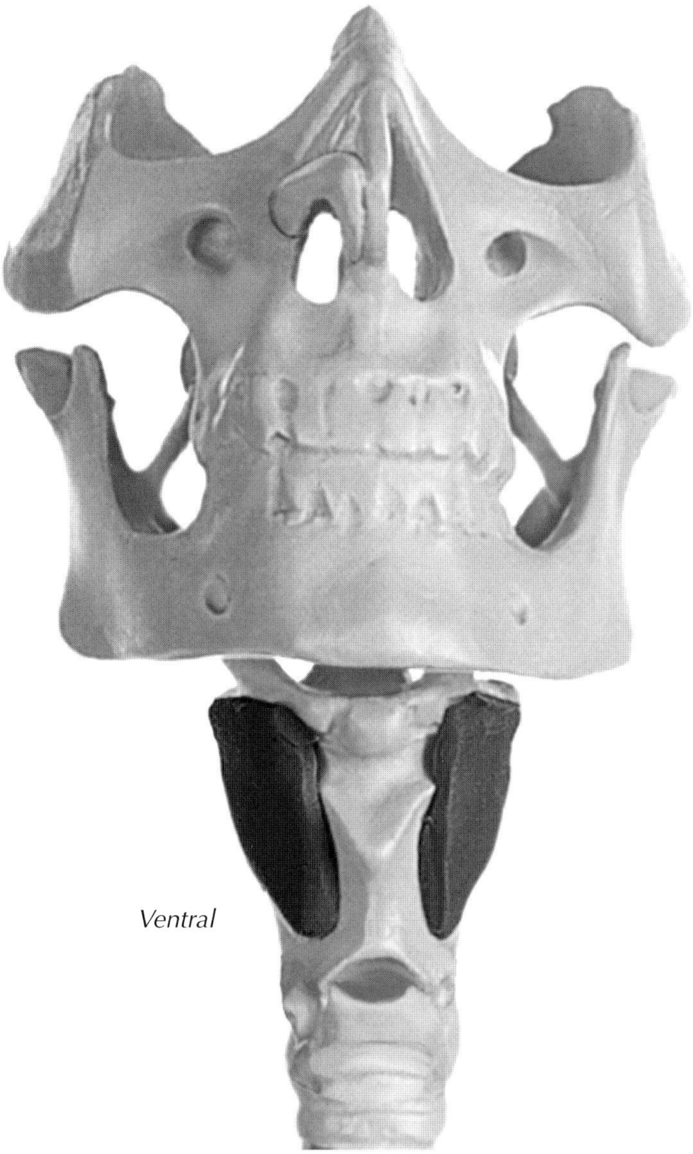

Ventral

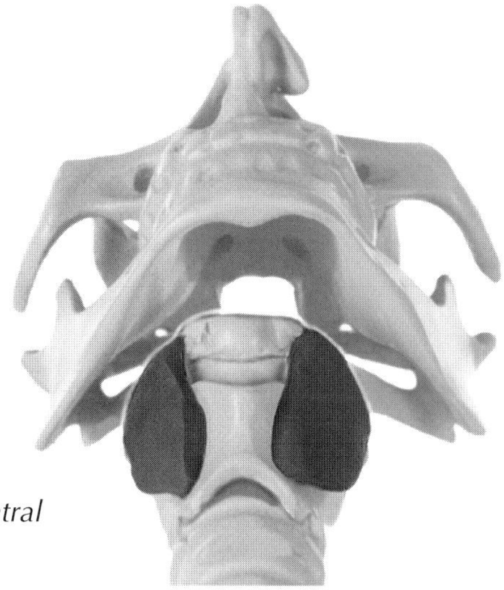

Caudoventral

Actions

Depresses hyoid bone, elevates larynx.

Attachments

Caudal

Oblique line on lateral surface of the thyroid cartilage.

Cranial

Caudal border and greater cornu of the hyoid bone.

*thy•ro•hy•***oid***•ee•us*

From the Greek, “thyra,” meaning door, and, “hyo”+“eidos,” meaning U-shaped.

Transversus linguae

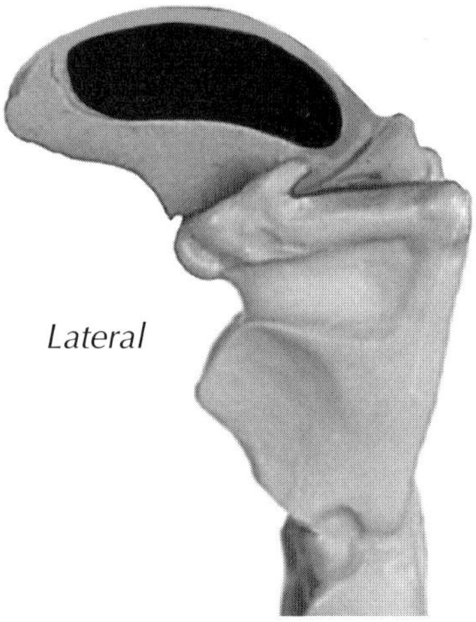

Lateral

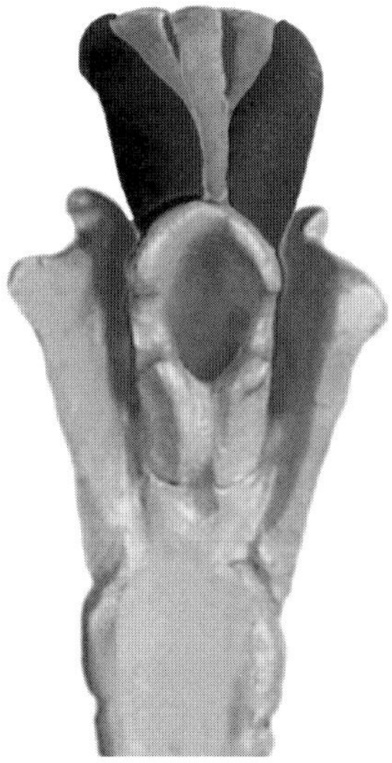

Dorsal

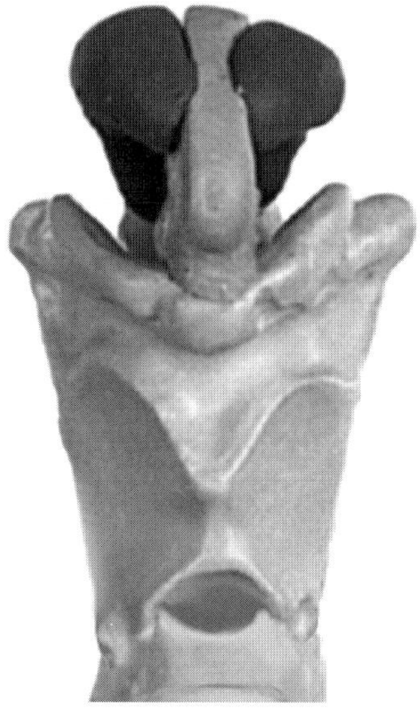

ventral

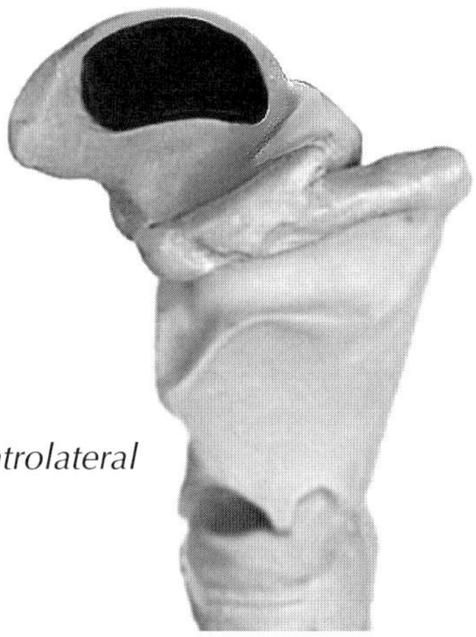

Ventrolateral

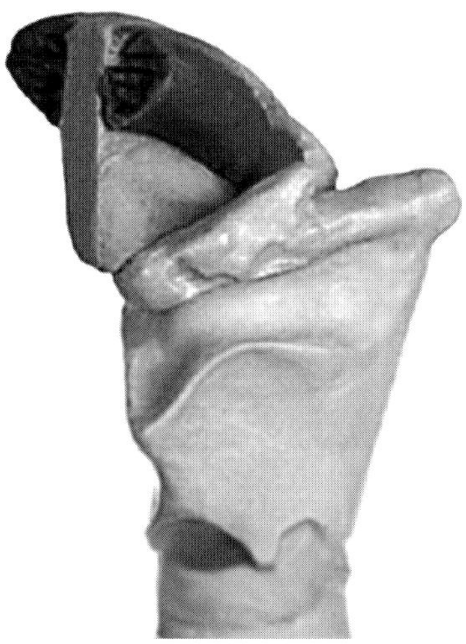

Ventrolateral

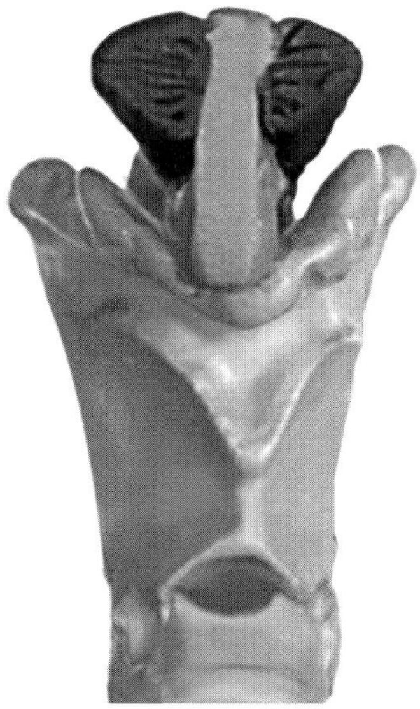

Ventral

Craniolateral

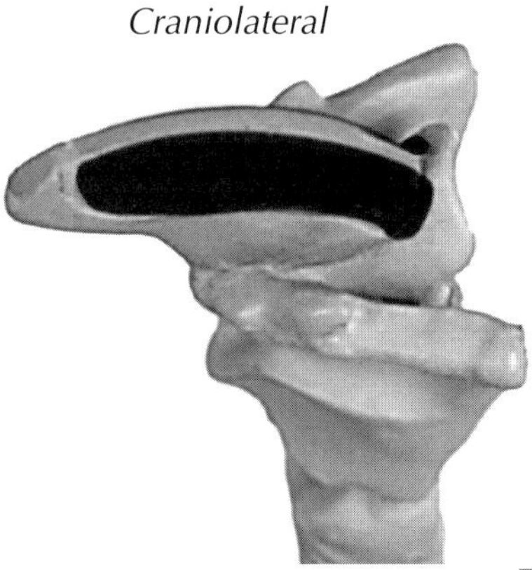

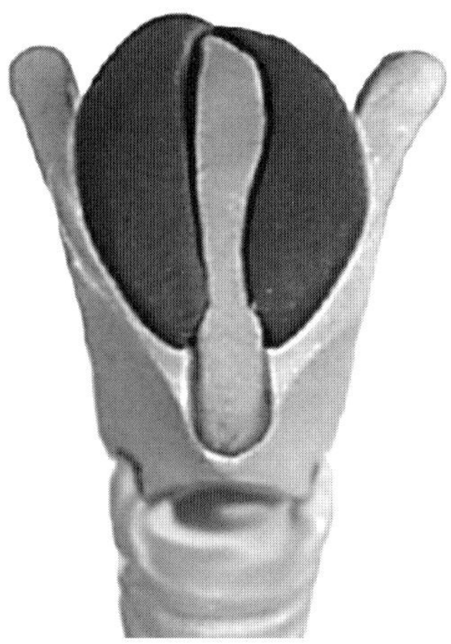

Cranioventral

Transversus linguae

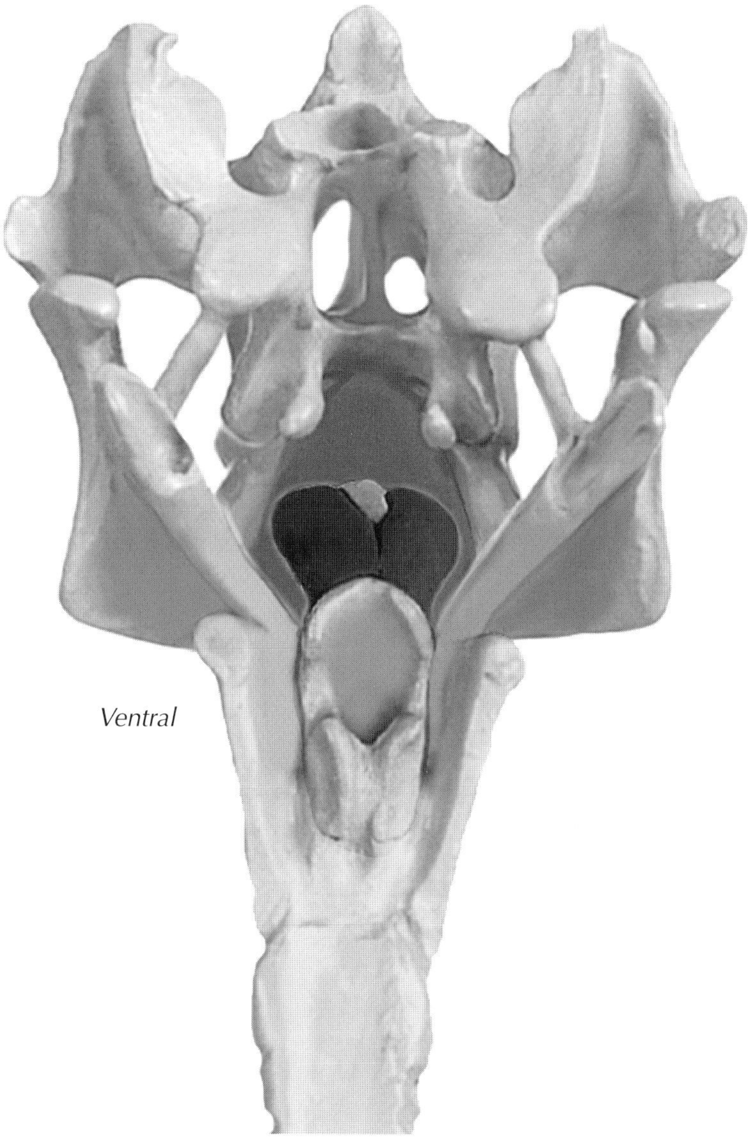

Ventral

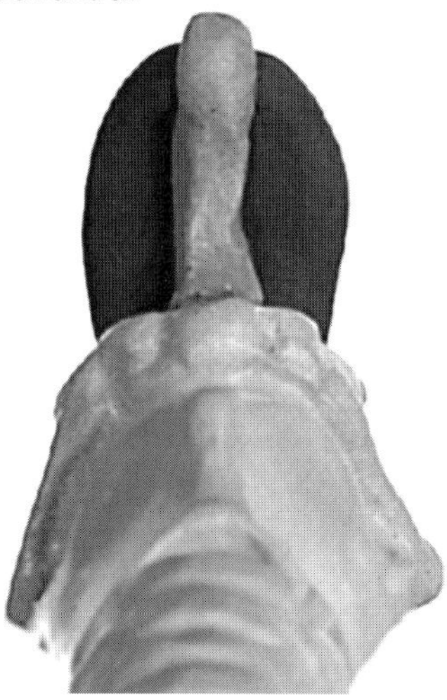

Caudoventral

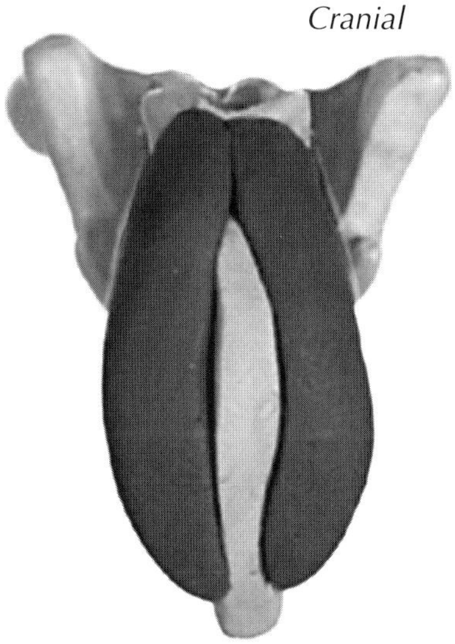

Cranial

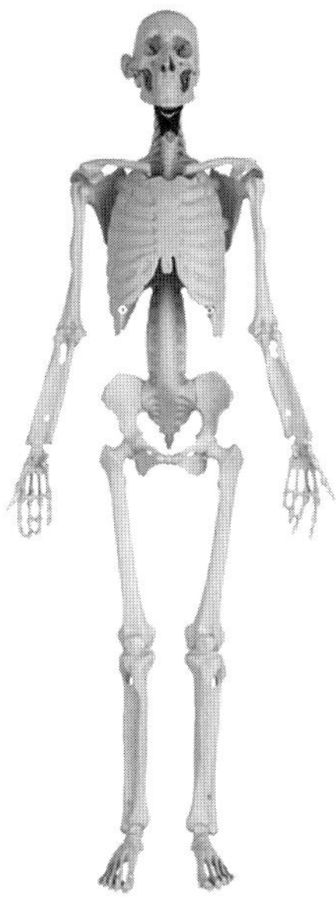

Actions

Narrows and elongates tongue.

Attachments

Medial

Median septum of the tongue.

Lateral

Submucous fibrous capsule of the side of the tongue.

Note

The two illustrations in the middle of the page 56 are coronal *cross-sections*

*trans•**ver**•sus*

Latin, meaning crossing.

ling•*gwee*

Latin, meaning belonging to the tongue.

Verticalis linguae

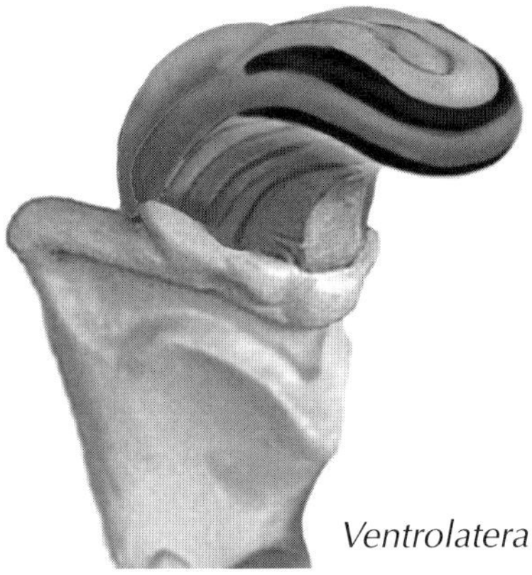

Ventrolateral

Ventral

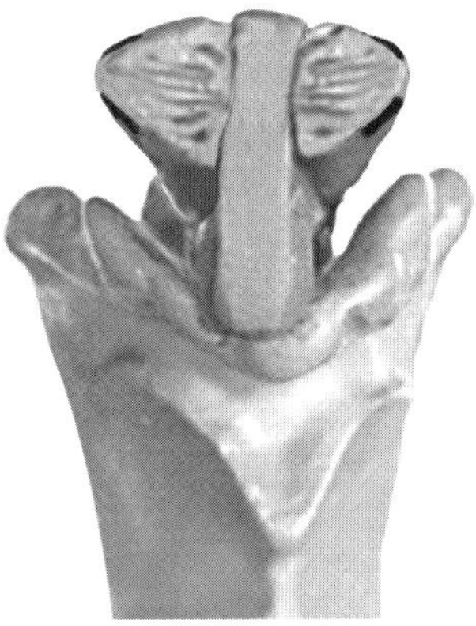

Cranial

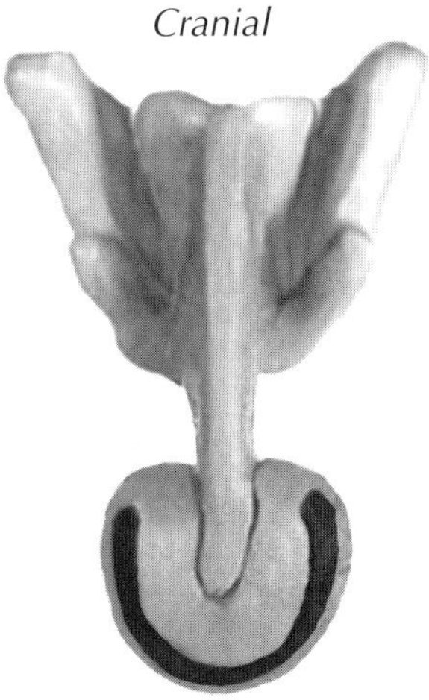

Ventrolateral

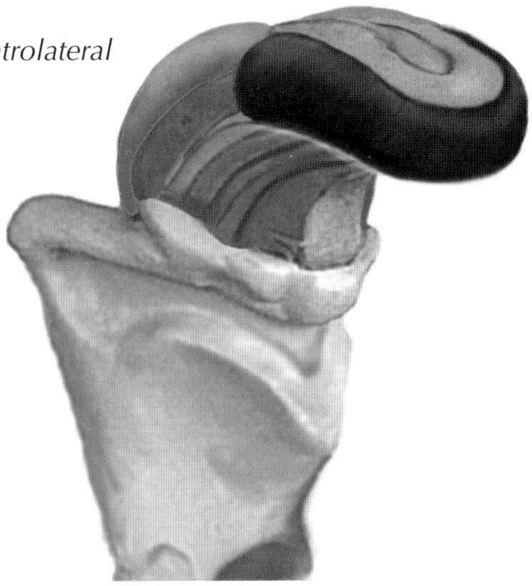

Lateral

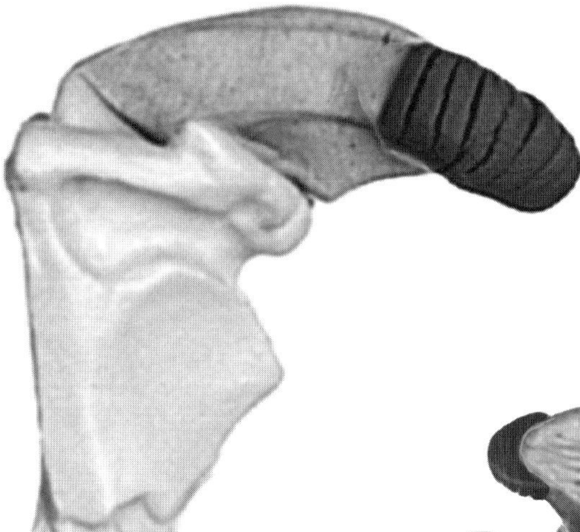

Ventral

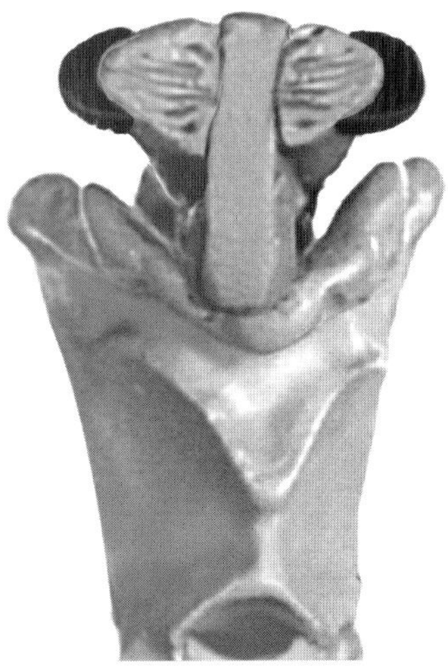

Dorsal

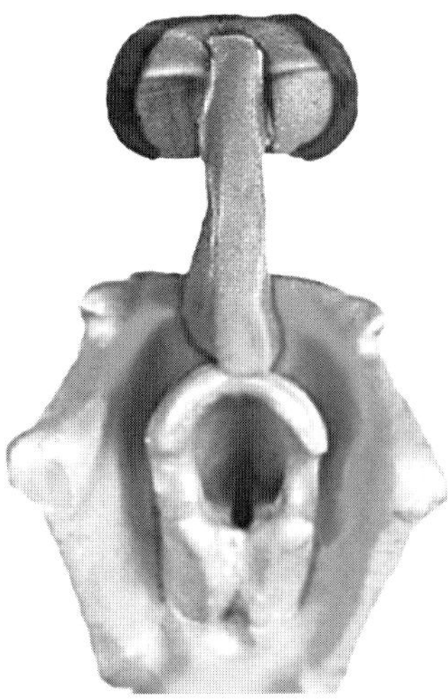

Verticalis **linguae**

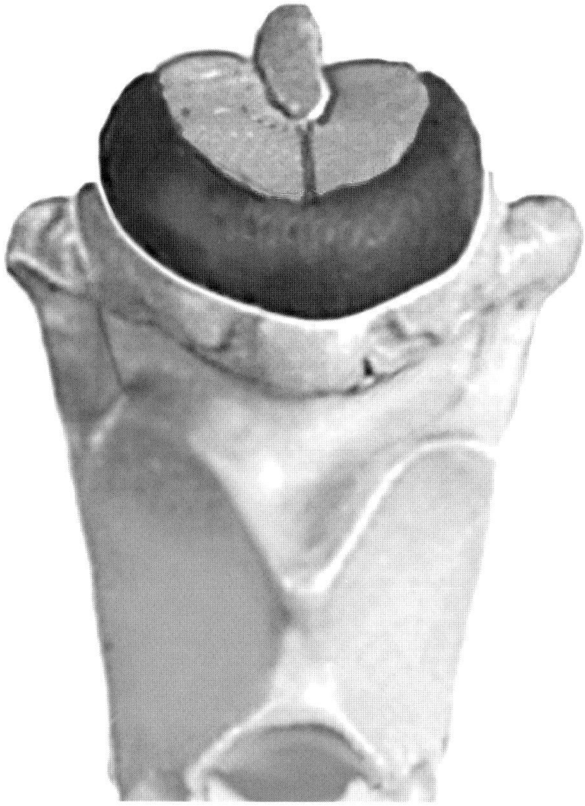

Ventral

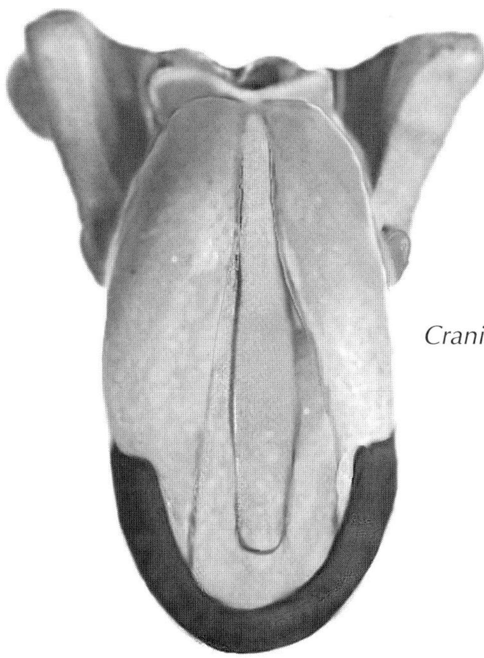

Cranial

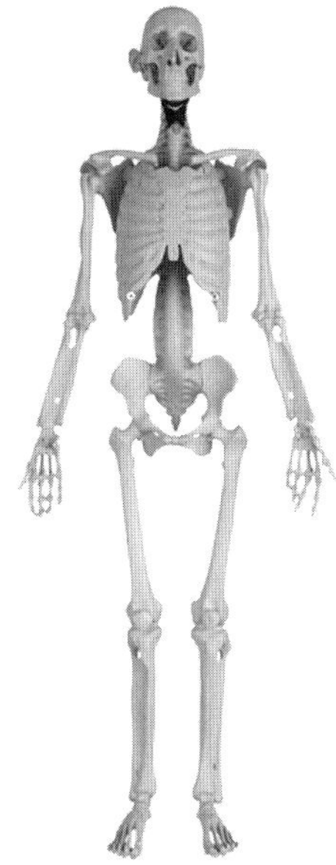

Actions

Flattens and broadens the tongue.

Attachments

Dorsal

Dorsal fibrous capsule of the tongue.

Ventral

Ventral fibrous capsule of the tongue.

*ver•tuh•**kal**•us*

Latin, meaning related to the peak (pointing upwards, vertical).

ling•*gwee*

Latin, meaning belonging to the tongue.

Longitudinalis linguae inferioris

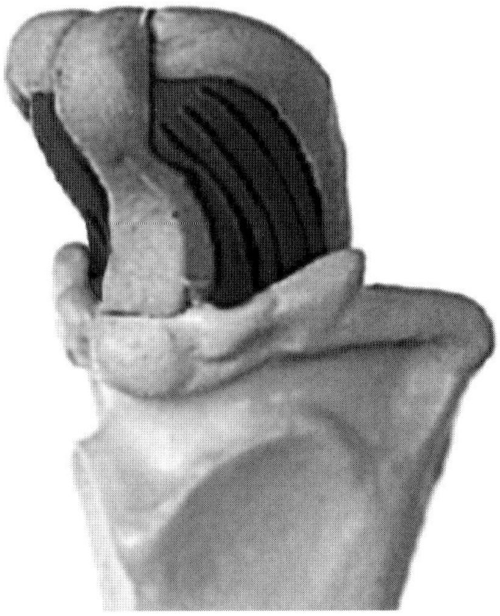

Ventrolateral

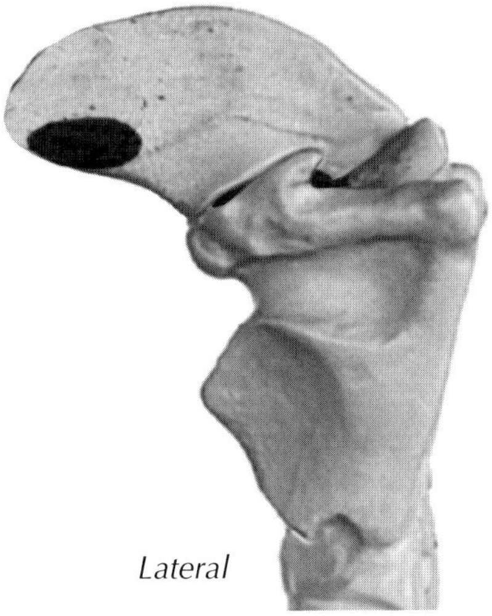

Lateral

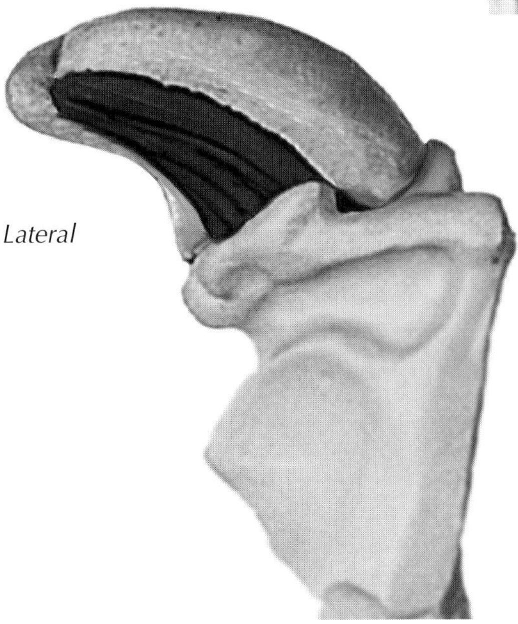

Lateral

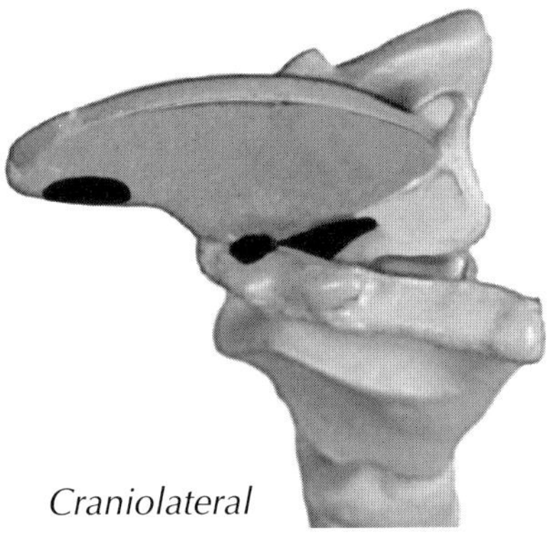

Craniolateral

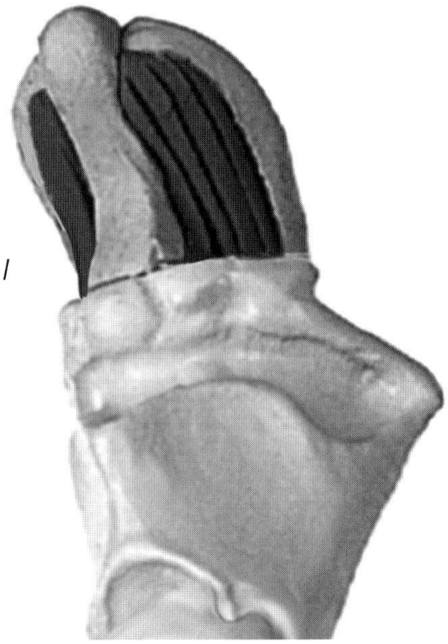

Caudoventrolateral

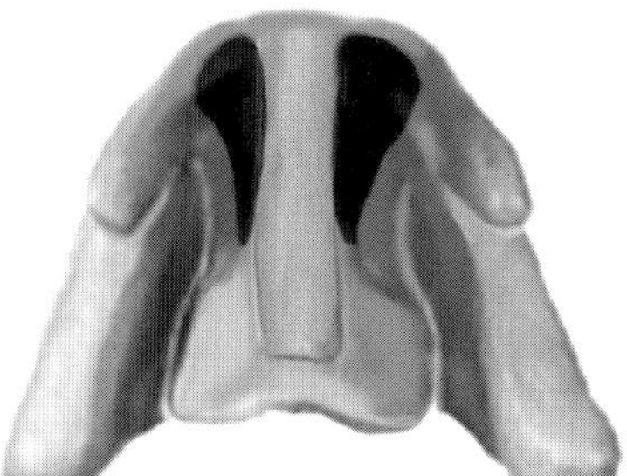

Cranial

Longitudinalis ***linguae*** *inferioris*

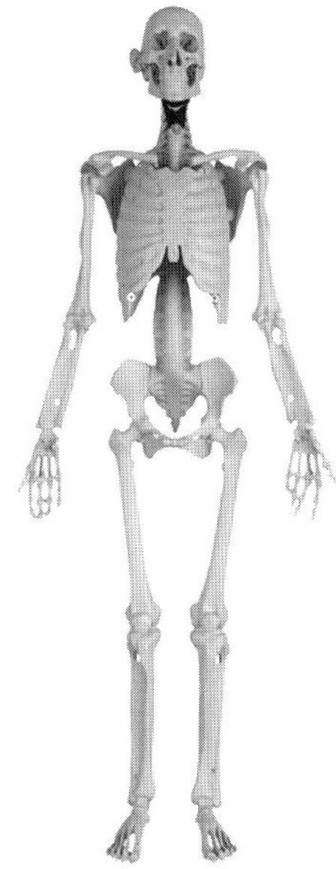

Actions

Shortens the tongue. Turns the tip of the tongue downward.

Attachments

Occipital

Undersurface of the posterior part of the tongue.

Rostral

Fibers from both sides converge at the tip of the tongue.

*lon•ji•**tood**•in•al•is*

From the Latin, "longitudinis"+"alis," meaning related to length.

ling•*gwee*

Latin, meaning belonging to the tongue.

*in•feer•ee•**or**•us*

From the Latin, "inferior," meaning lower or beneath.

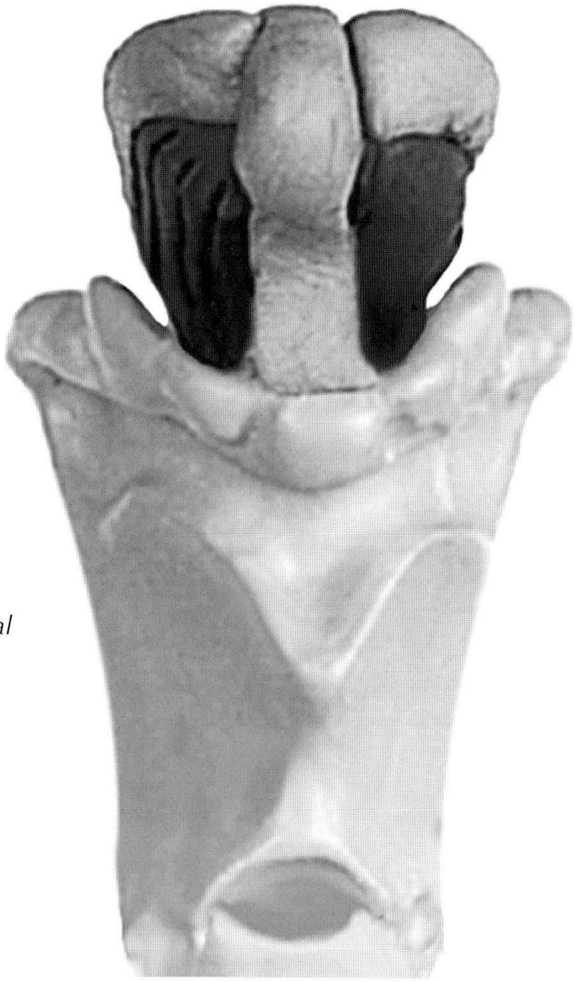

Ventral

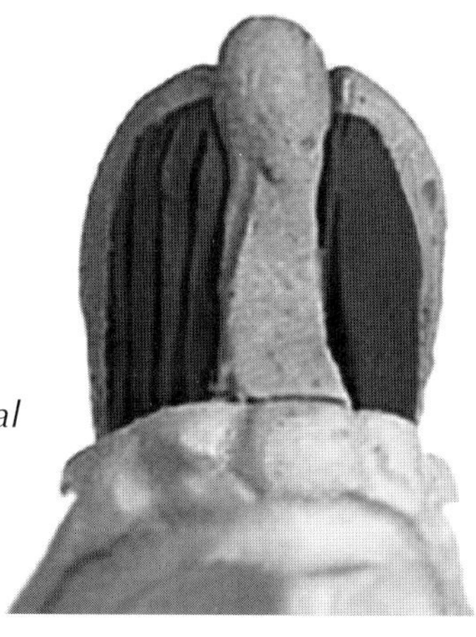

Caudoventral

Longitudinalis ***linguae*** ***superioris***

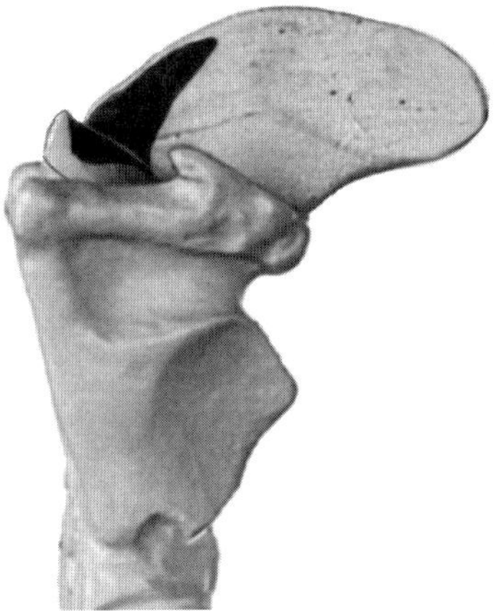

Lateral

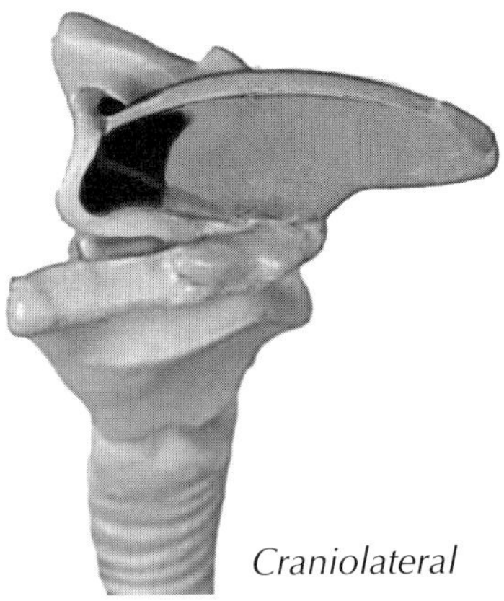

Craniolateral

Cranial

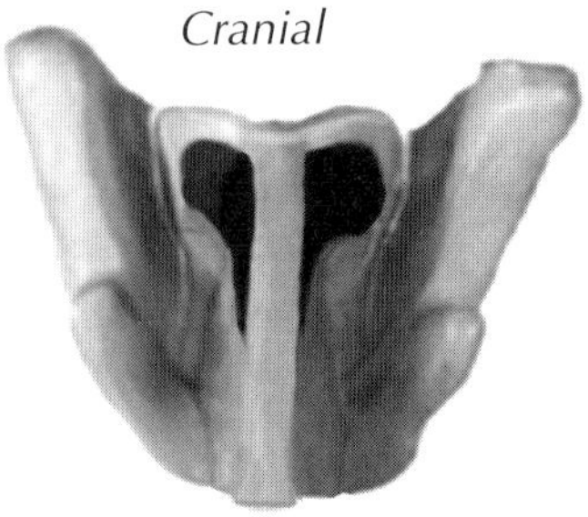

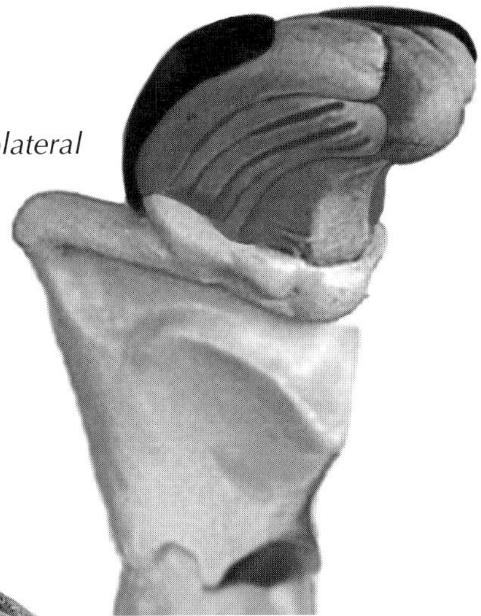

Cranioventrolateral

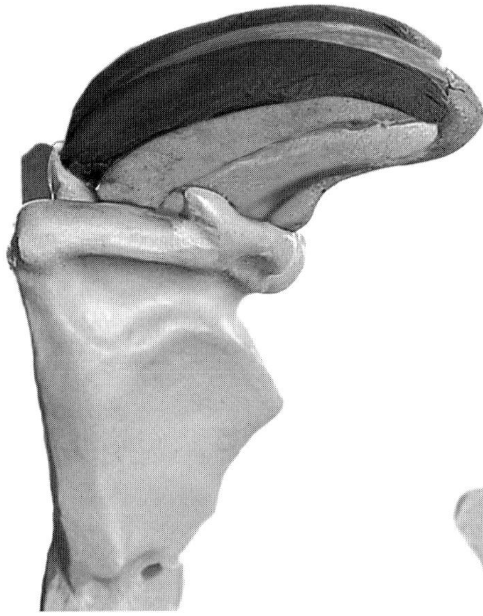

Lateral

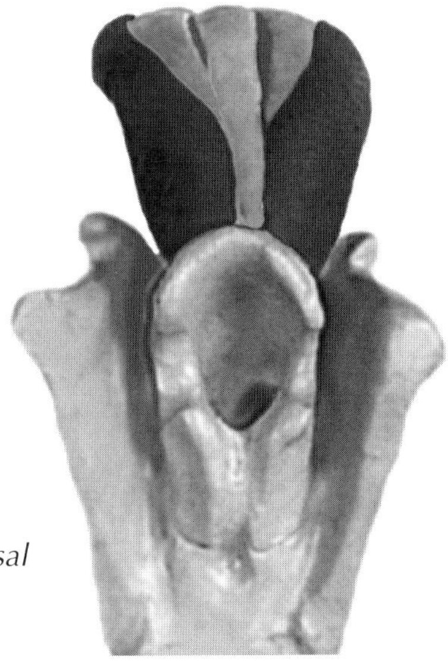

Dorsal

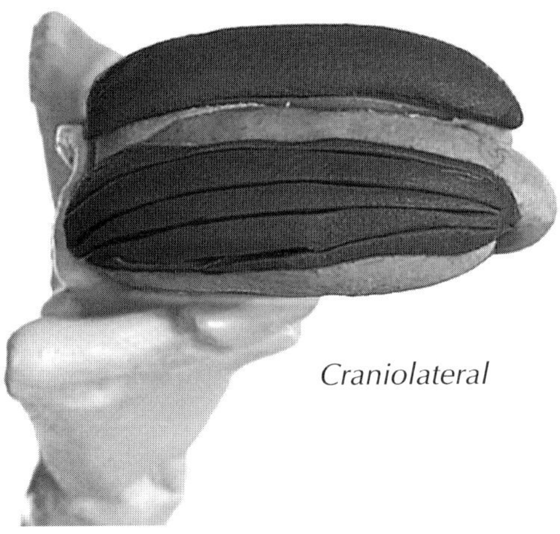

Craniolateral

Longitudinalis *linguae* *superioris*

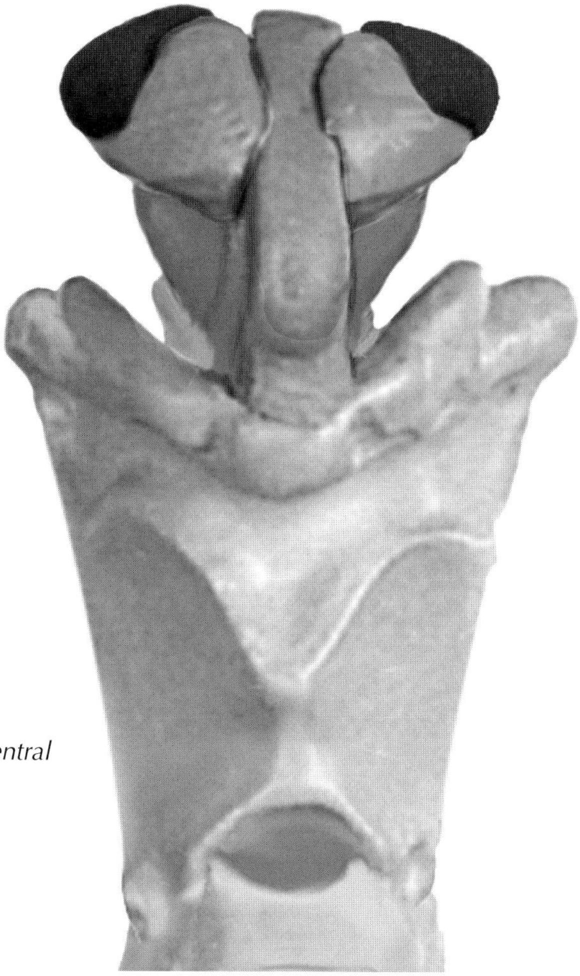

Ventral

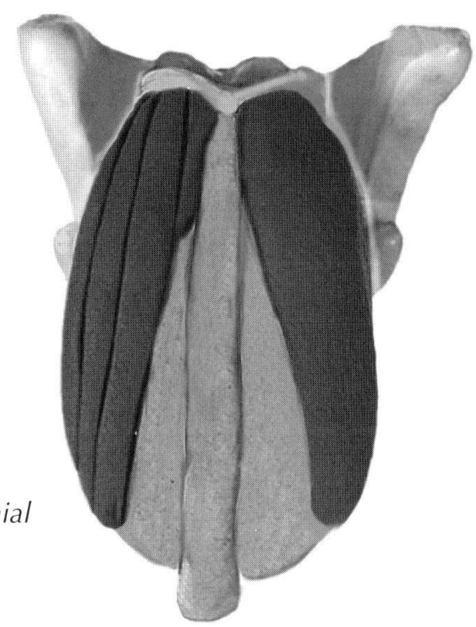

Cranial

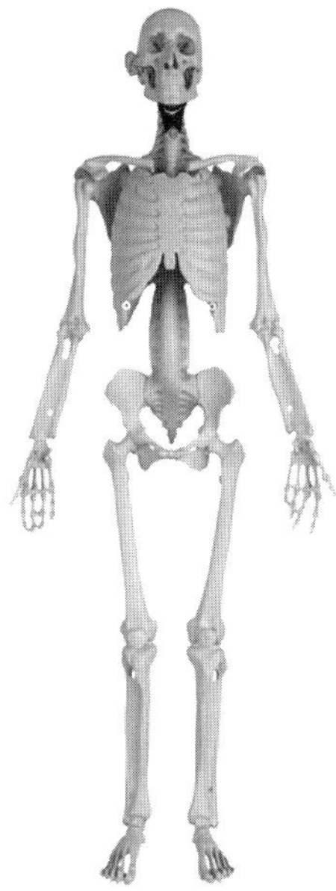

Actions

Shortens the tongue. Turns the tip of the tongue upward. Each bundle alone curls tongue dorsolaterally.

Attachments

Posterior

Fibrous capsule of the tongue near the epiglottis.

Anterior

Fibers from both sides converge at the tip of the tongue.

*lon•ji•**tood**•in•al•is*

From the Latin, "longitudinis"+"alis," meaning related to length.

ling•*gwee*

Latin, meaning belonging to the tongue.

*sue•peer•ee•**or**•us*

From the Latin, "superior," meaning above or higher.

Genioglossus

Craniolateral

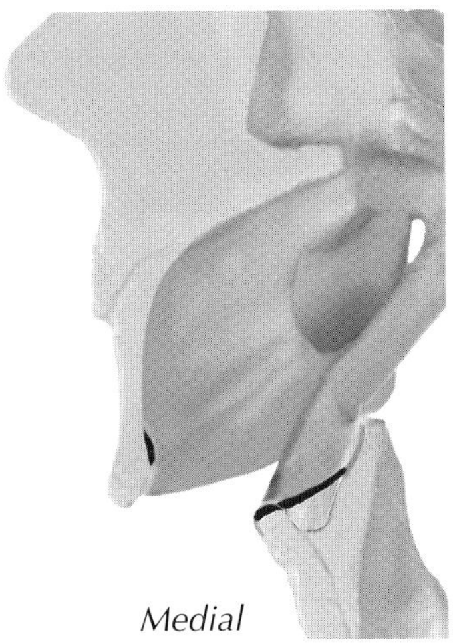

Medial

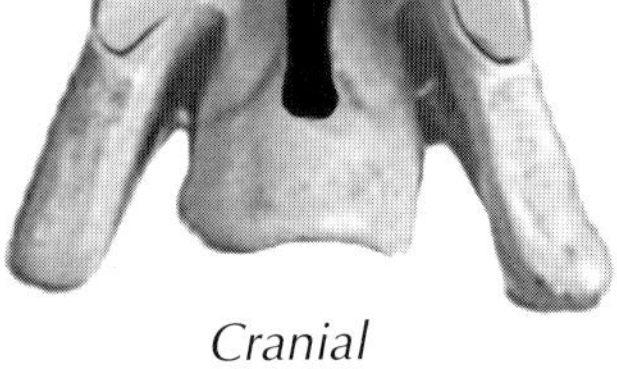

Cranial

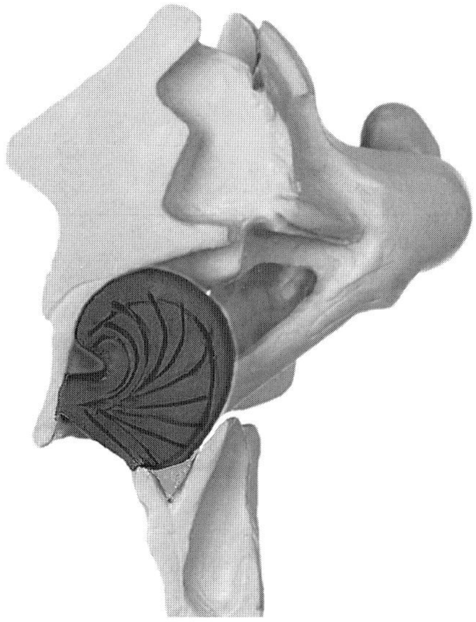

Medial

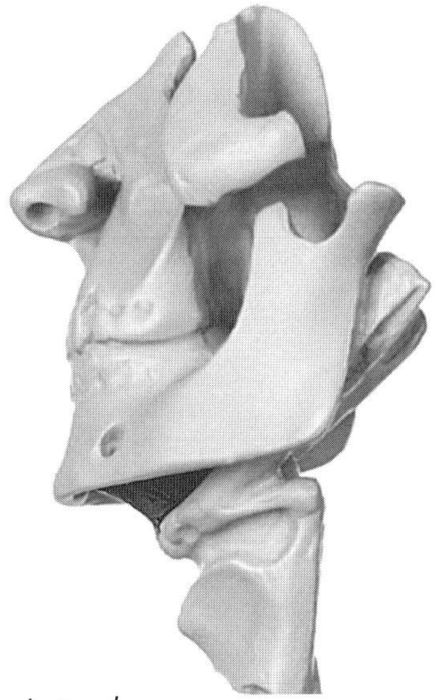

Lateral

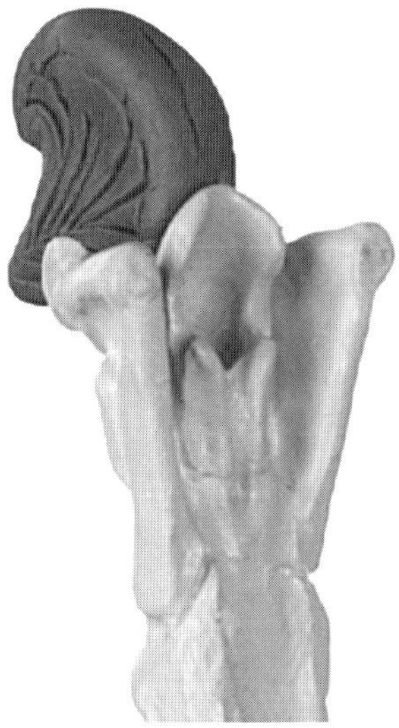

Dorsolateral

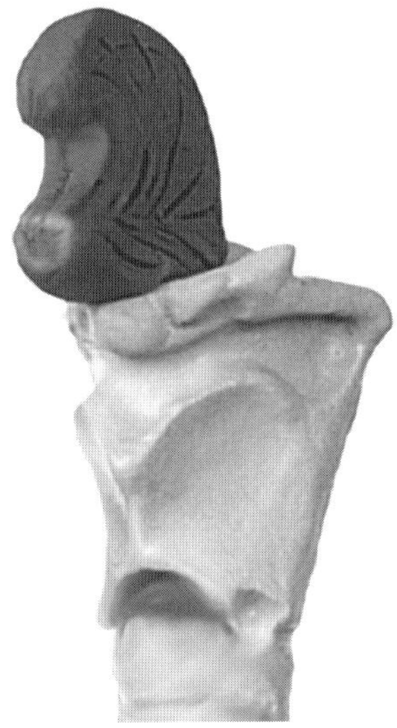

Ventrolateral

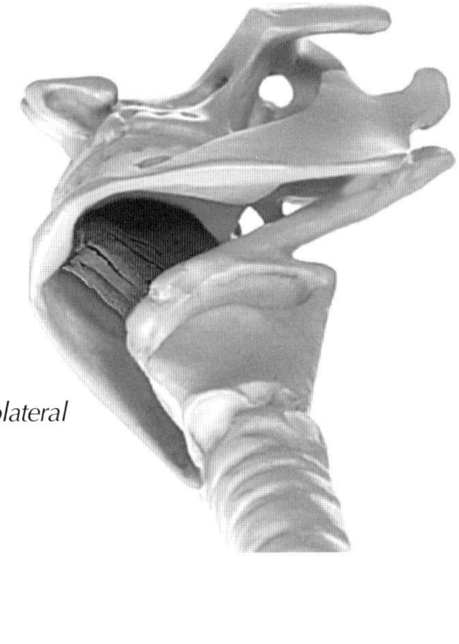

Caudoventrolateral

Genioglossus

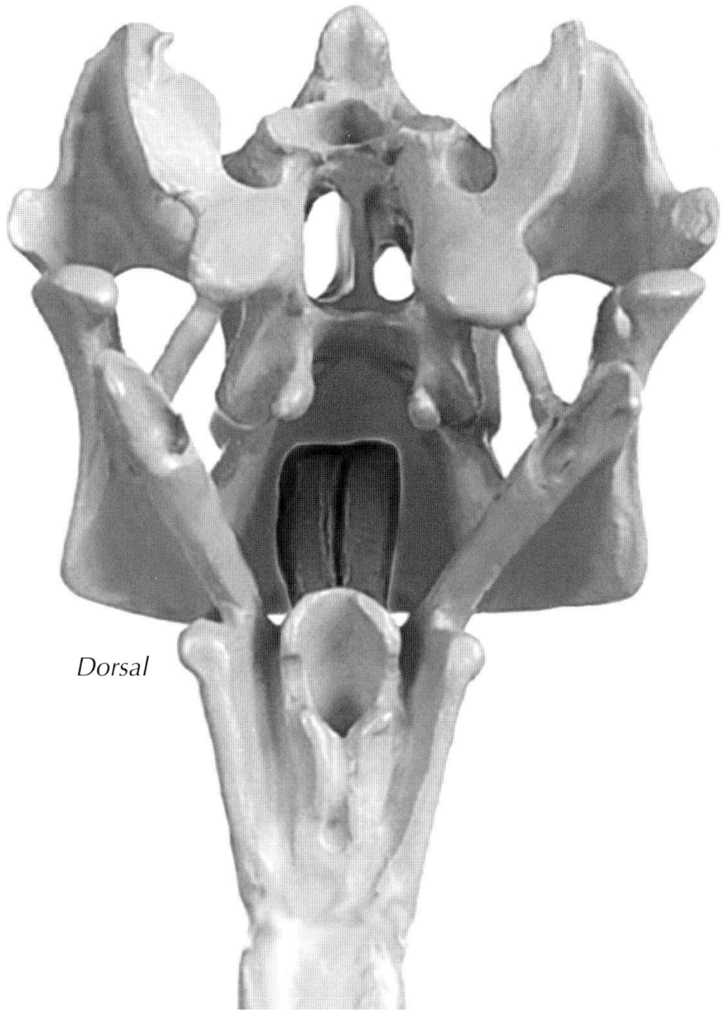

Dorsal

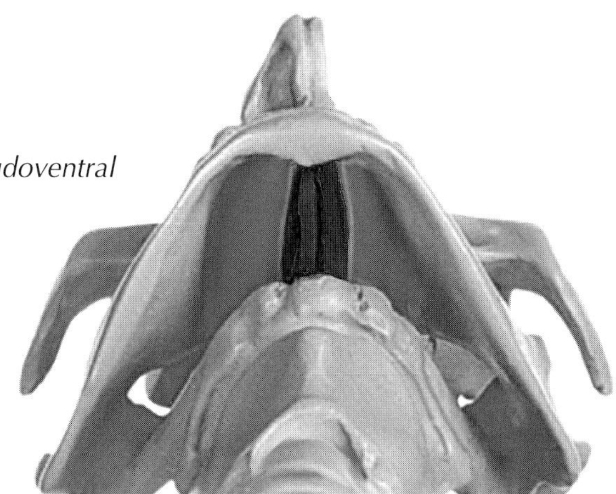

Caudoventral

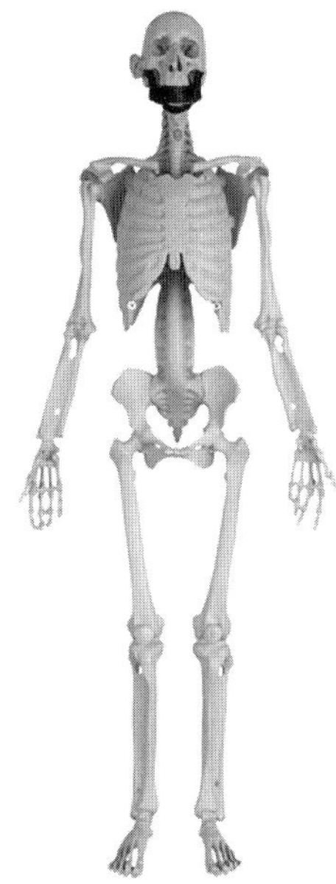

Actions

Protrudes tongue.

Attachments

Caudal

Superior tubercle on the inner surface of the symphysis of the mandible.

Cranial

Body of hyoid bone, undersurface of the tongue adjacent to the septum and tip of the tongue.

*jean•ee•oh•**glos**•us*

From the Greek, "geneion," meaning chin, and from the Greek, "glossa," meaning tongue.

Chondroglossus

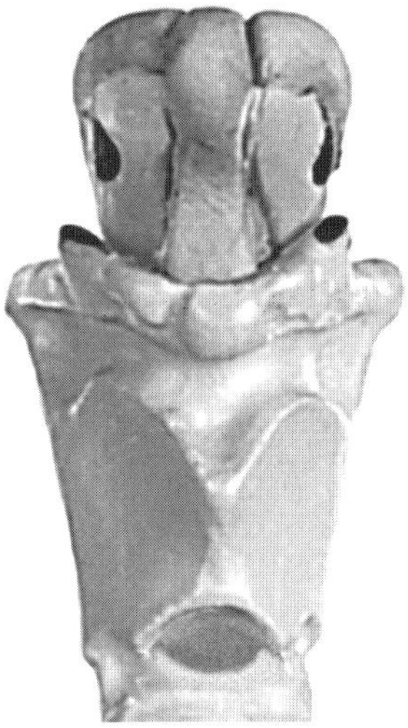

Ventral

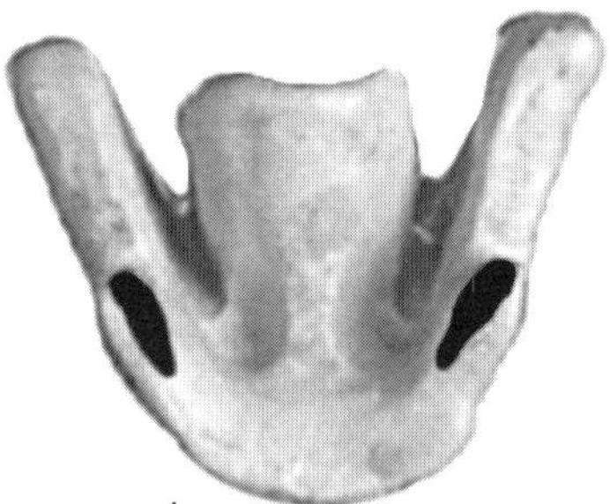

Cranial

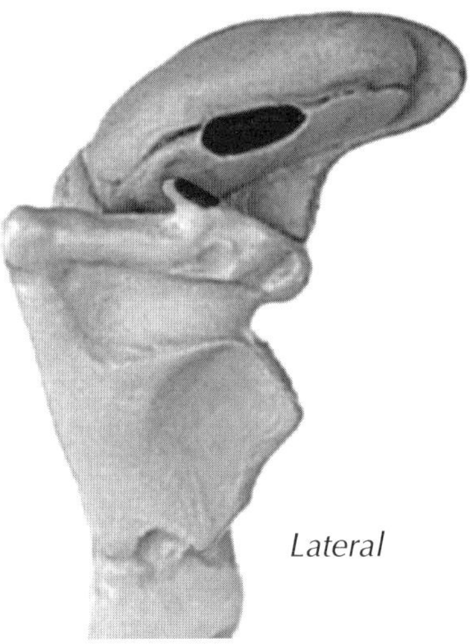

Lateral

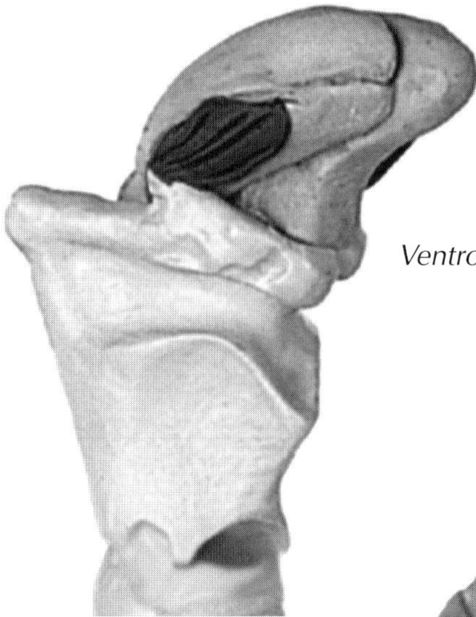

Ventrolateral

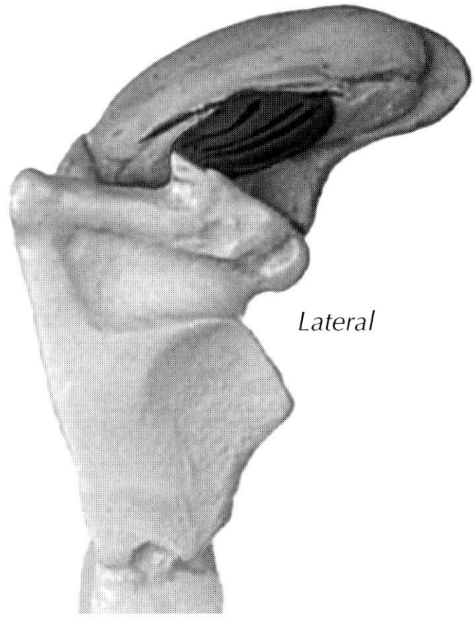

Lateral

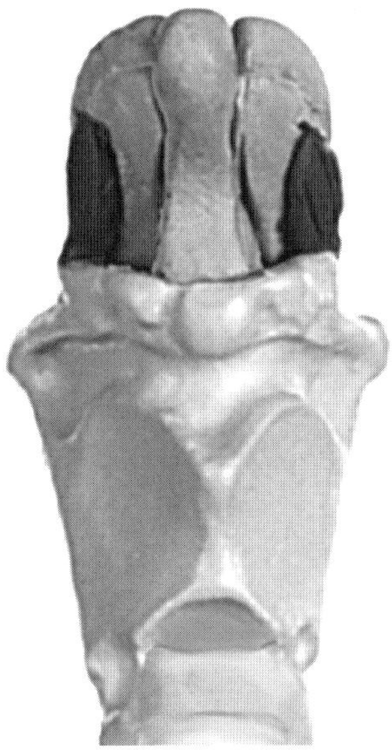

Ventral

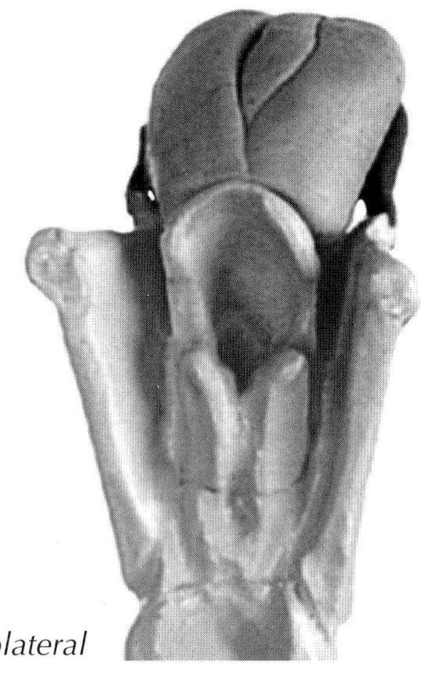

Dorsolateral

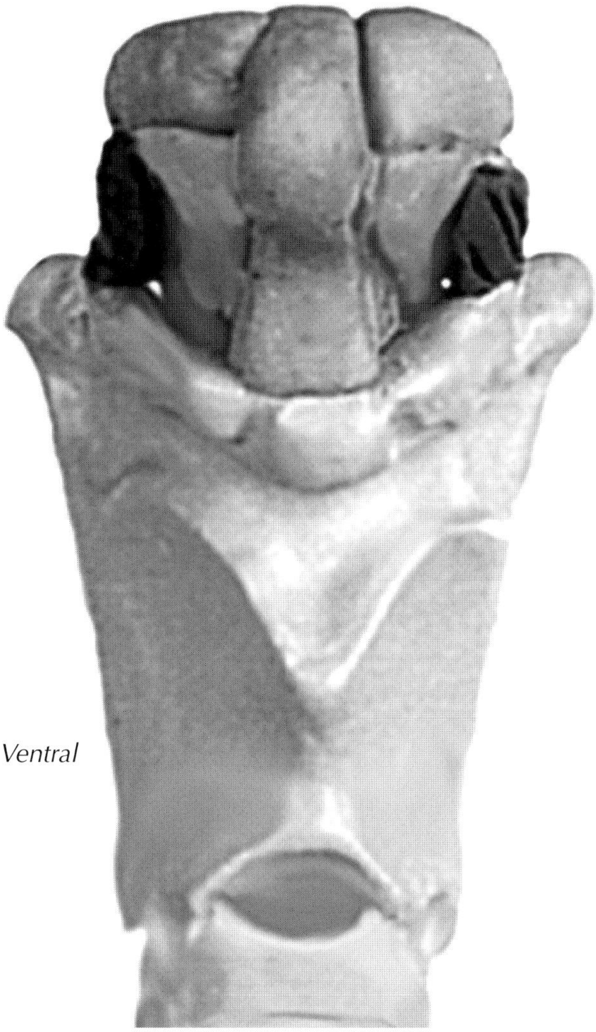

Ventral

Caudal

Chondroglossus

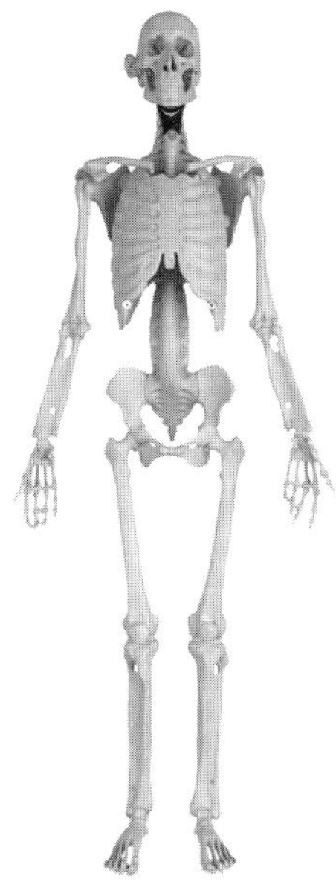

Actions

Depresses tongue.

Attachments

Caudal

Lesser cornu of the hyoid bone.

Cranial

Blends with intrinsic muscles of the tongue between the genioglossus and hyoglossus.

Note

This muscle is often considered to be a deep layer of the hyoglossus.

*kon•dro•***glos***•us*

From the Greek, "chondros," meaning cartilage, and from the Greek, "glossa," meaning tongue.

Hyoglossus

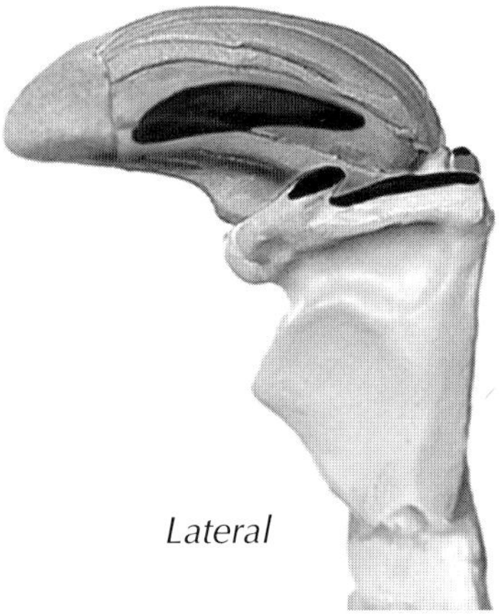

Lateral

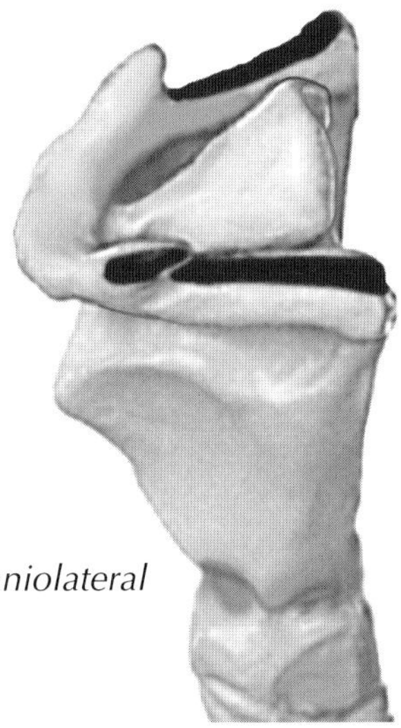

Craniolateral

Cranial

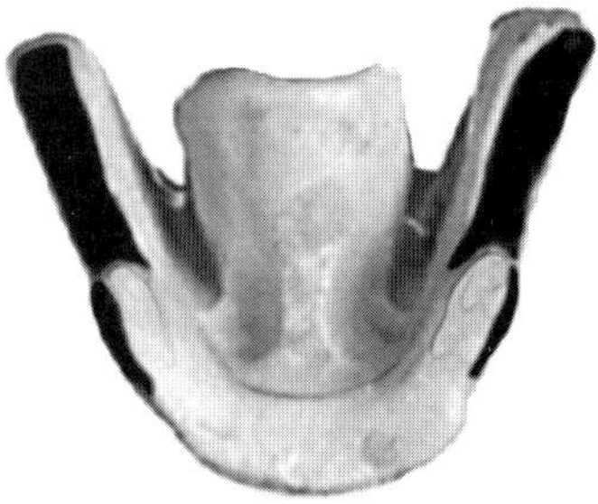

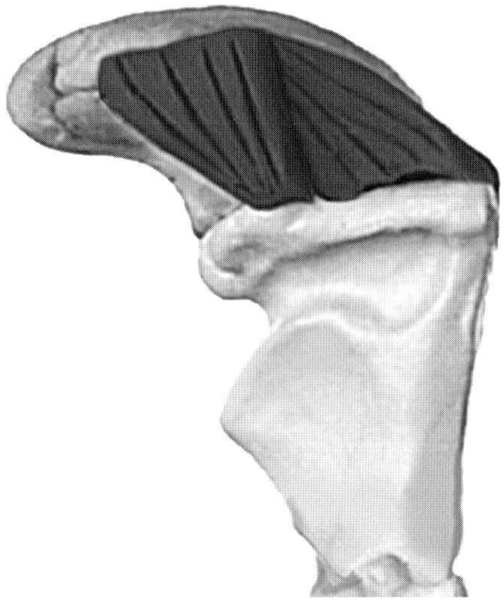

Lateral

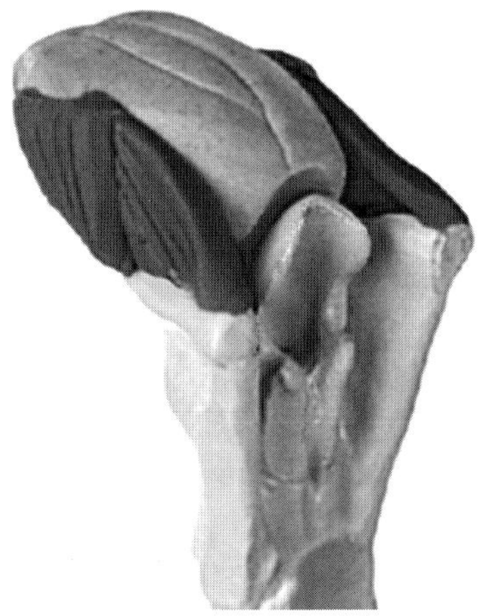

Dorsolateral

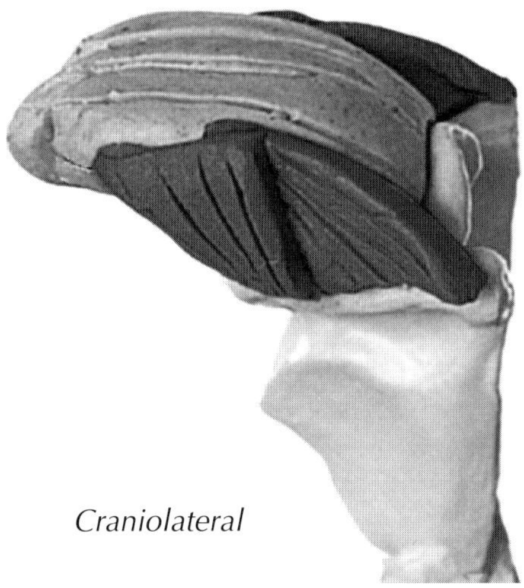

Craniolateral

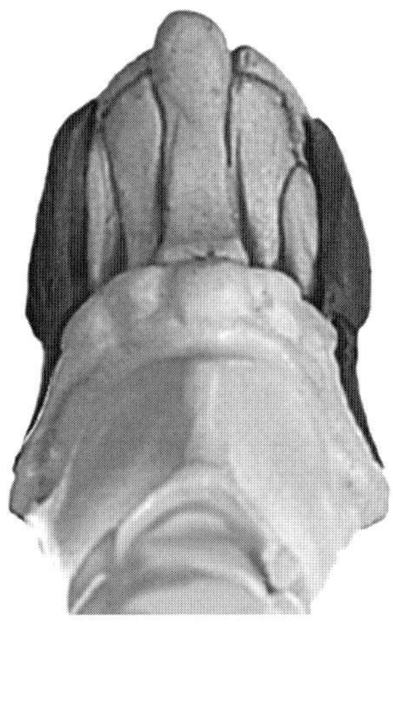

Caudal

Hyoglossus

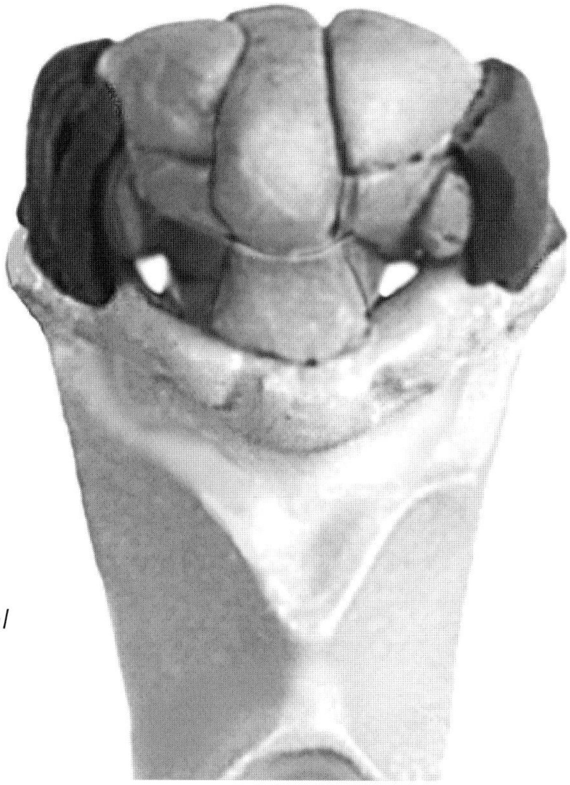

Ventral

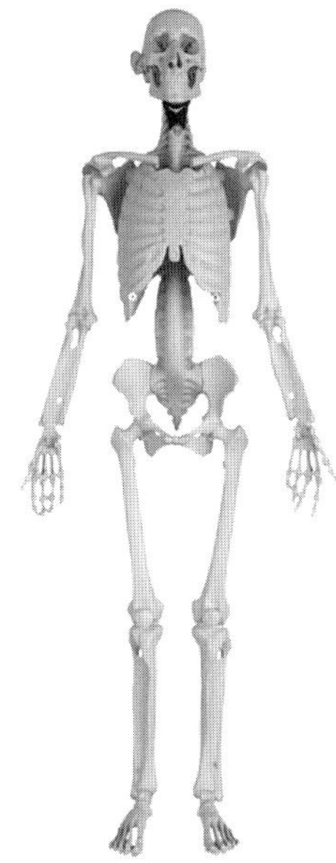

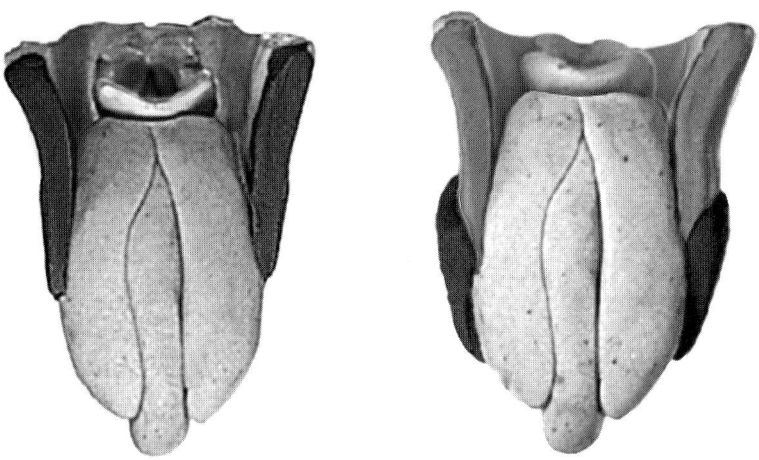

Cranial

Actions

Depresses tongue.

Attachments

Caudal

Body and greater cornu of the hyoid bone.

Cranial

Undersurface of the tongue interlacing with the longitudinalis inferior and styloglossus.

*hy•oh•***glos***•us*

From the Greek, "hyo"+"eidos," meaning U-shaped, and from the Greek, "glossa," meaning tongue.

Styloglossus

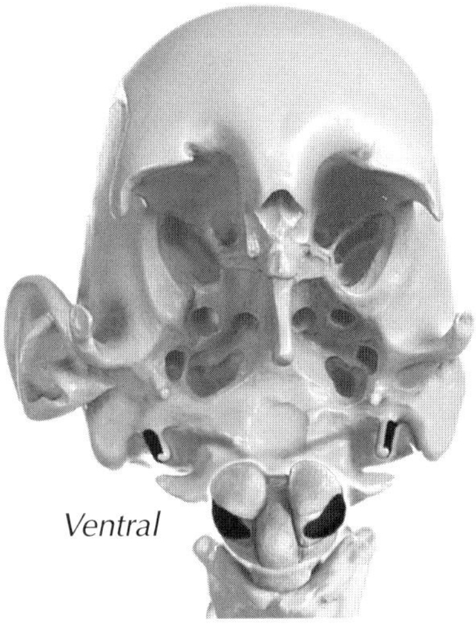

Ventral

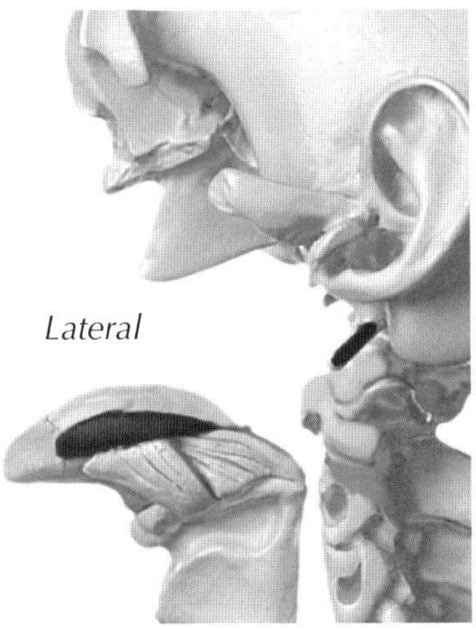

Lateral

Caudoventral

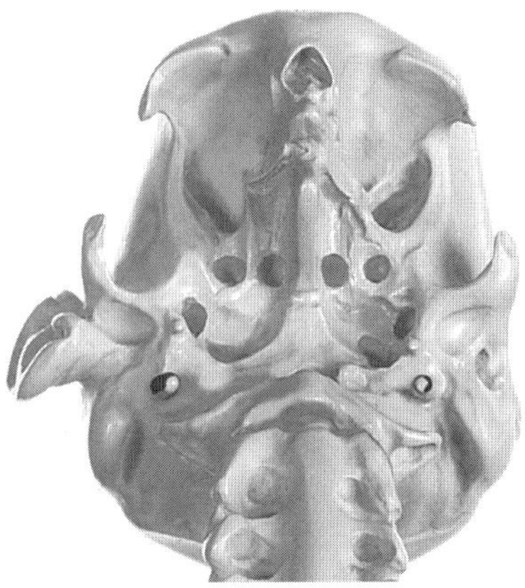

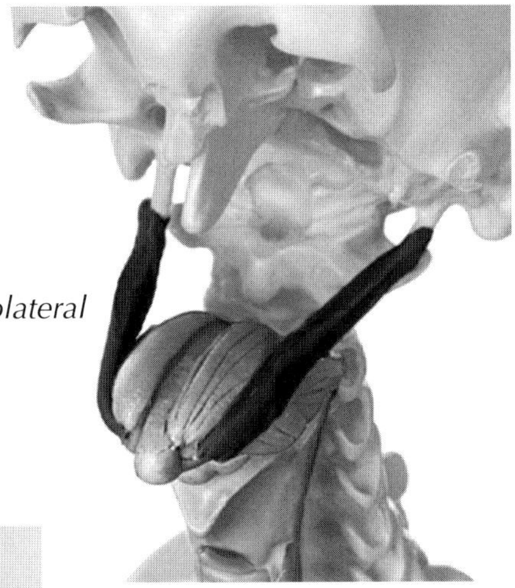

Ventrolateral

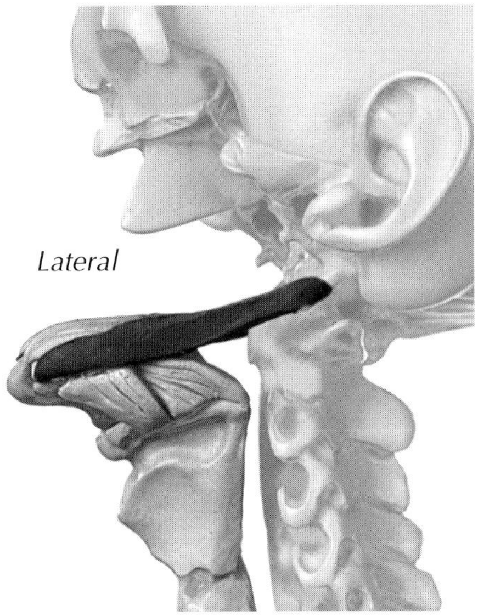

Lateral

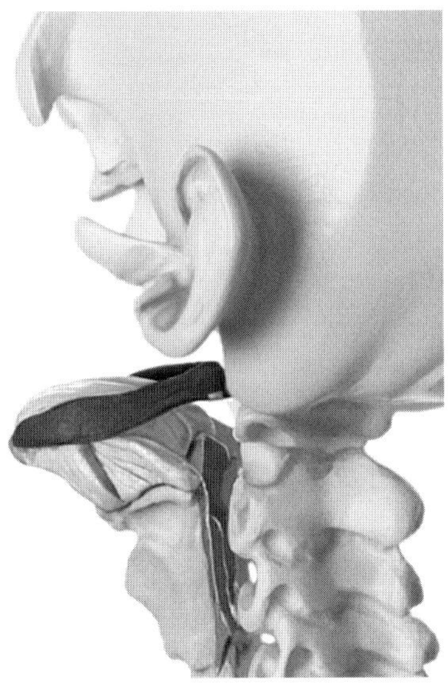

Dorsolateral

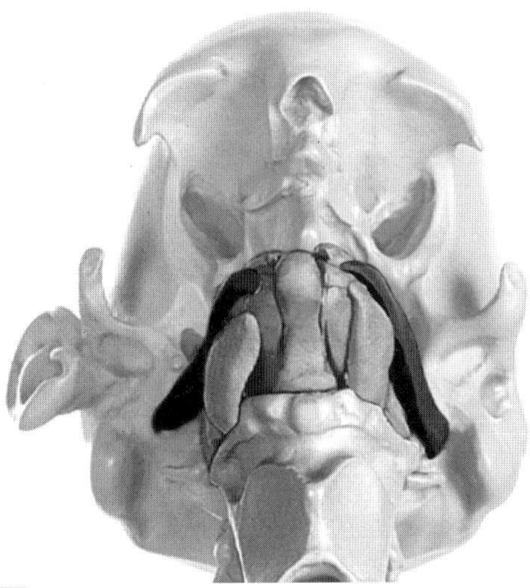

Caudoventral

Styloglossus

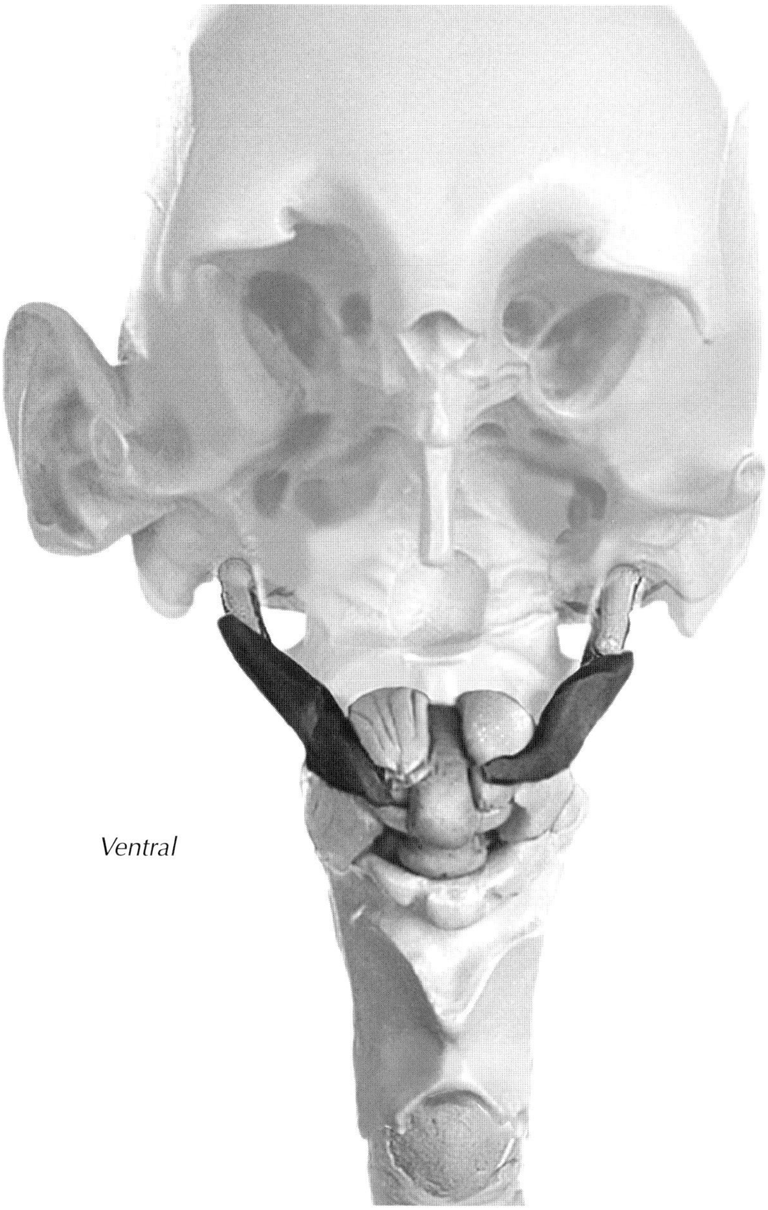

Ventral

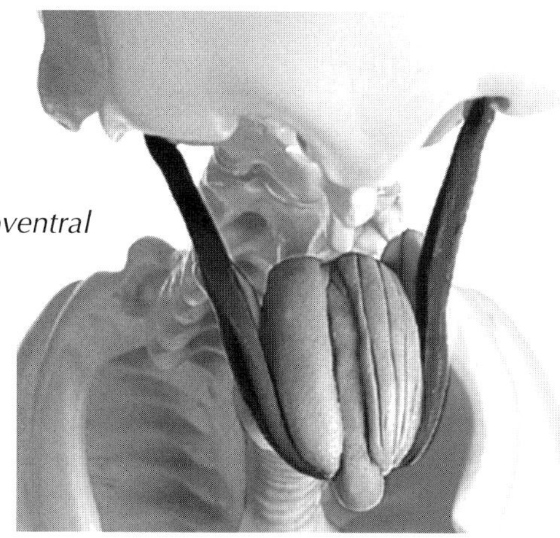

Cranioventral

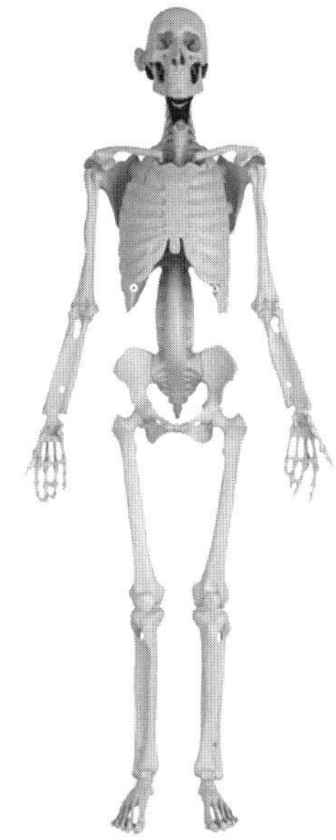

Actions

Elevates and retracts tongue.

Attachments

Dorsocranial

Anterior styloid process and upper part of the stylomandibular ligament.

Caudoventral

Side of tongue.

*sty•low•***glos***•us*

From the Latin, "stylus," meaning pen, and from the Greek, "glossa," meaning tongue.

Geniohyoideus

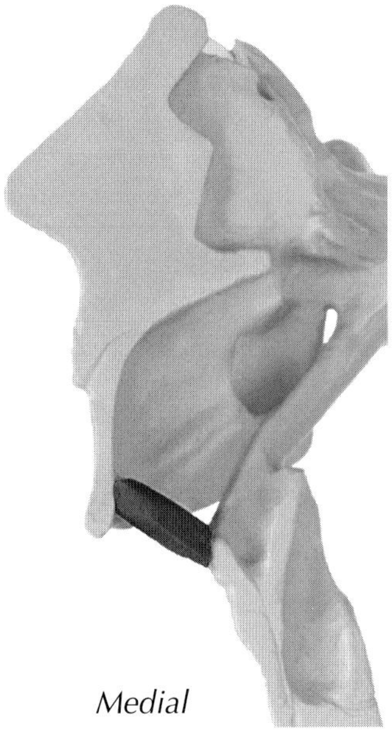

Medial

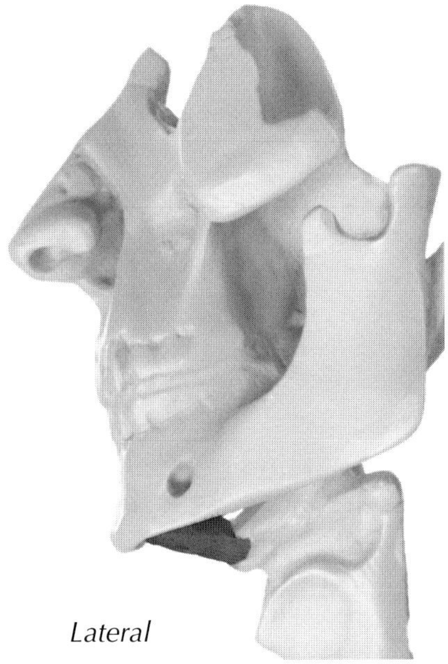

Lateral

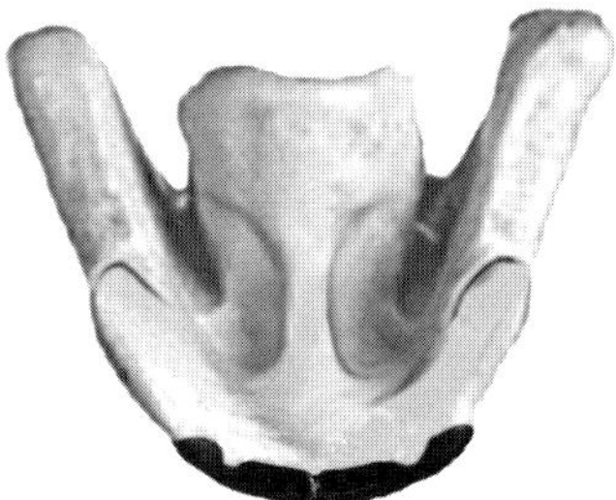

Cranial

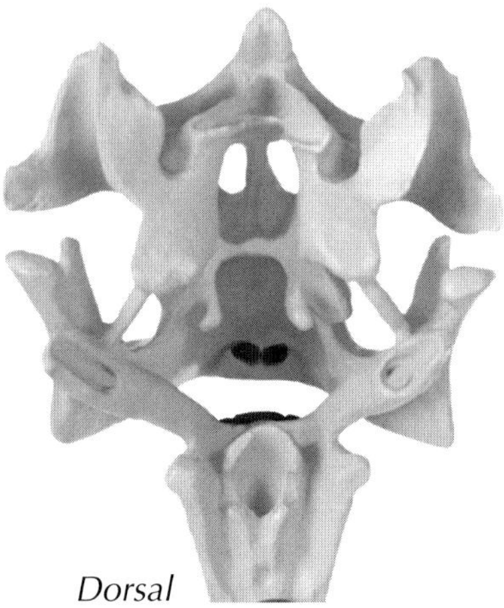

Dorsal

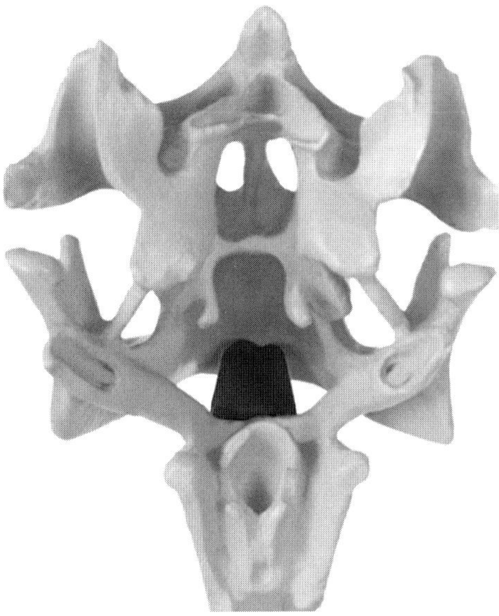

Dorsal

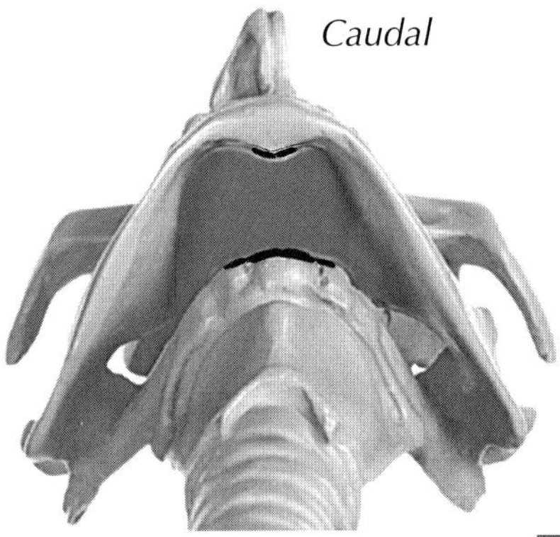

Caudal

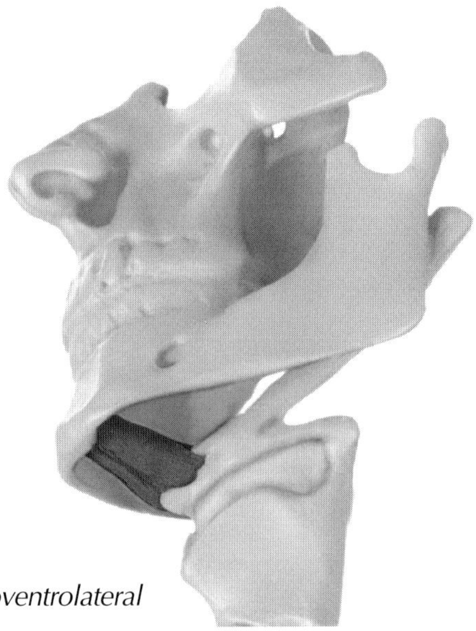

Caudoventrolateral

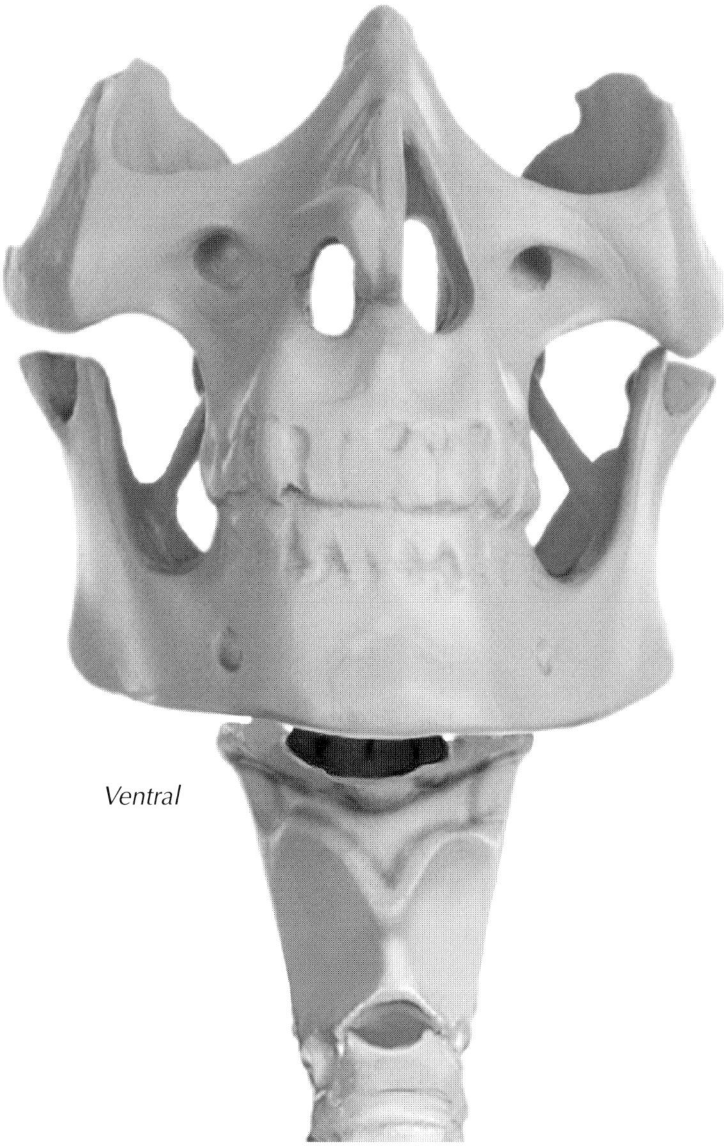

Ventral

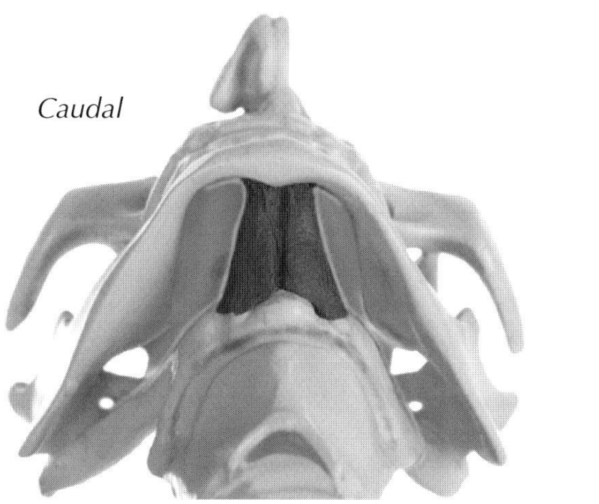

Caudal

Geniohyoideus

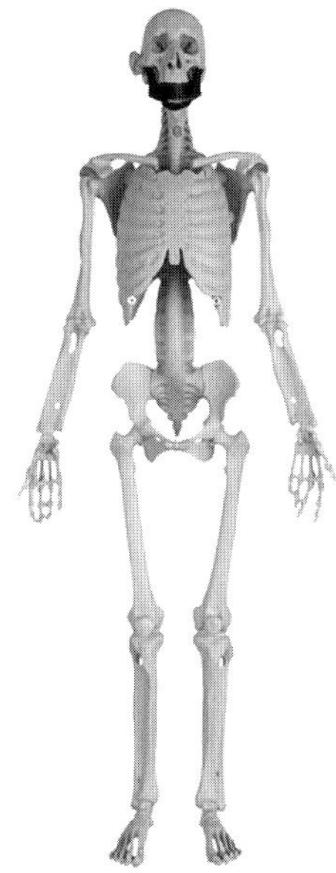

Actions

Elevates and pulls hyoid bone forward.

Attachments

Ventral

Inferior tubercle on the inner surface of the symphysis of the mandible.

Dorsal

Anterior body of the hyoid bone.

*jean•ee•oh•hy•**oid**•ee•us*

From the Greek, "geneion," meaning chin, and "hyo"+"eidos," meaning U-shaped.

Mylohyoideus

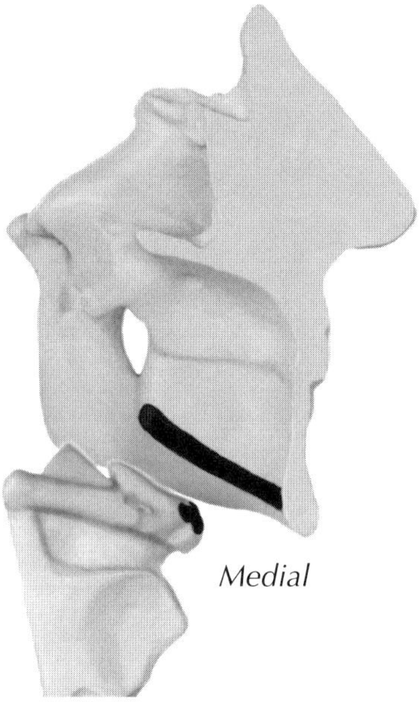

Medial

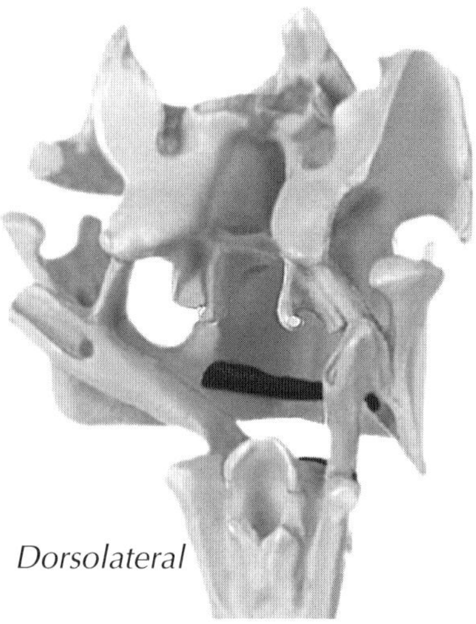

Dorsolateral

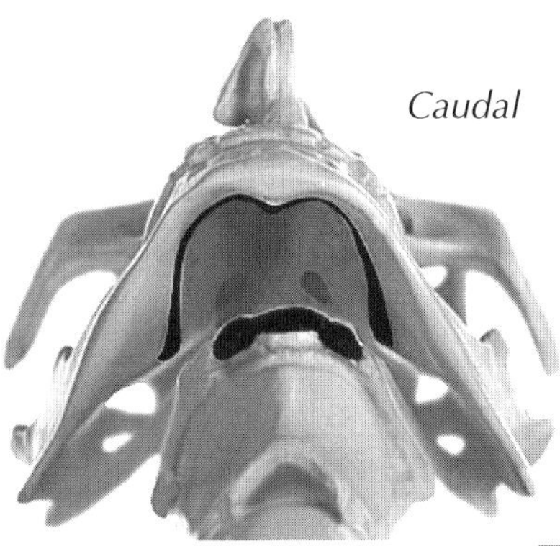

Caudal

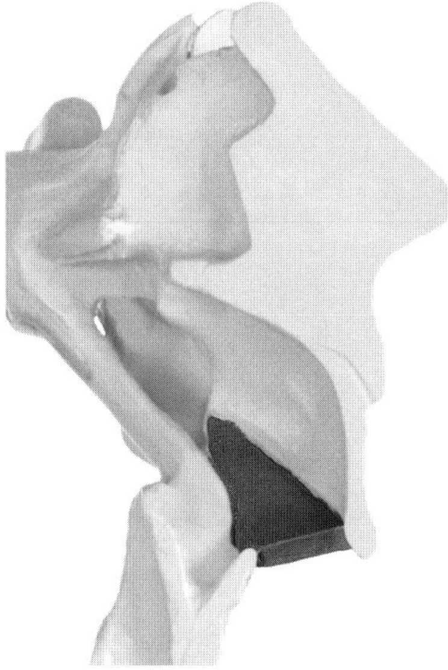

Medial

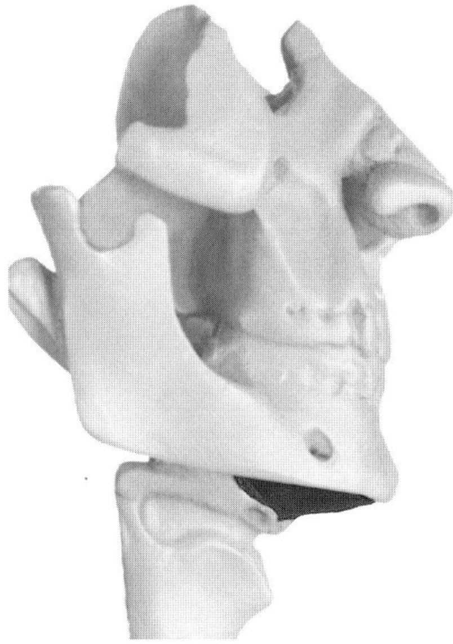

Lateral

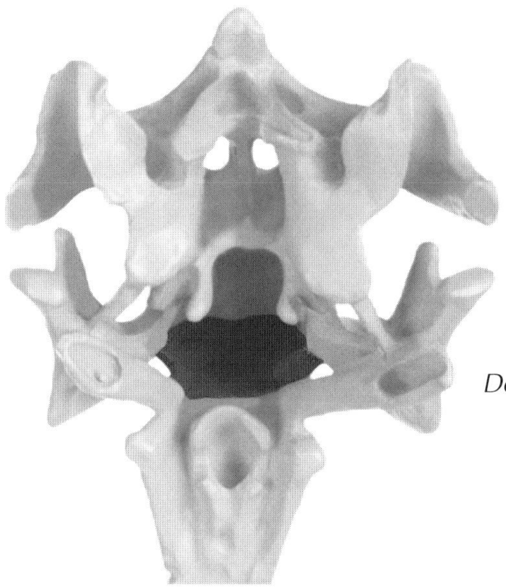

Dorsal

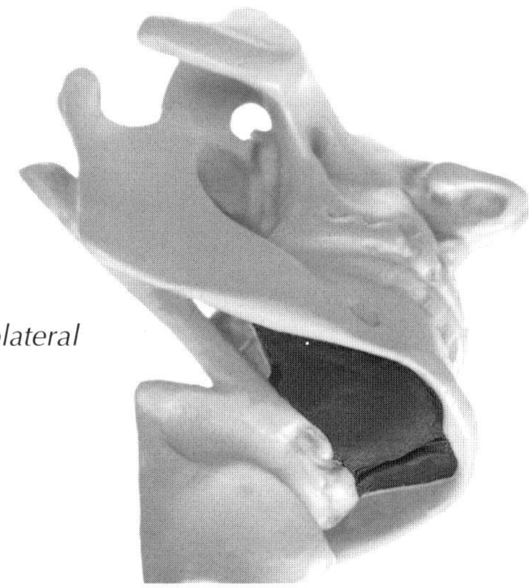

Caudoventrolateral

Mylohyoideus

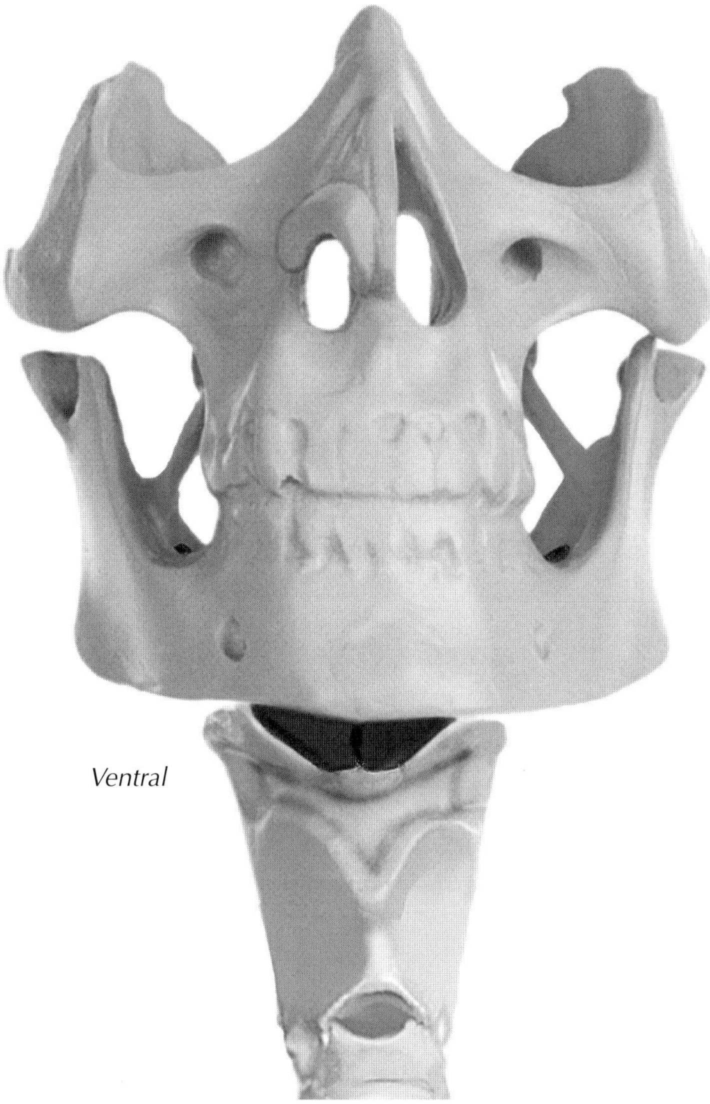

Ventral

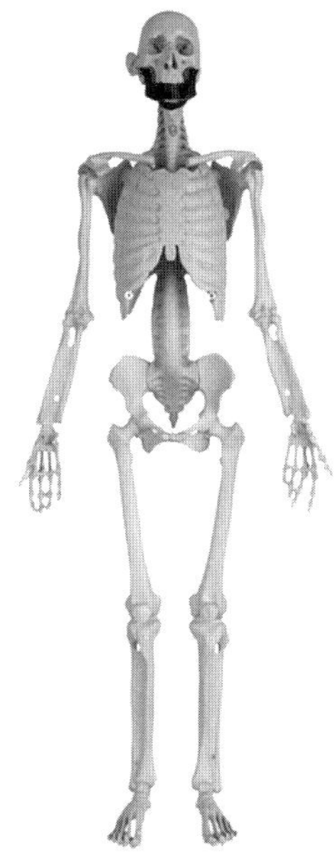

Actions

Elevates floor of mouth in swallowing. Elevates hyoid bone.

Attachments

Lateral

Mylohyoid line on the lingual or medial surface of the mandible.

Medial

Midline raphé that stretches from the symphysis of the mandible to the hyoid bone, and the body of the hyoid bone.

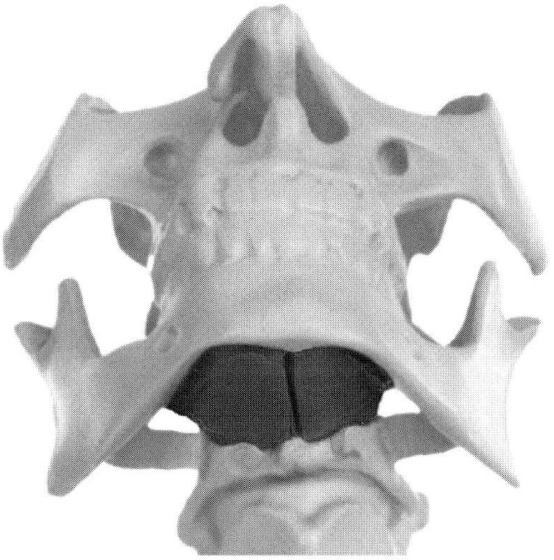

Caudoventral

*my•low•hy•**oid**•ee•us*

From the Greek, "myle," meaning mill or molar, and "hyo"+"eidos," meaning U-shaped.

Stylohoyoideus

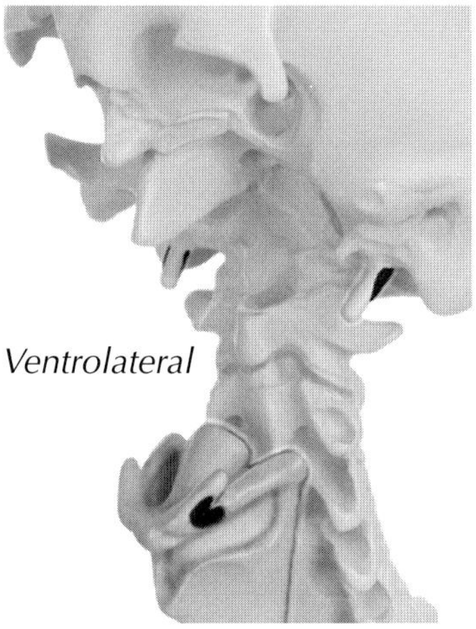

Ventrolateral

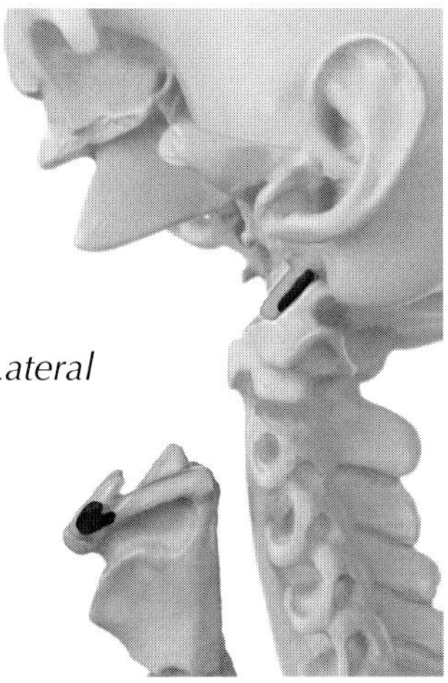

Lateral

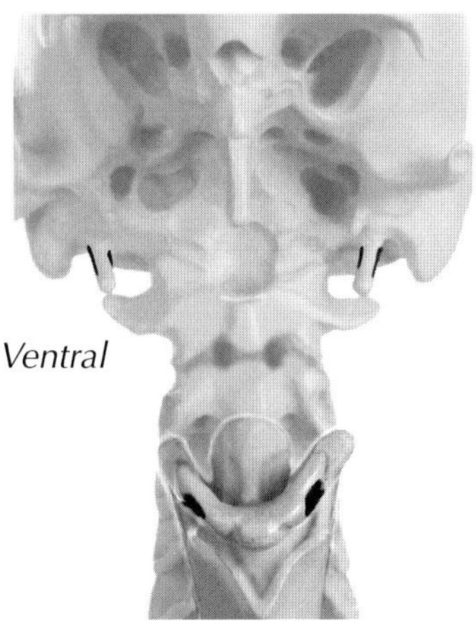

Ventral

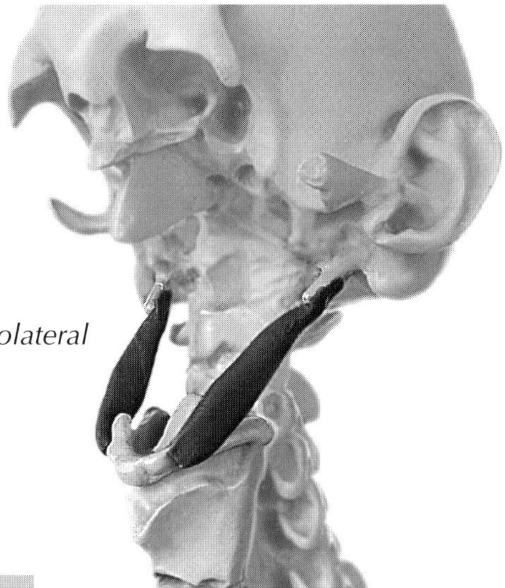

Ventrolateral

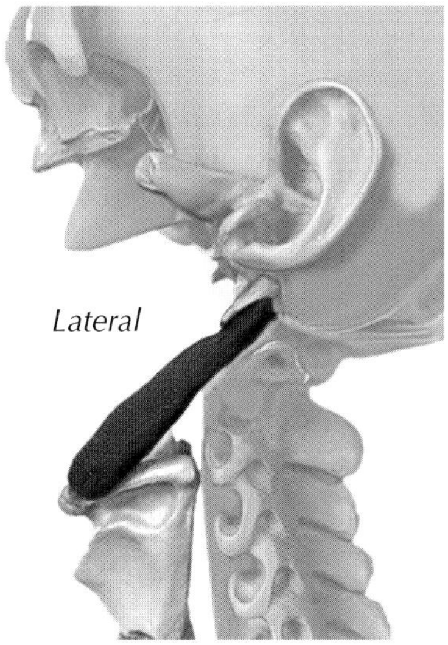

Lateral

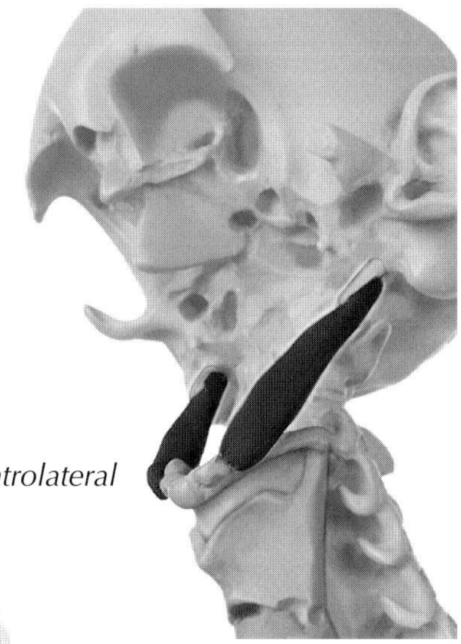

Caudoventrolateral

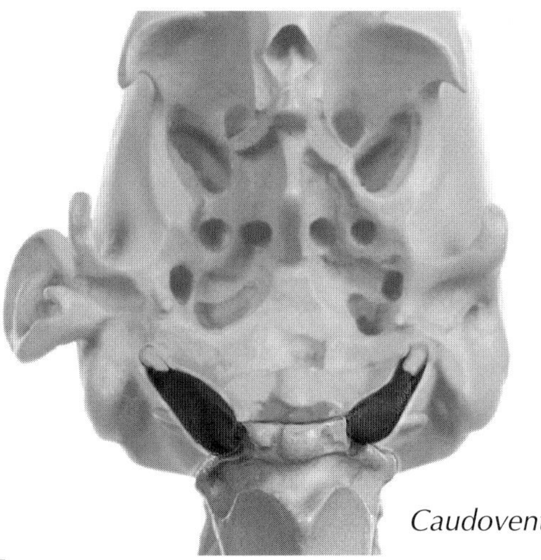

Caudoventral

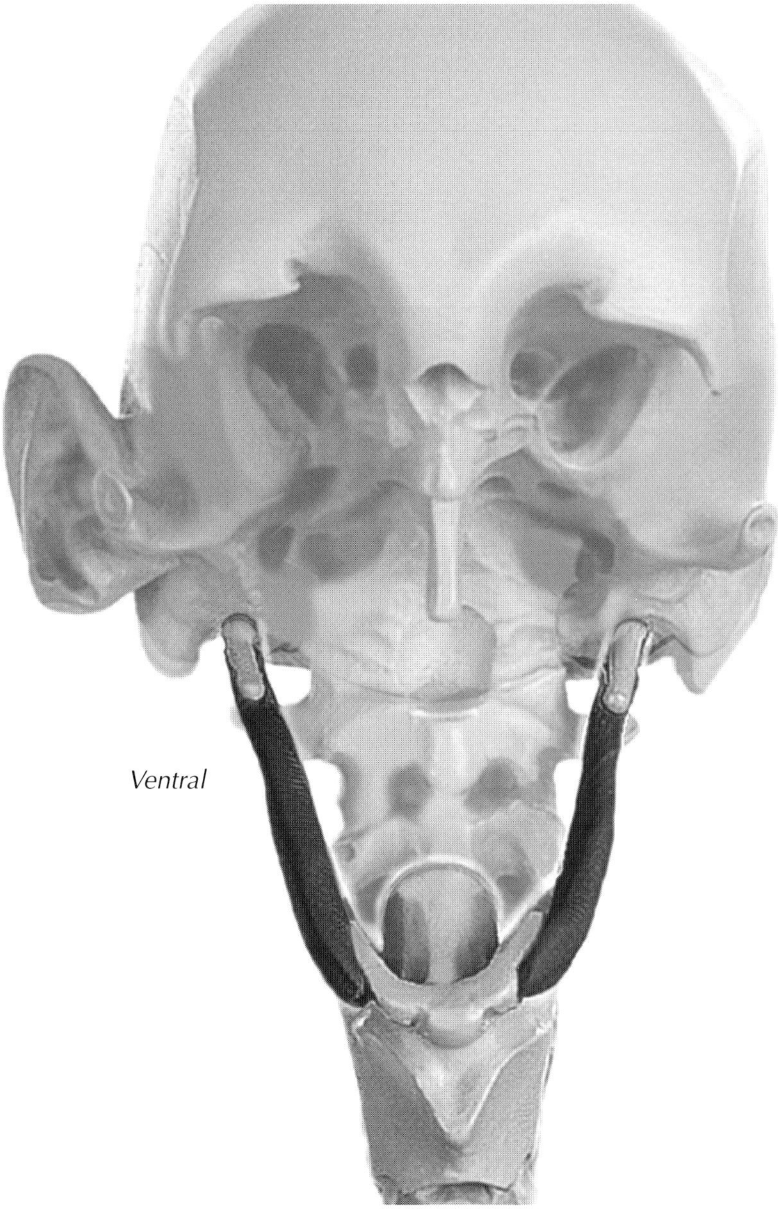

Ventral

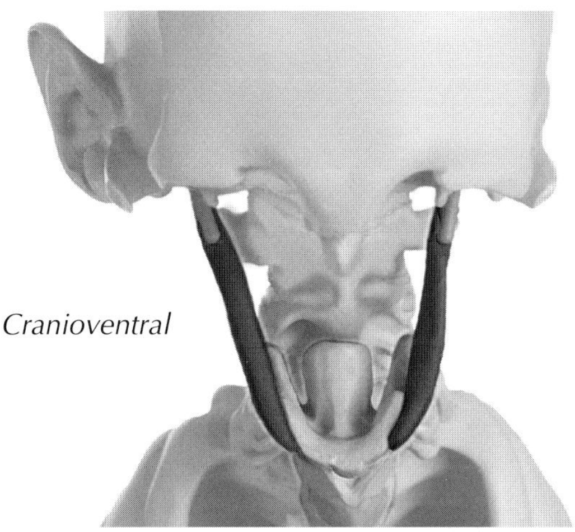

Cranioventral

Stylohoyoideus

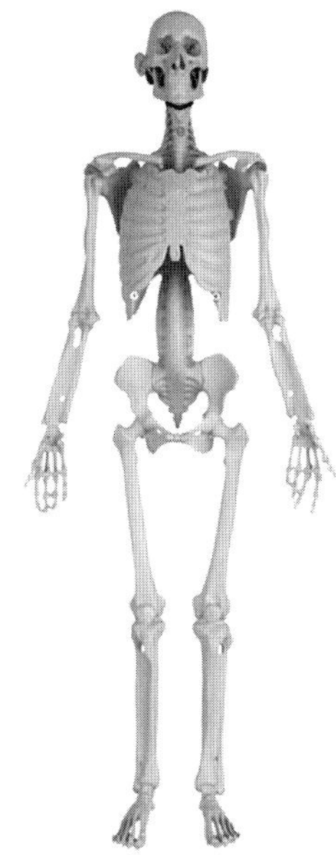

Actions

Elevates hyoid bone.

Attachments

Dorsocranial

Dorsal border of styloid process.

Ventrolateral

Body of hyoid bone at junction with greater cornu.

*sty•low•hy•**oid**•ee•us*

From the Latin, "stylus," meaning pen and from the Greek, "hyo"+"eidos," meaning U-shaped.

Pterygoideus lateralis caput superior

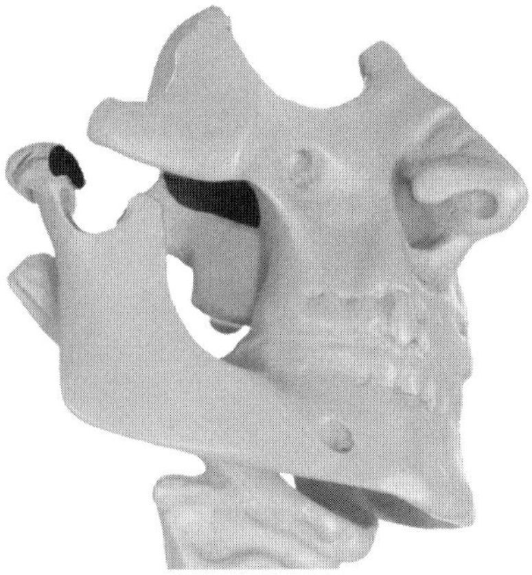

Ventrolateral

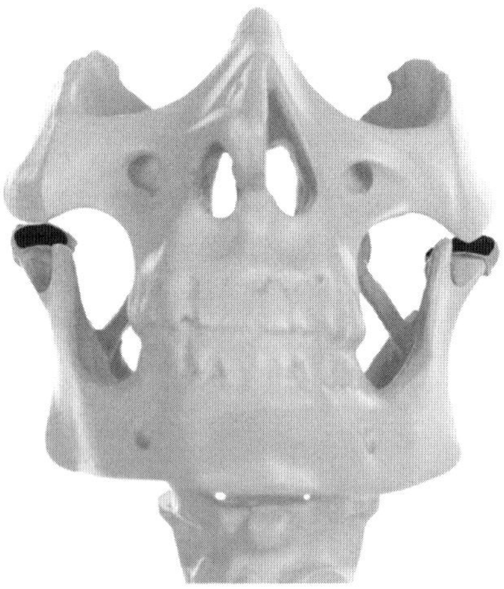

Ventral

Crainial

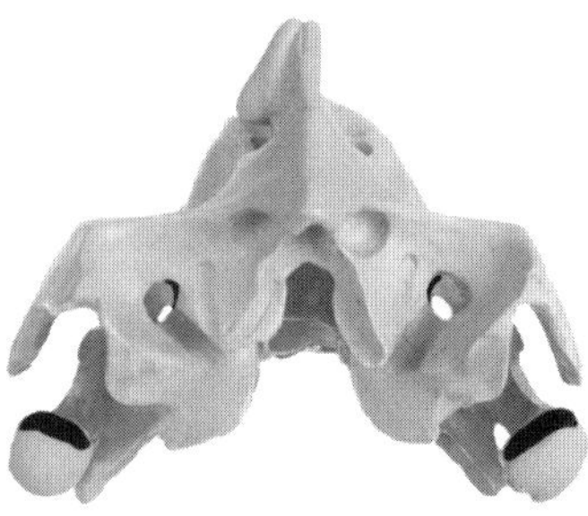

Caudoventrolateral

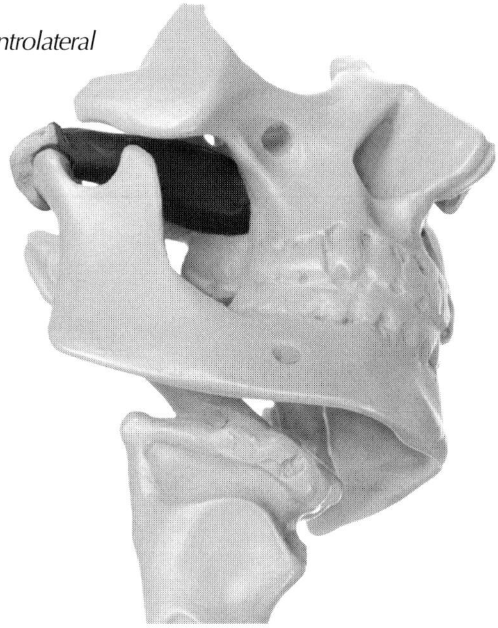

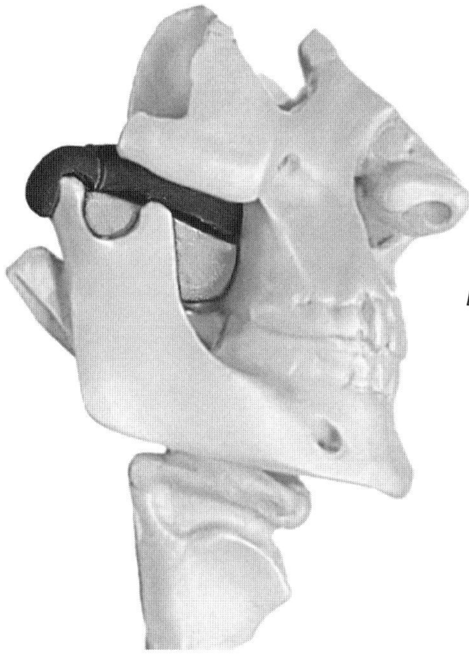

Lateral

Caudoventral

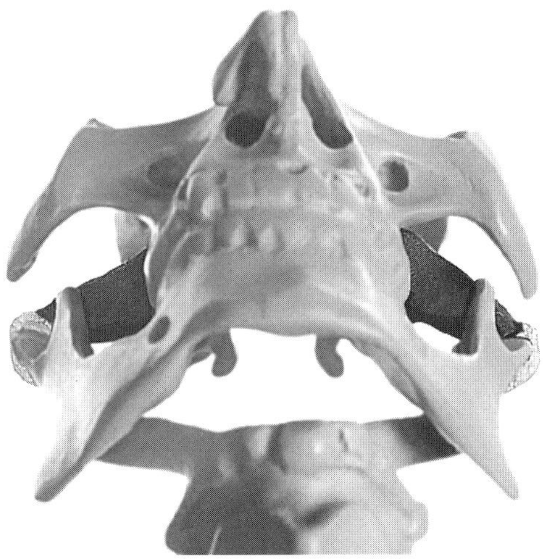

Pterygoideus lateralis caput superior

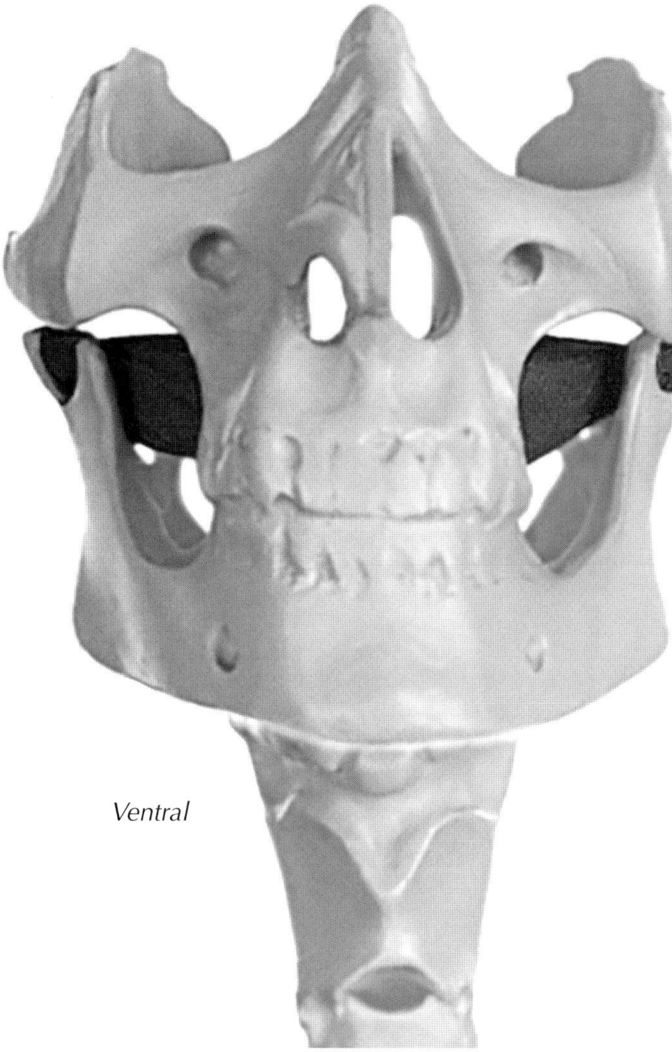

Ventral

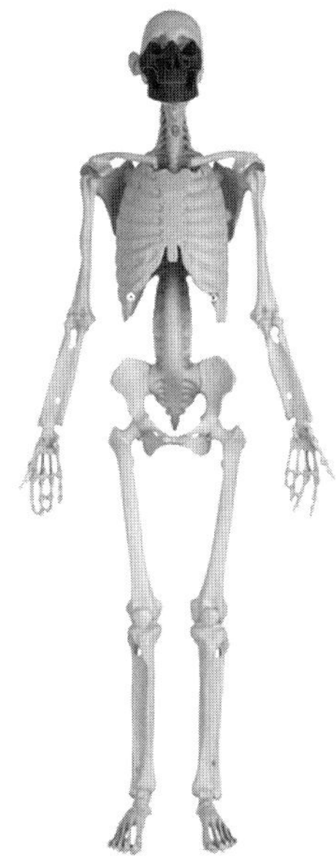

Actions

Pulls articular disc of temporomandibular joint forward in opening jaw.

Attachments

Ventral

Infratemporal surface and crest of the greater wing of the sphenoid bone.

Dorsal

Joint capsule and articular disc of the temporomandibular joint (TMJ).

Dorsal

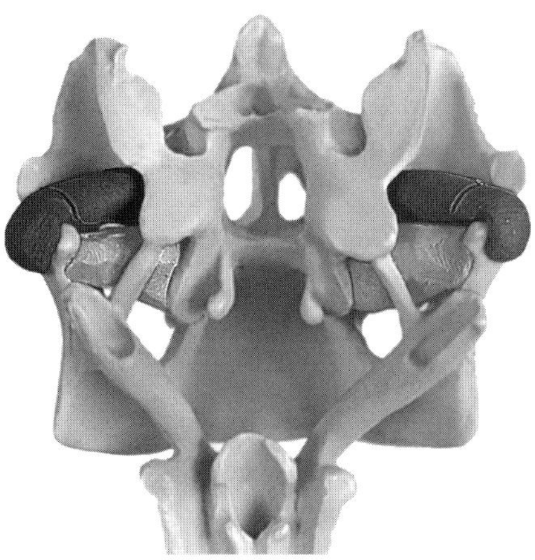

Ventral

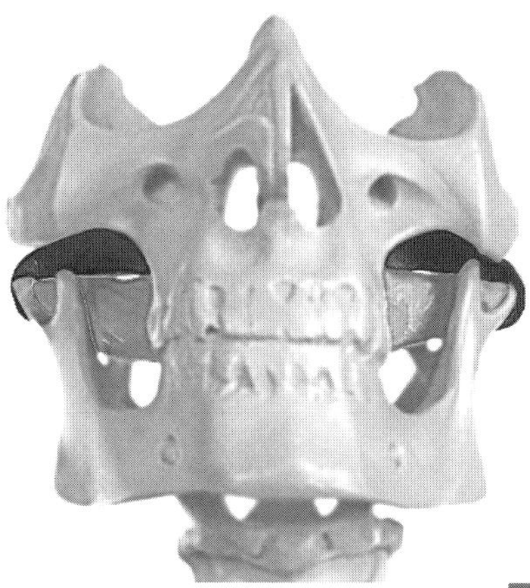

tare•uh•**goid**•ee•us

From the Greek, "pteron"+"eidos", meaning wing shaped.

lat•er•**al**•us

From the Latin, "latus"+"alis", meaning related to the side.

ka•**put**

Latin, meaning head.

sue•**peer**•ee•or

Latin, meaning above or higher.

Pterygoideus
lateralis
caput inferior

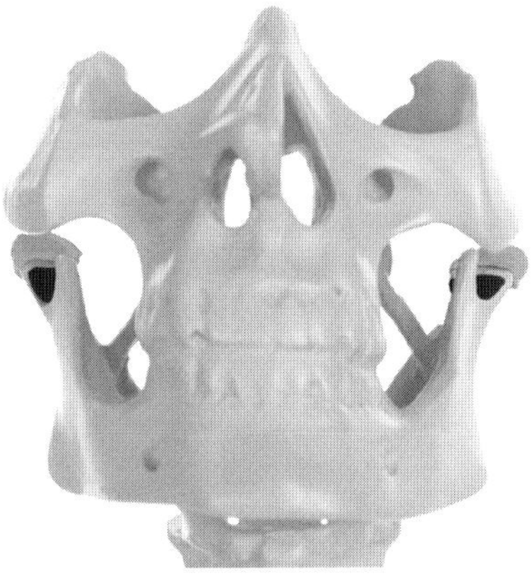

Ventral

Caudoventrolateral

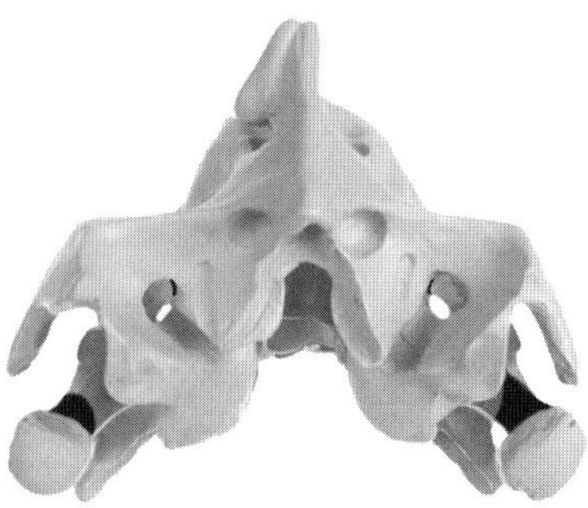

Cranial

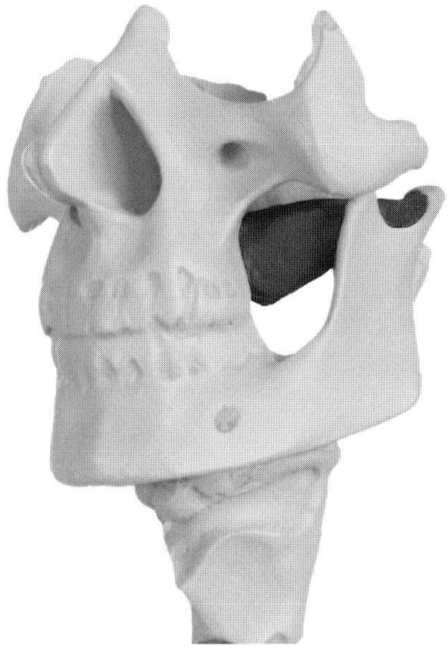

Ventrolateral

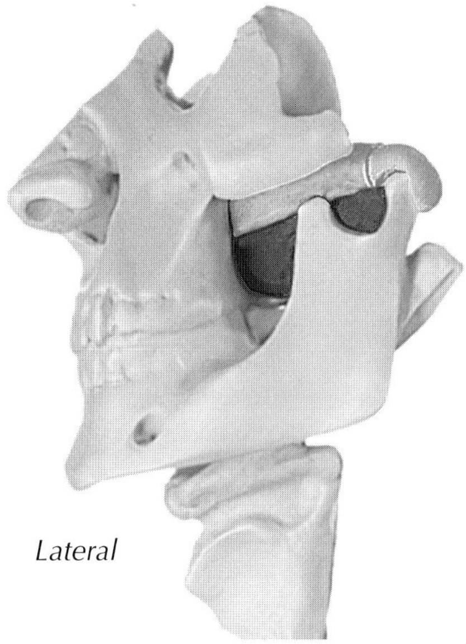

Lateral

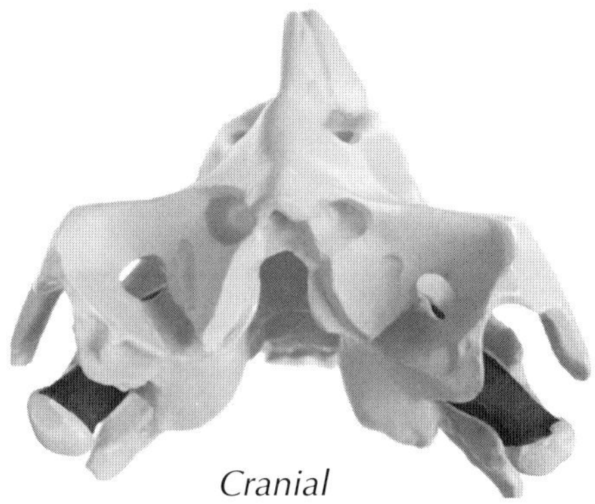

Cranial

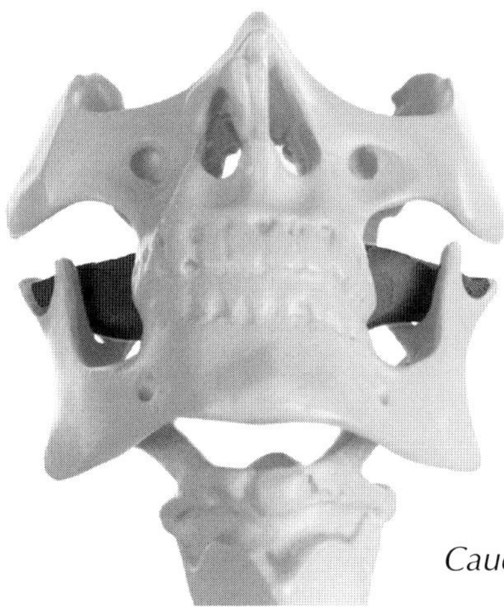

Caudoventral

Pterygoideus lateralis caput inferior

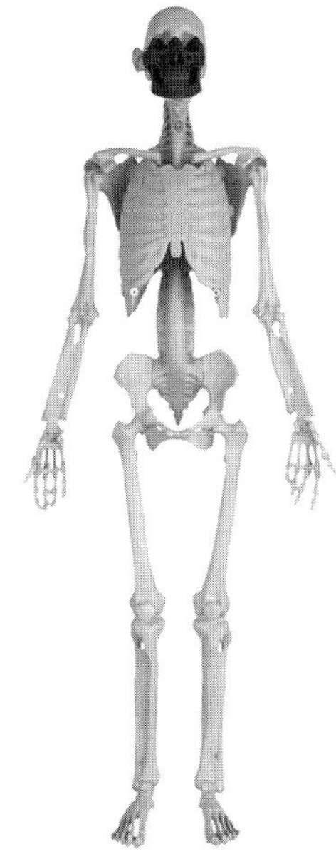

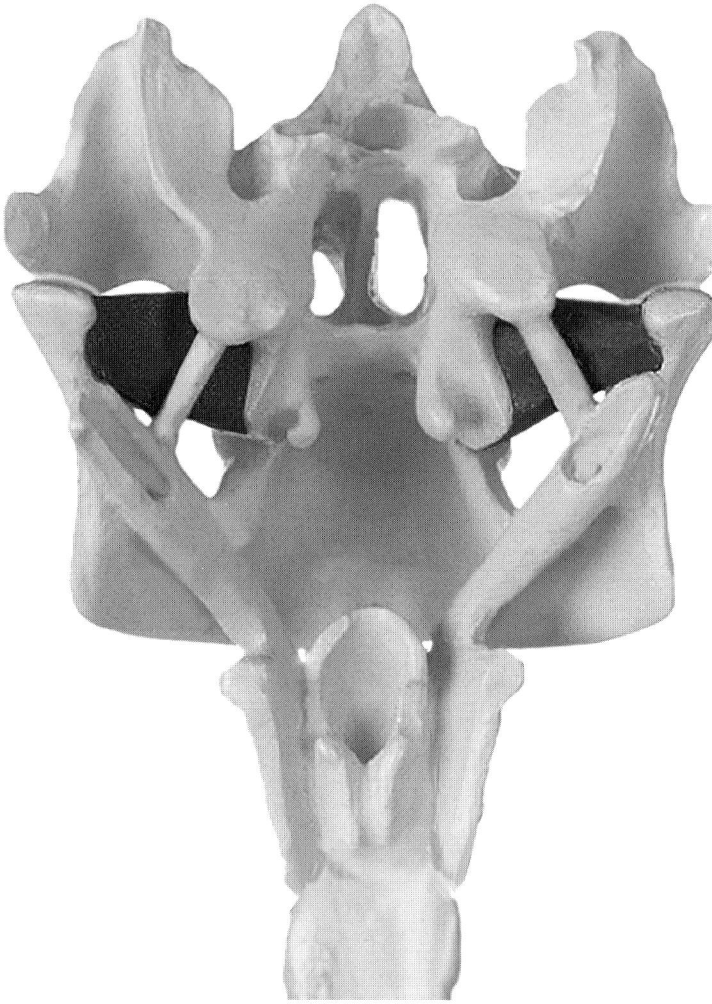

Dorsal

Ventral

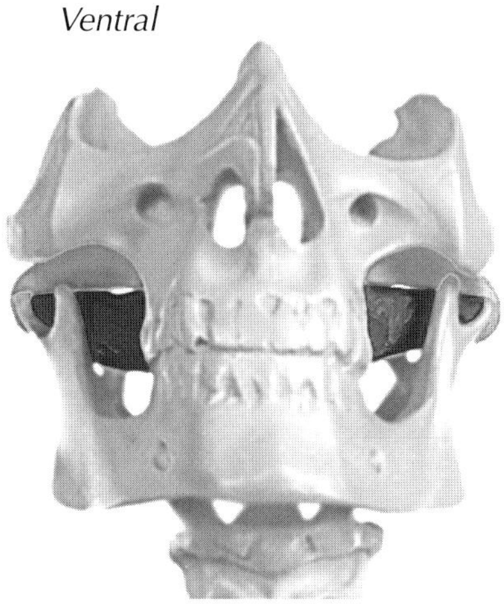

Caudodorsal

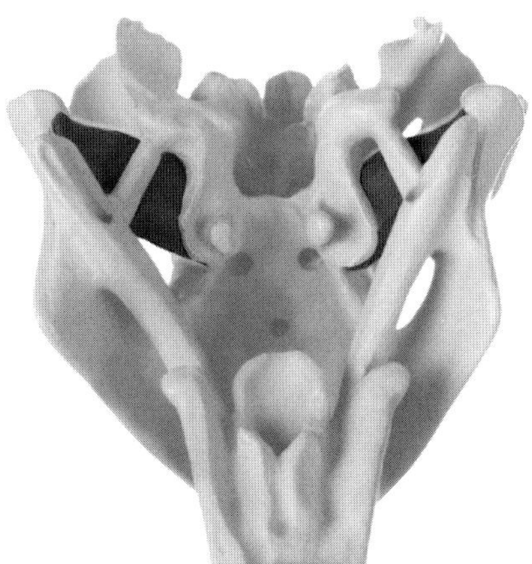

Actions

Protracts, assists in depression of the mandible in opening the jaw. Opposite side rotator of the mandible.

Attachments

Ventral

Lateral surface of the lateral pterygoid plate.

Dorsal

Anterior surface of the neck of the mandible.

*tare•uh•**goid**•ee•us*

From the Greek, "pteron"+"eidos", meaning wing shaped.

*lat•er•**al**•us*

From the Latin, "latus"+"alis," meaning related to the side.

*ka•**put***

Latin, meaning head.

*in•**feer**•ee•or*

Latin, meaning lower or beneath.

Pterygoideus medialis

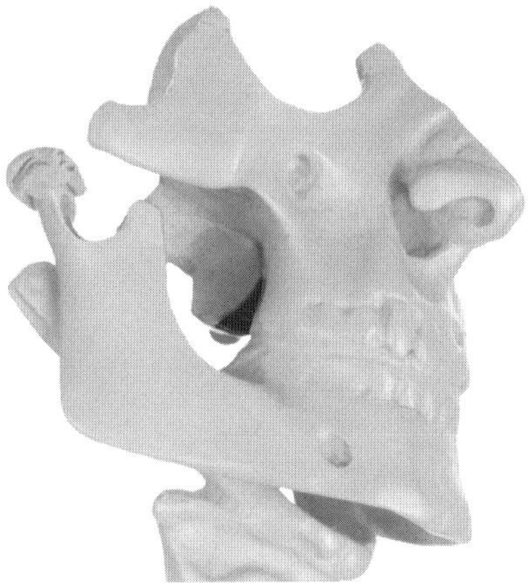

Lateral

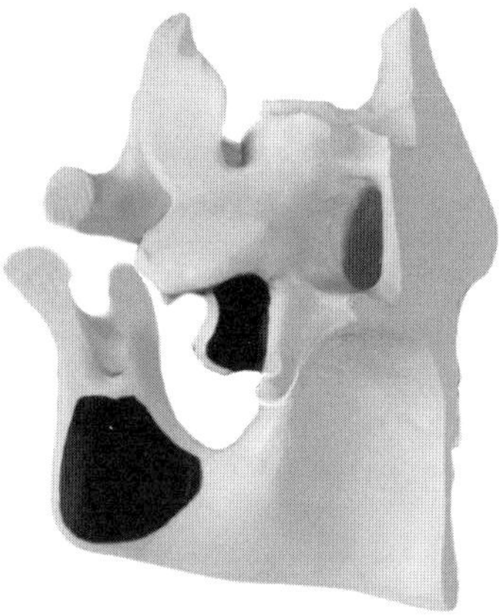

Dorsomedial

Dorsal

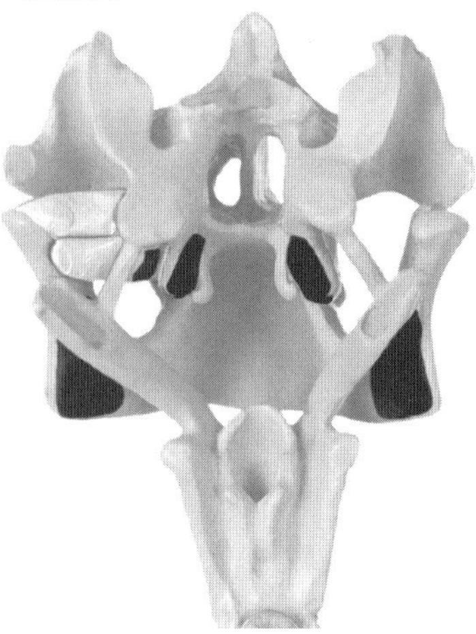

Ventral

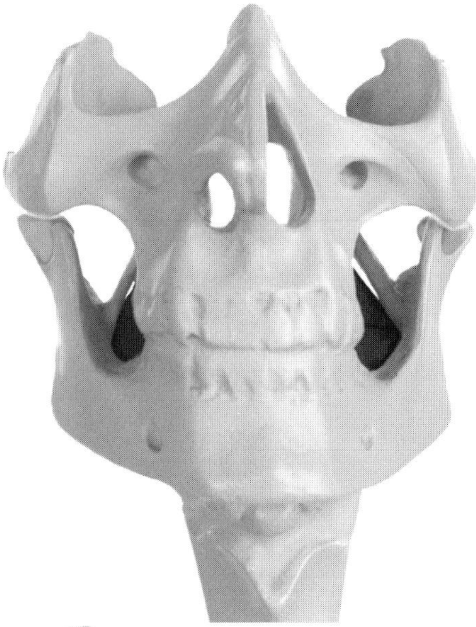

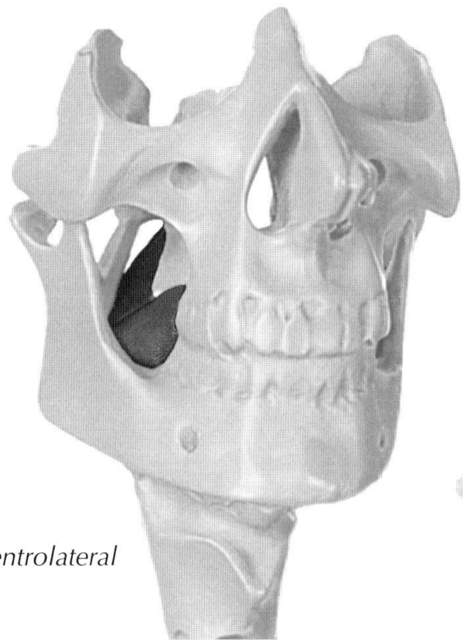

Ventrolateral

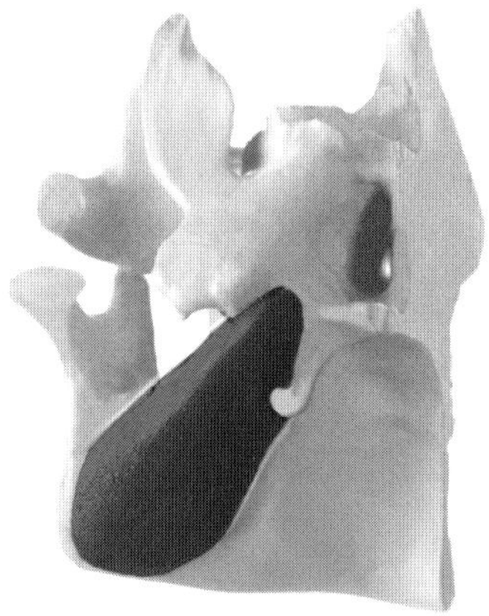

Dorsomedial

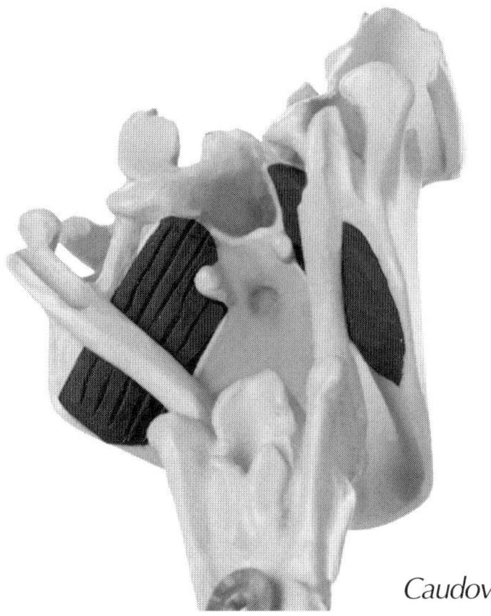

Caudoventrolateral

Pterygoideus medialis

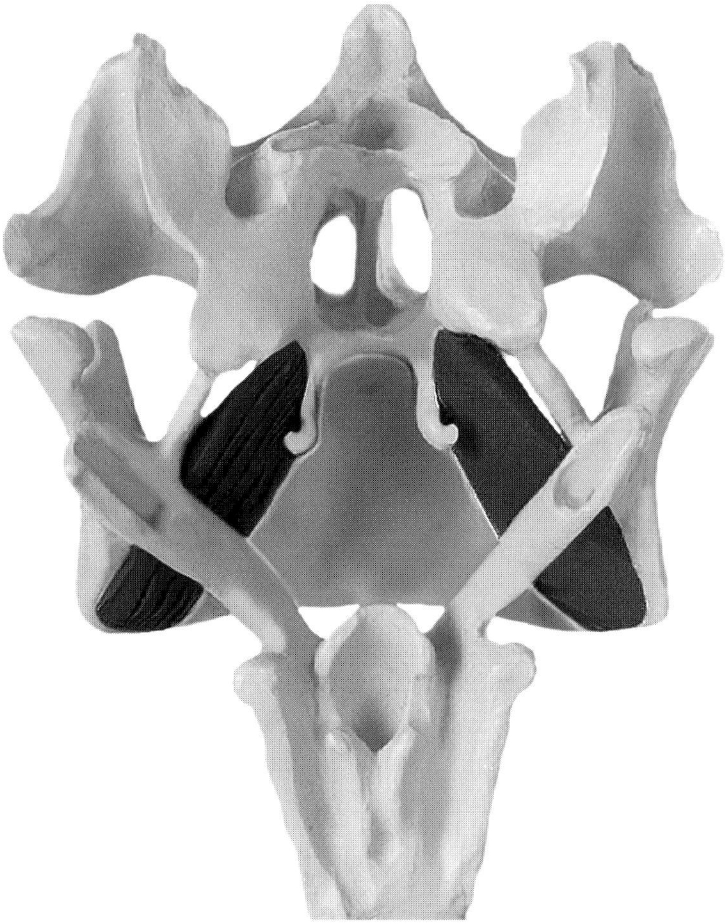

Dorsal

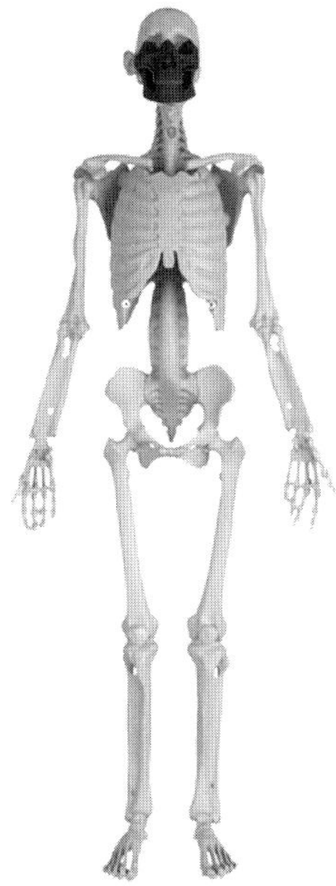

Actions

Elevates and protracts mandible to close the jaw. Opposite side rotator of the mandible.

Attachments

Cranial

Medial surface of lateral pterygoid plate and pyramidal process of the palatine bone.

Caudal

Medial surface of ramus and angle of the mandible.

Note

With attachments on both sides of the lateral pterygoid plate, this muscle splits around the inferior lateral pterygoid.

*tare•uh•**goid**•ee•us*

From the Greek, "pteron"+"eidos", meaning wing shaped.

*mead•ee•**al**•us*

From the Latin, "medius"+"alis", meaning related to the middle.

Craniodorsal

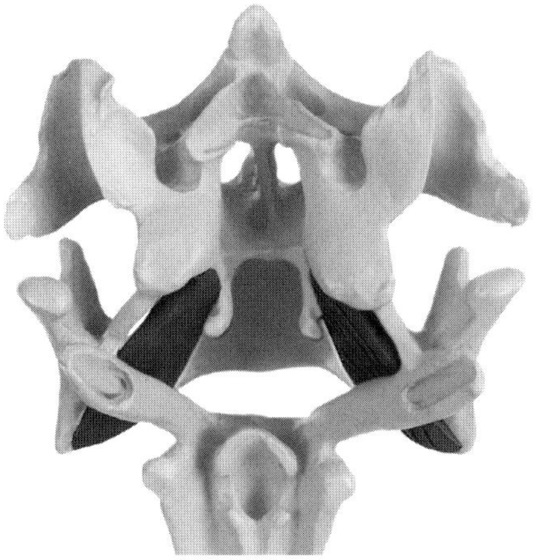

Caudodorsal

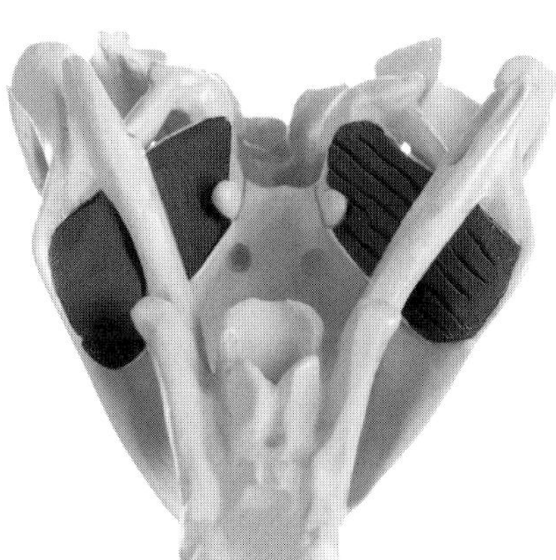

Tensor veli palatini

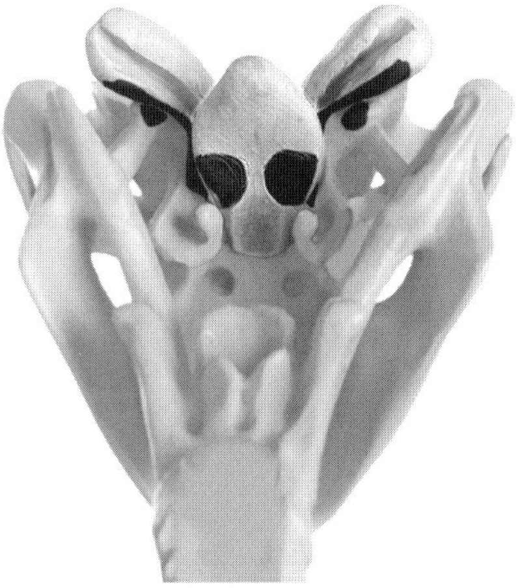

Caudodorsal

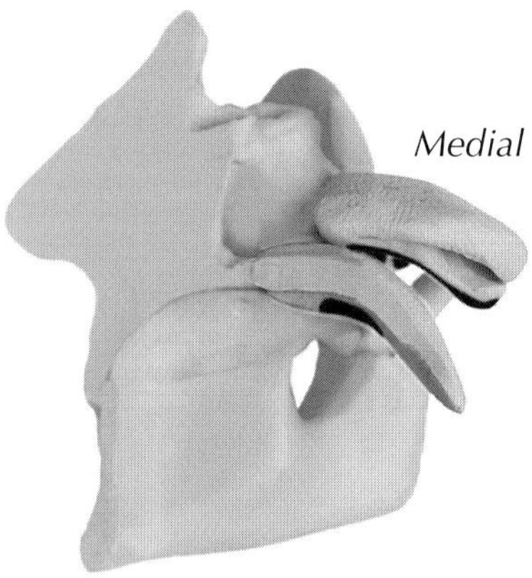

Medial

Caudodorsolateral

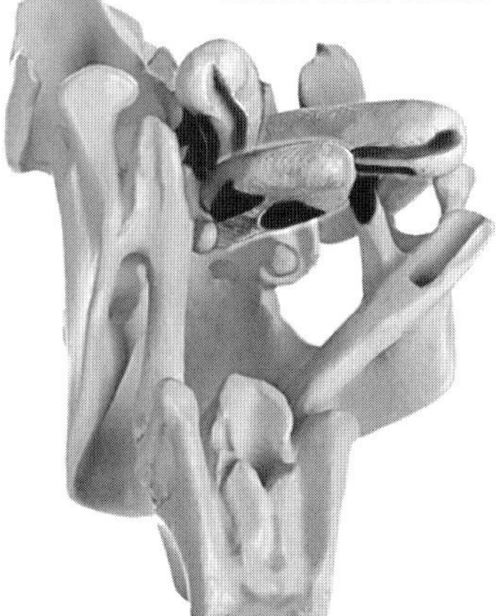

Caudodorsal

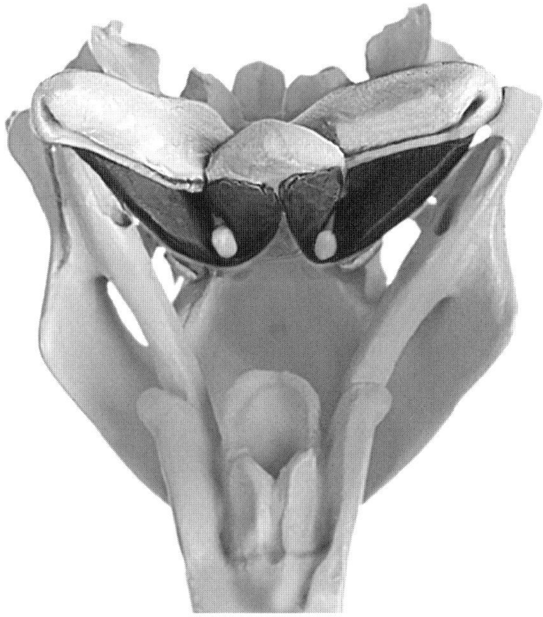

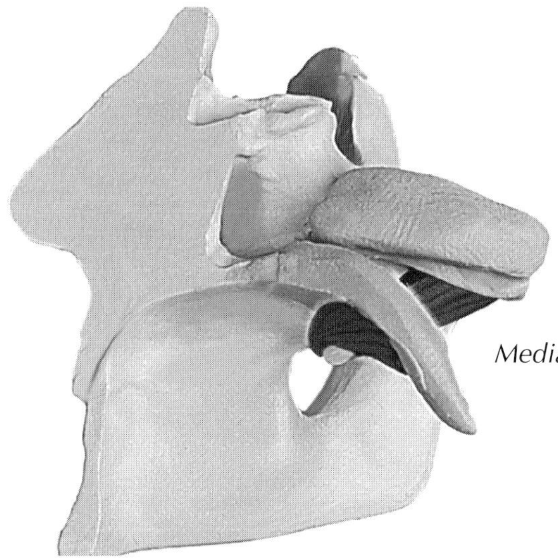

Medial

Caudodorsolateral

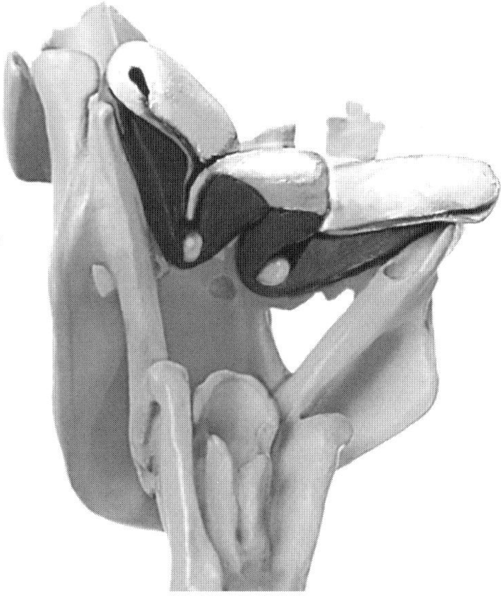

Tensor veli palatini

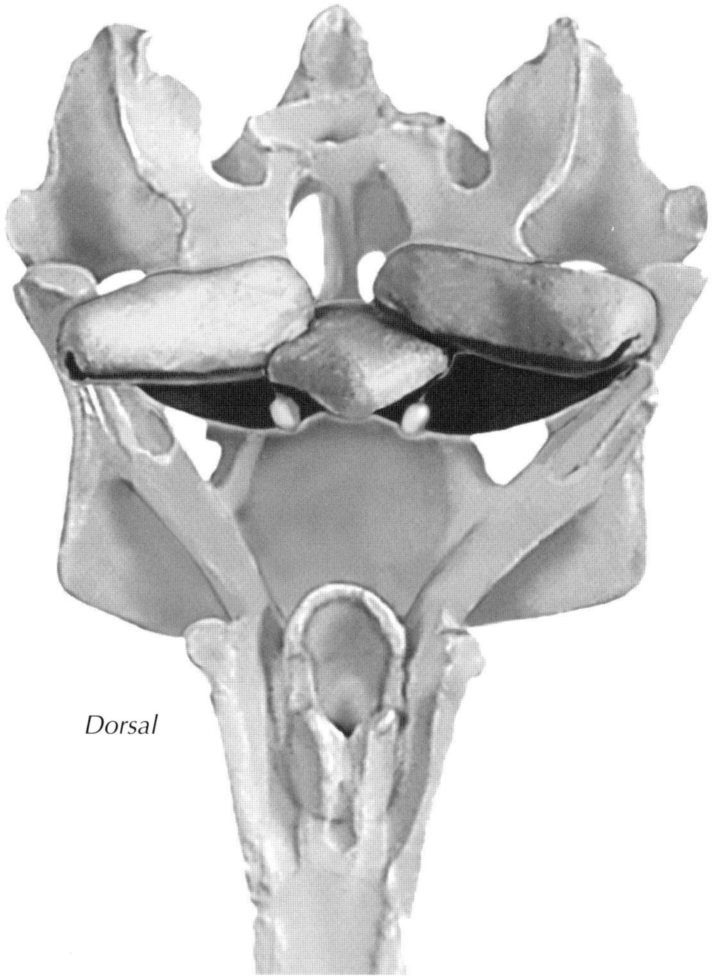

Dorsal

Caudodorsal

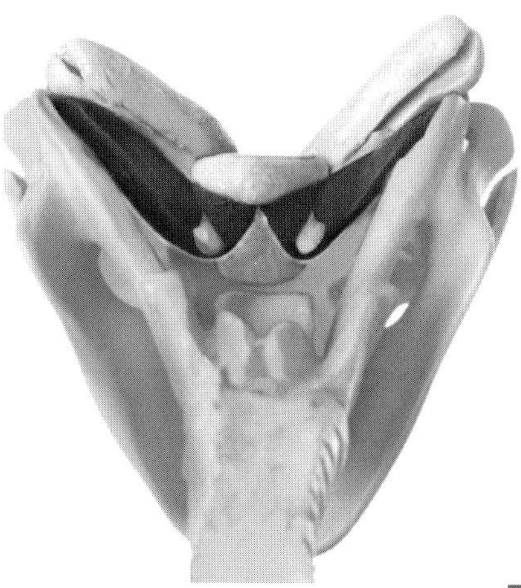

Caudodorsal

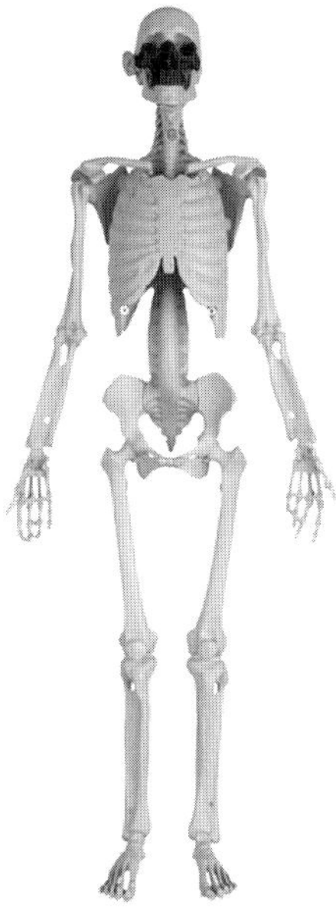

Actions

Tenses and depresses soft palate.

Attachments

Cranial

Spine and scaphoid foss of the sphenoid bone, lateral side of eustachian (auditory) tube.

Caudal

Tendon wraps around hamulus of the medial pterygoid plate and into the aponeurosis of the soft palate.

Note

The soft palate and auditory tube are shown in light clay.

ten•*ser*

From the Latin, “tensus”+“or”, meaning one that stretches.

vee•*lee*

Latin, meaning belonging to an awning or covering.

pal•*uh*•***tee***•*nee*

Latin, meaning belonging to the palate.

Palatoglossus

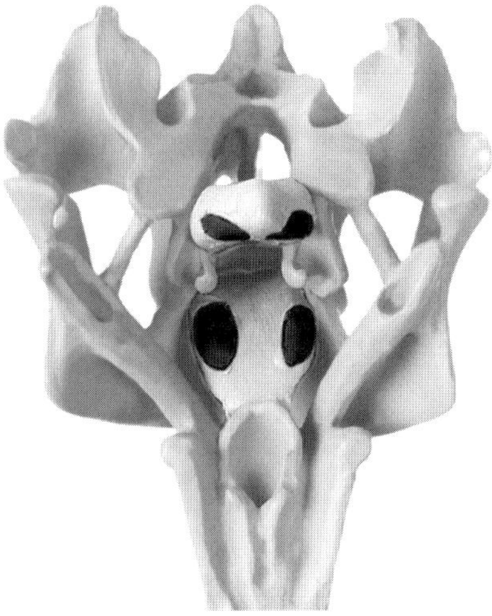

Dorsal

Dorsolateral

Caudodorsal

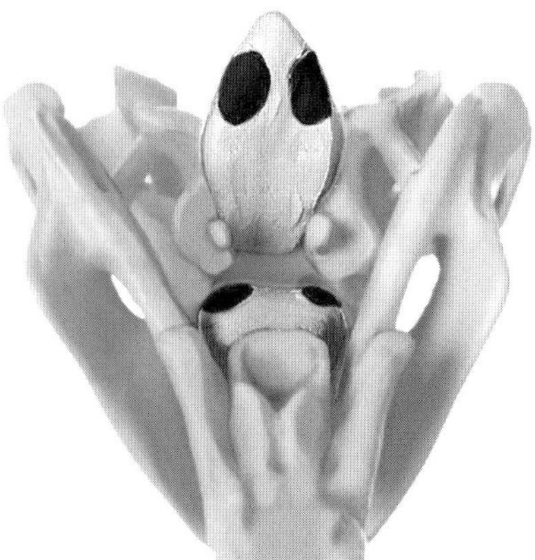

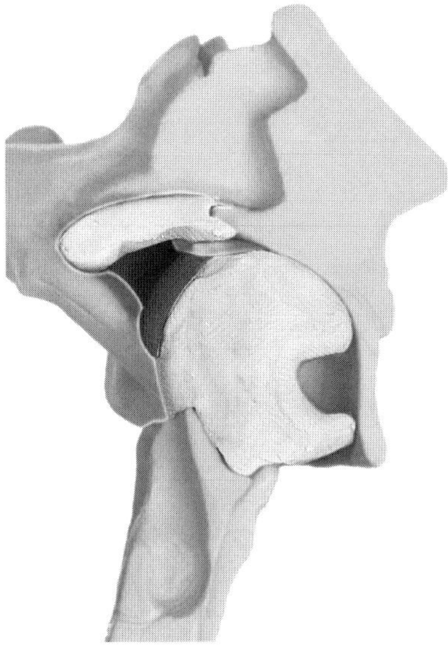

Medial

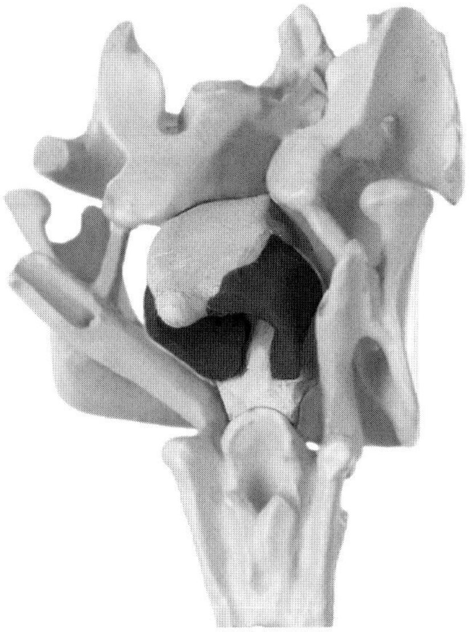

Dorsolateral

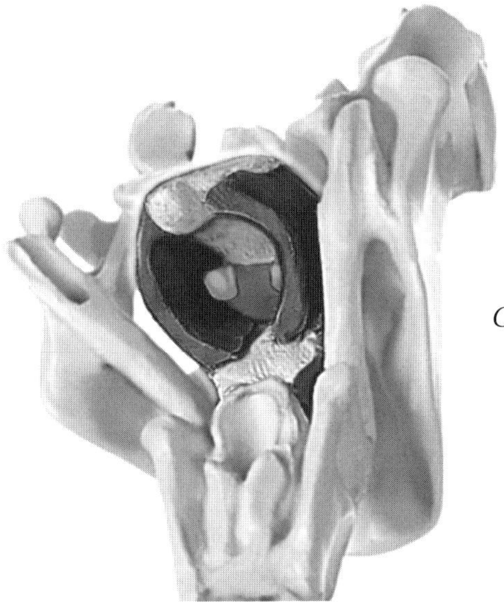

Caudodorsolateral

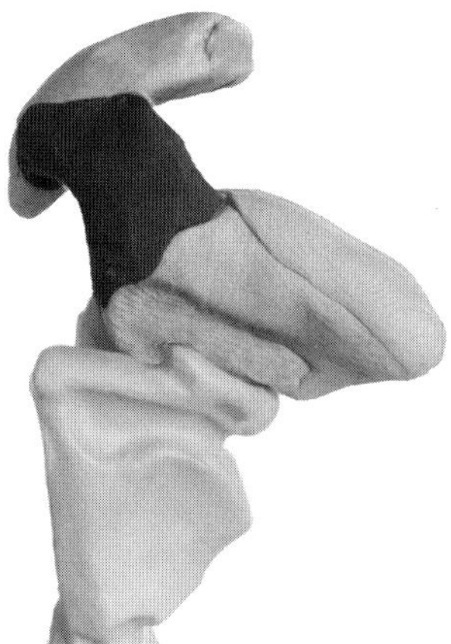

Lateral

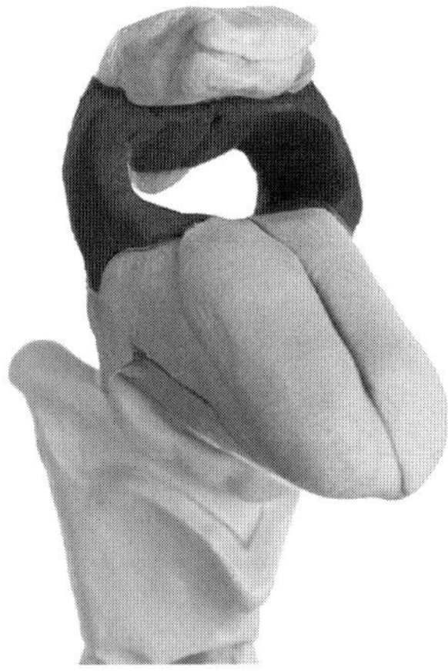

Ventrolateral

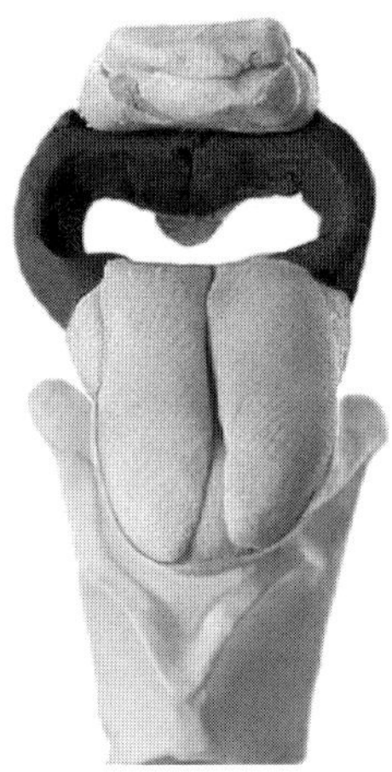

Ventral

Palatoglossus

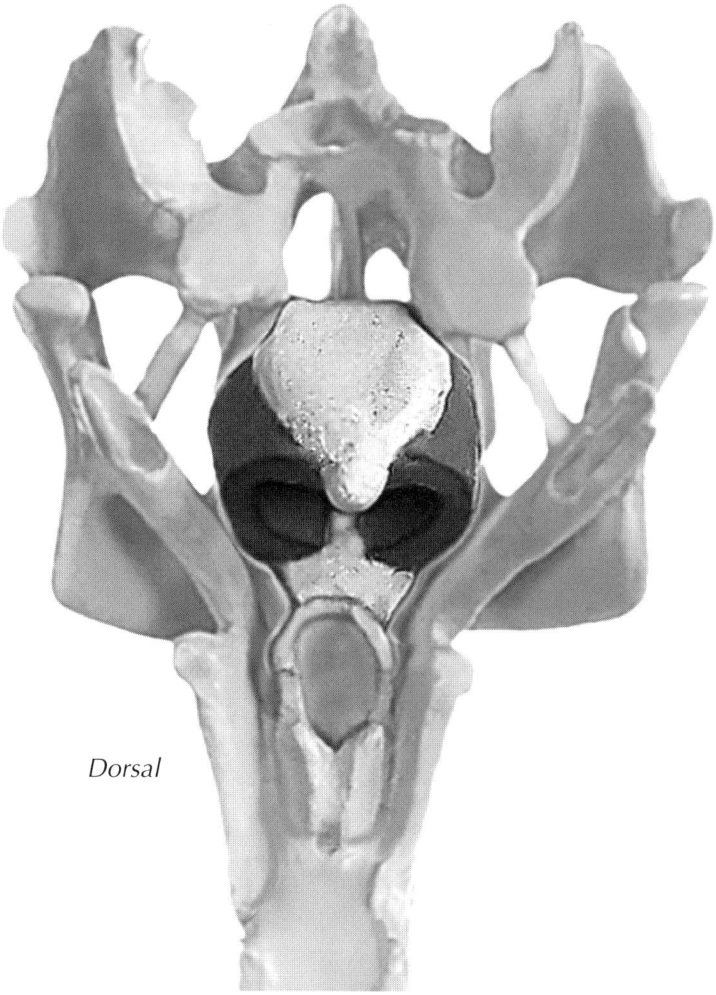

Dorsal

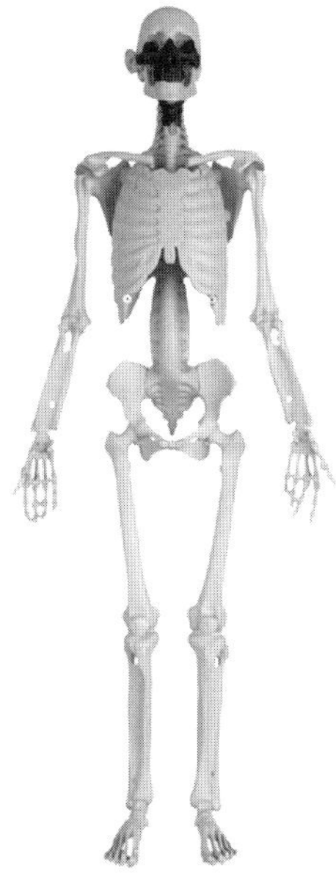

Actions

Elevates tongue, closes palatoglossal arch.

Attachments

Cranial

Aponeurosis of the soft palate.

Caudal

Side and dorsal surface of the tongue.

Caudodorsal

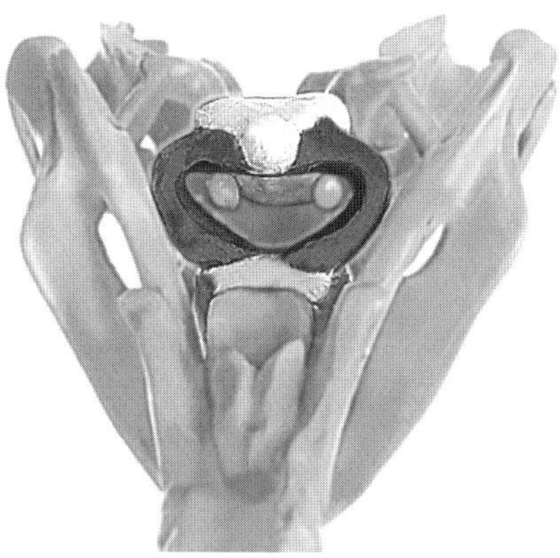

Caudodorsal

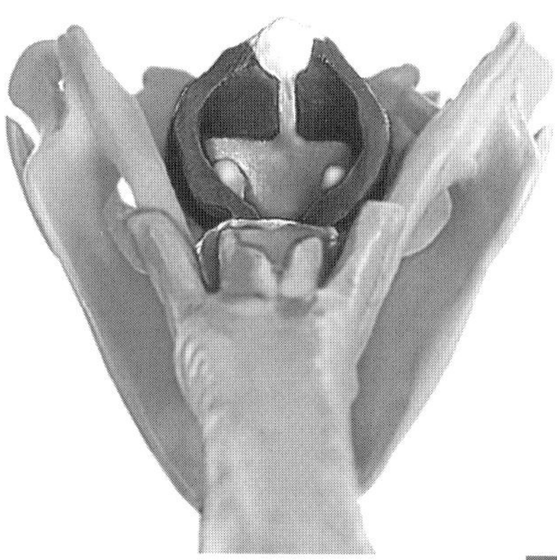

*pal•ah•toh•**glos**•us*

From the Latin, "palatum," meaning palate, and from the Greek, "glossa," meaning tongue.

Palatopharyngeus

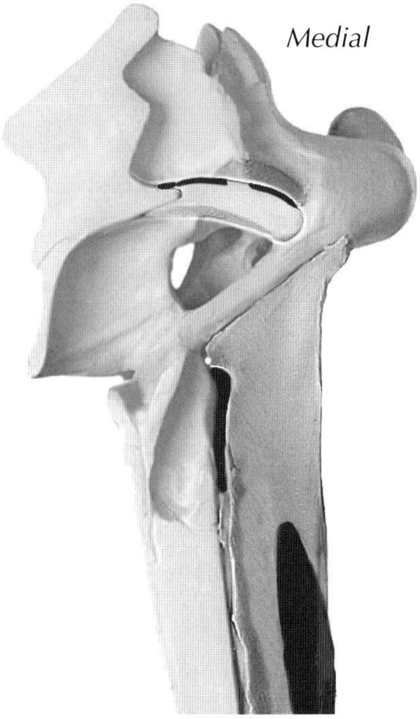

Medial

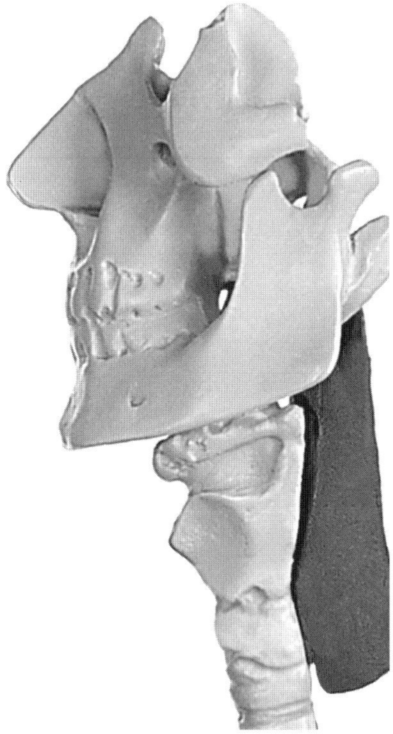

Lateral

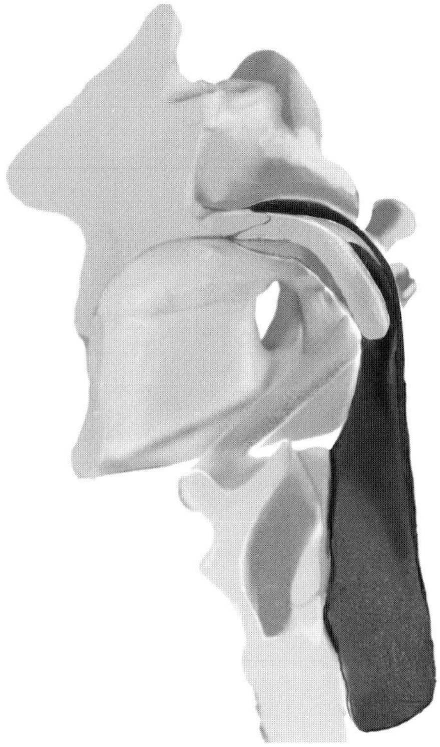

Medial

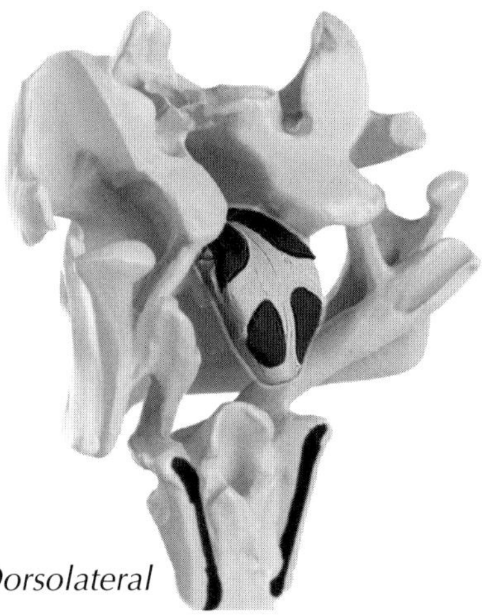

Dorsolateral

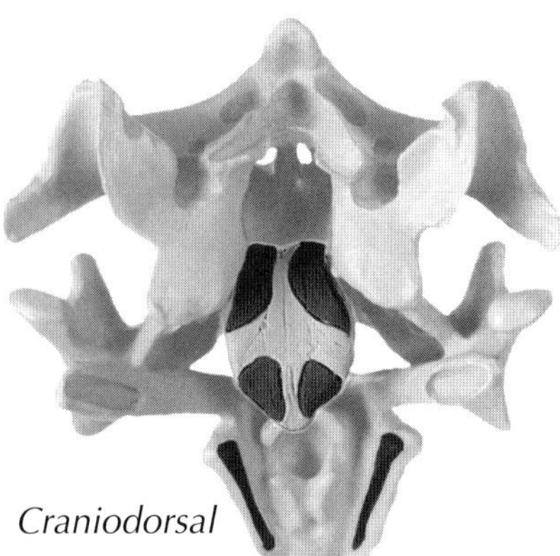

Craniodorsal

Caudodorsolateral

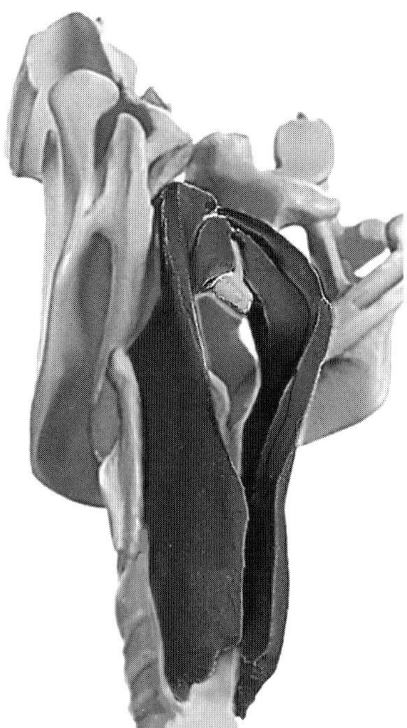

Dorsolateral

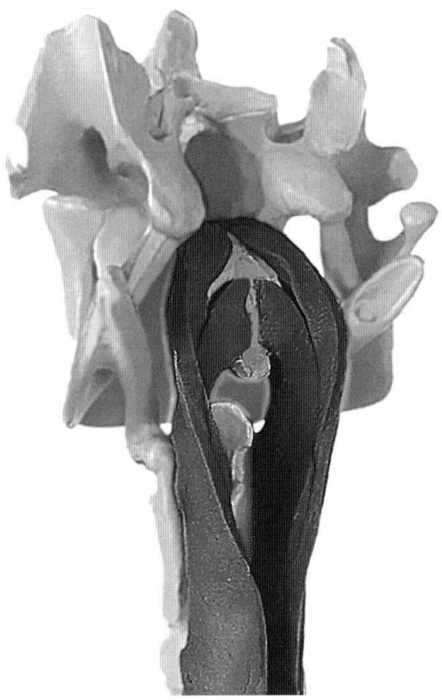

Palatopharyngeus

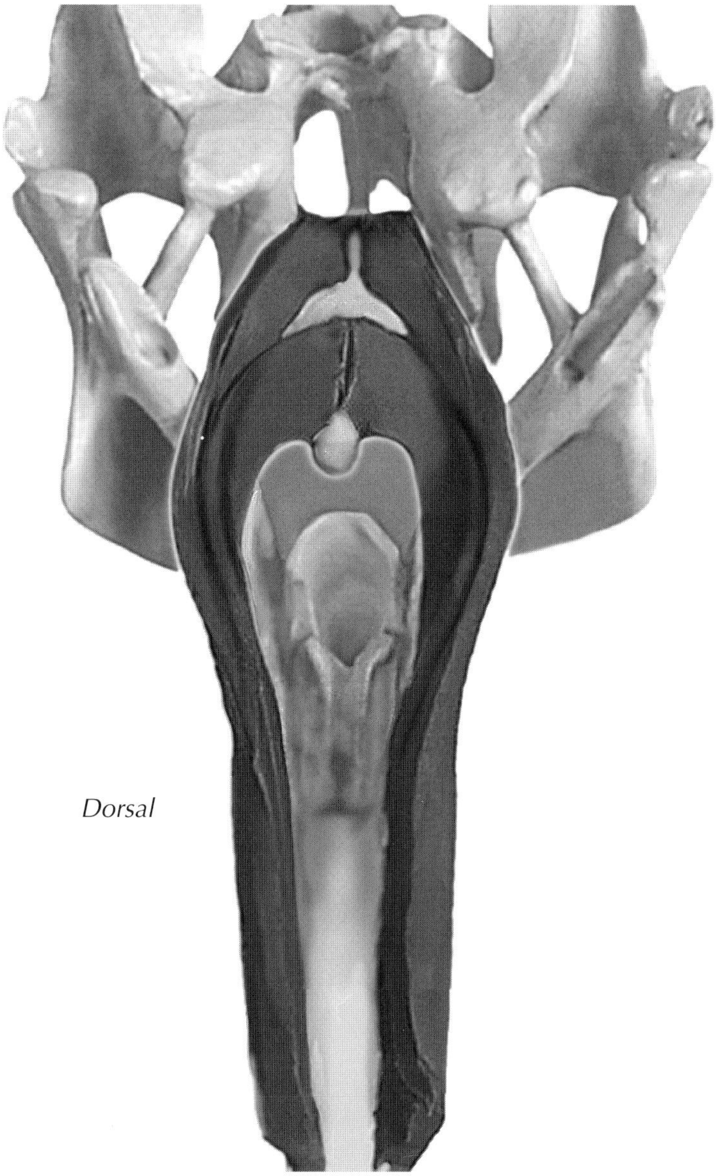

Dorsal

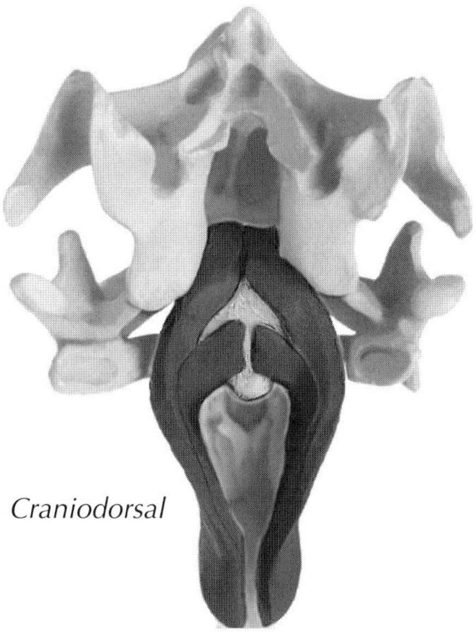

Craniodorsal

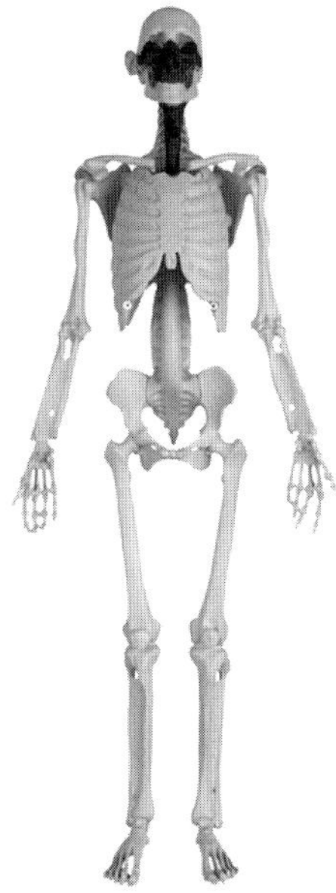

Actions

Elevates walls of pharynx, closes palatopharyngeal arch.

Attachments

Cranial

Ventral slip: Posterior part of the hard palate and aponeurosis and median raphé with contralateral muscle.
Dorsal slip: Median raphe with contralateral muscle and mucous membrane of the soft palate.

Caudal

Dorsal borders of thyroid cartilage.

*pal•lot•oh•fu•**rin**•jee•us*

From the Latin, "palatum," meaning palate, and from the Greek, "pharynx," meaning throat.

Salpingopharyngeus

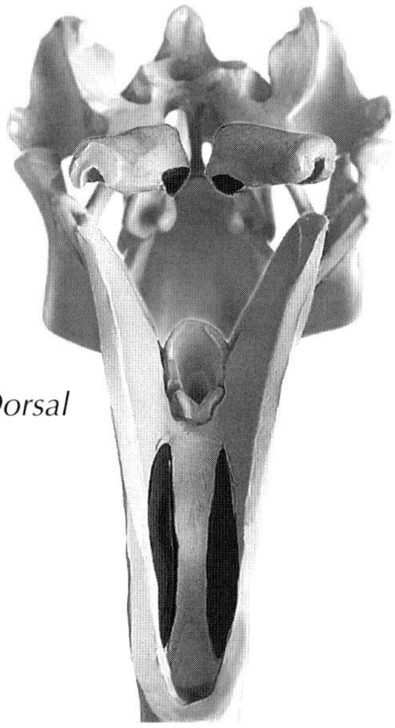

Dorsal

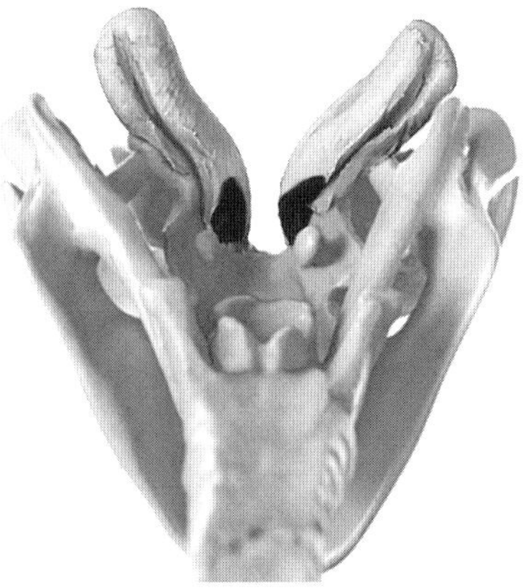

Caudodorsal

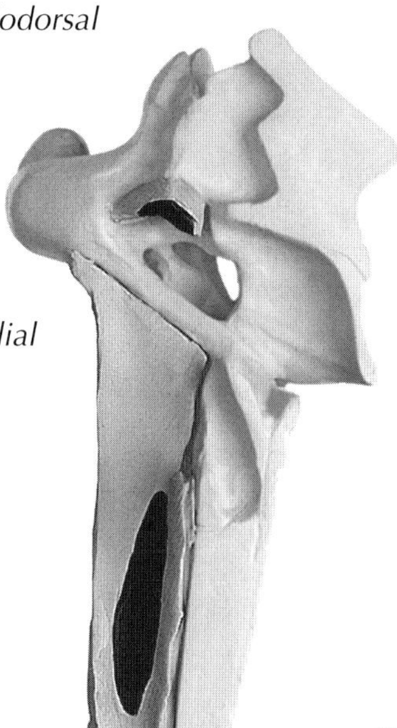

Medial

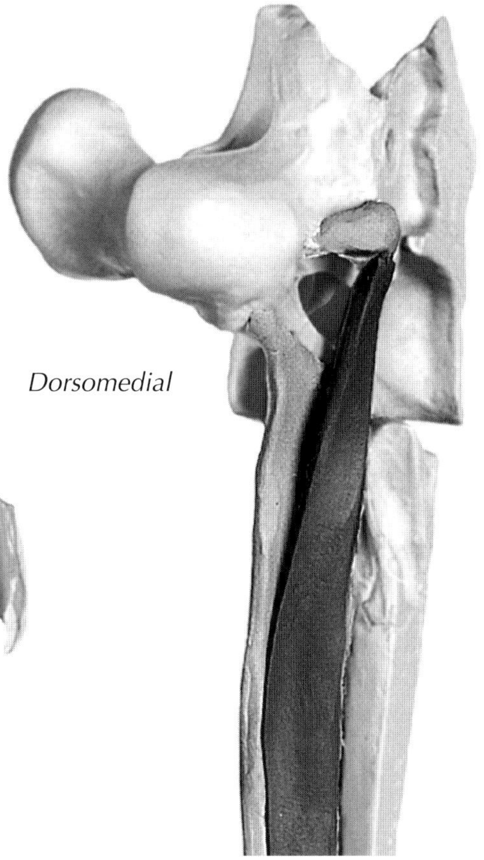

Dorsomedial

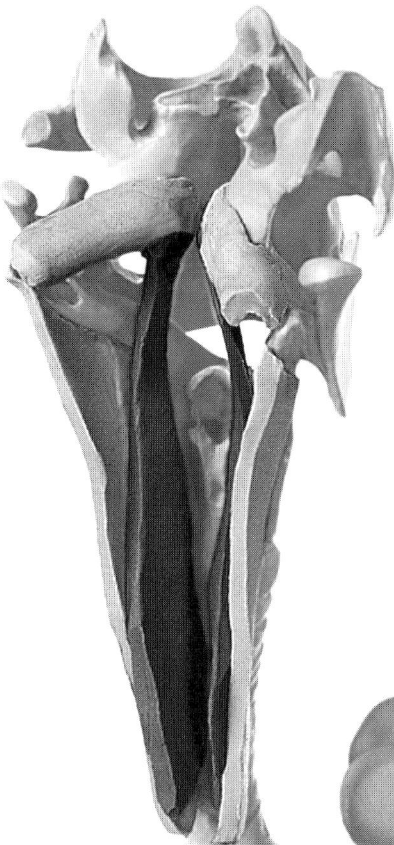

Craniodorsolateral

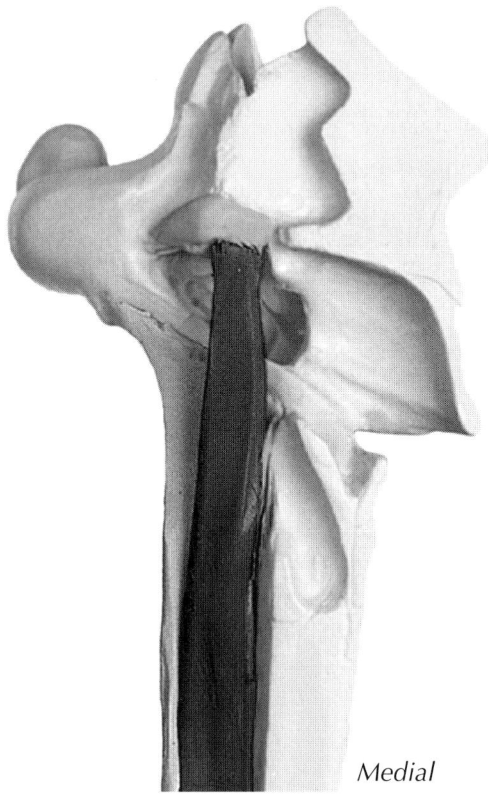

Medial

Salpingopharyngeus

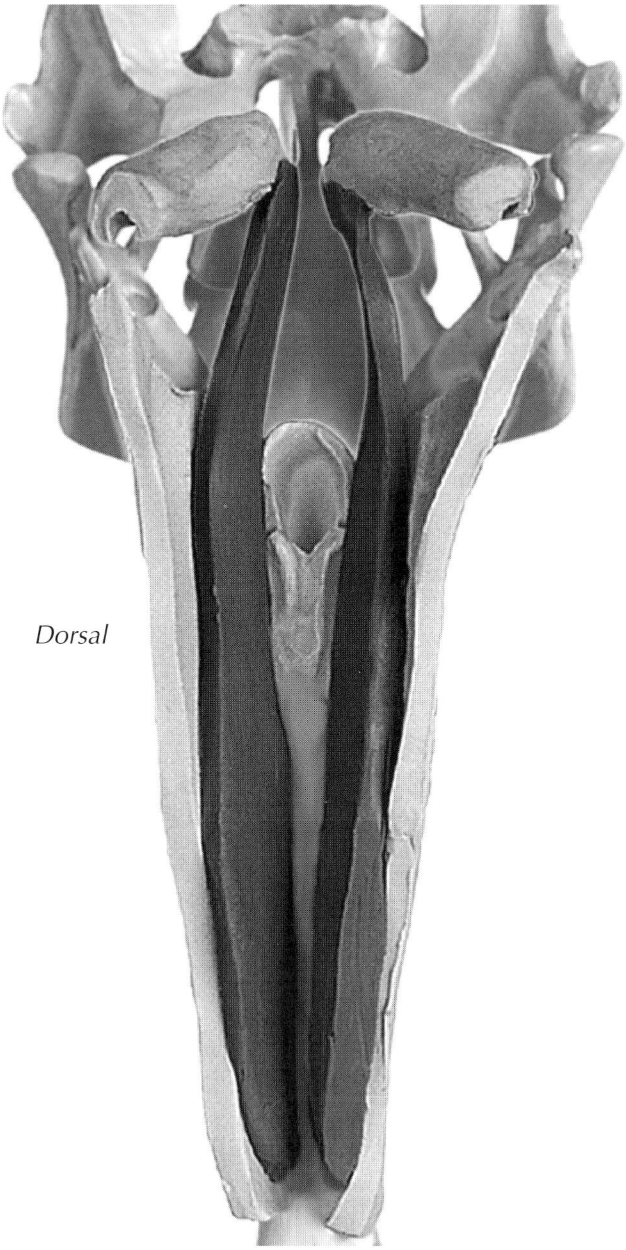

Dorsal

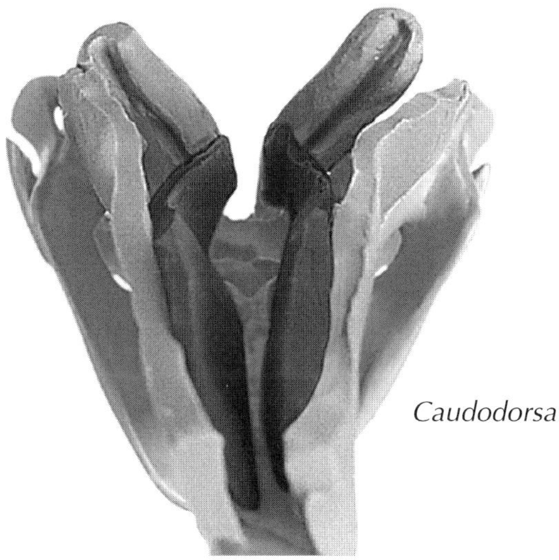

Caudodorsal

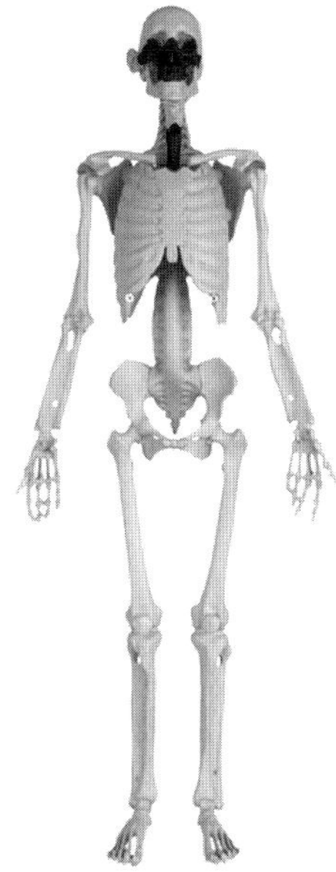

Actions

Raises the upper part of the pharyngeal wall in swallowing. Opens the nasal end of the eustachian tube in pressure equalization yawning.

Attachments

Cranial

Auditory tube near its pharyngeal opening.

Caudal

Blends with posterior part of palatopharyngeus.

*sal•**pin**•jo•fu•**rin**•jee•us*

From the Greek, "salpinx," meaning trumpet, and "pharynx," meaning throat.

Levator veli palatini

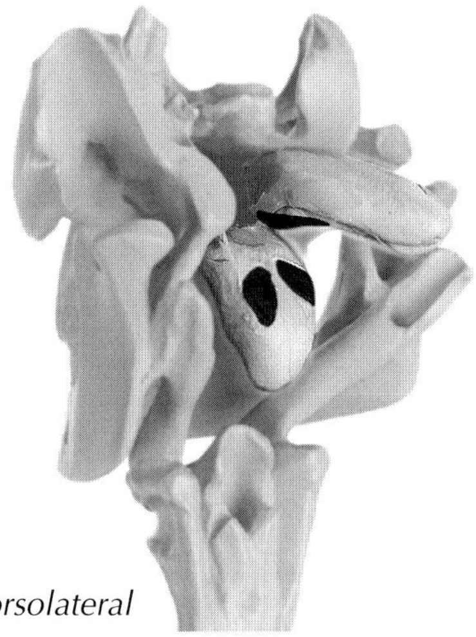

Dorsolateral

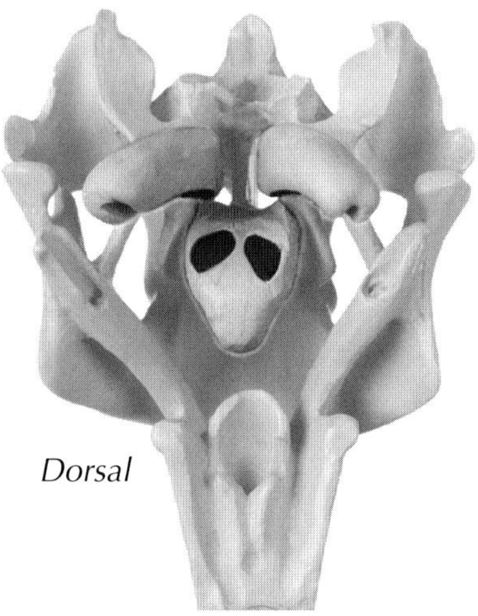

Dorsal

Caudodorsal

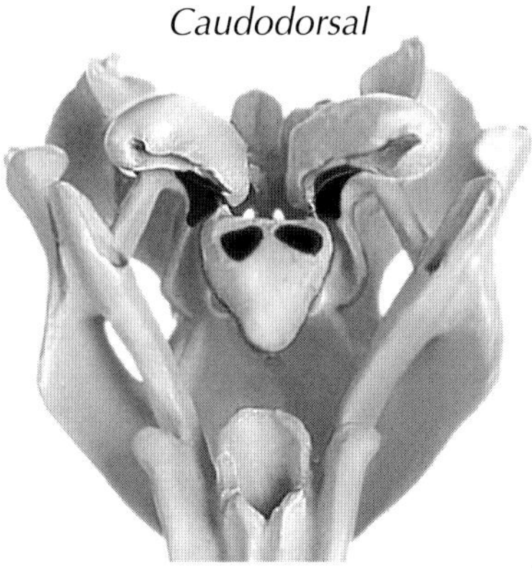

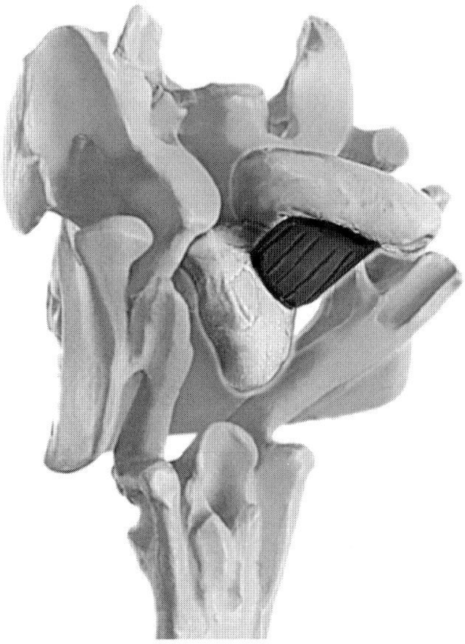

Dorsolateral

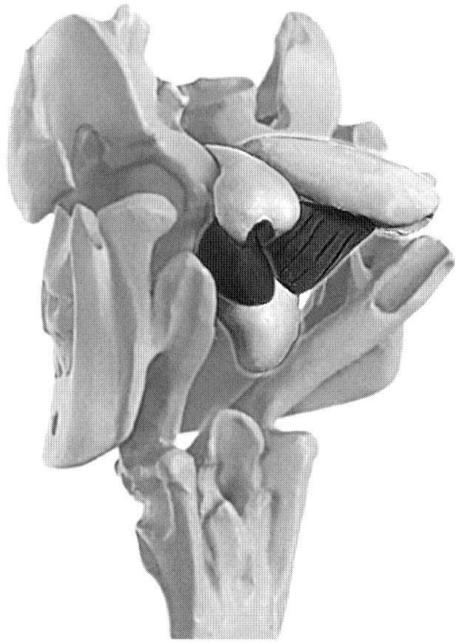

Dorsolateral

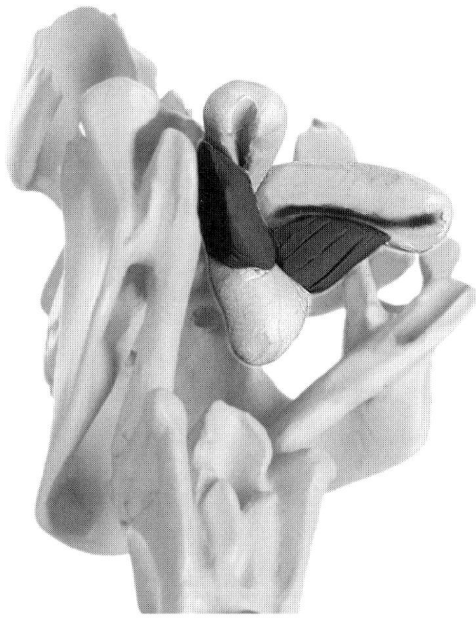

Caudodorsolateral

Caudodorsal

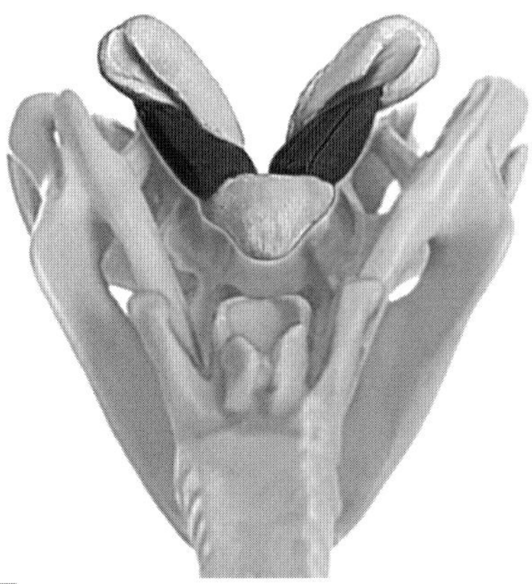

Caudodorsal

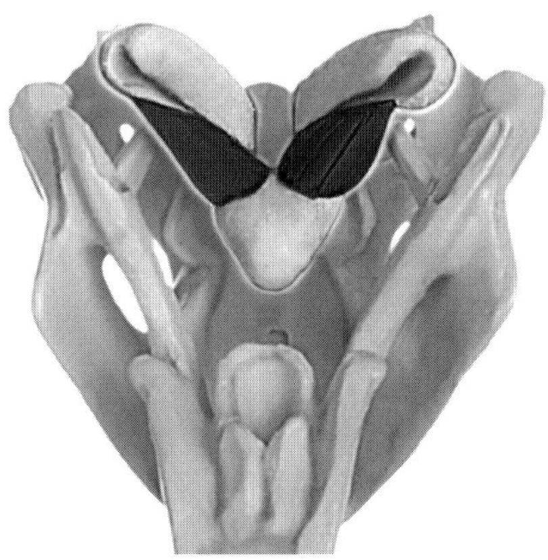

Levator veli palatini

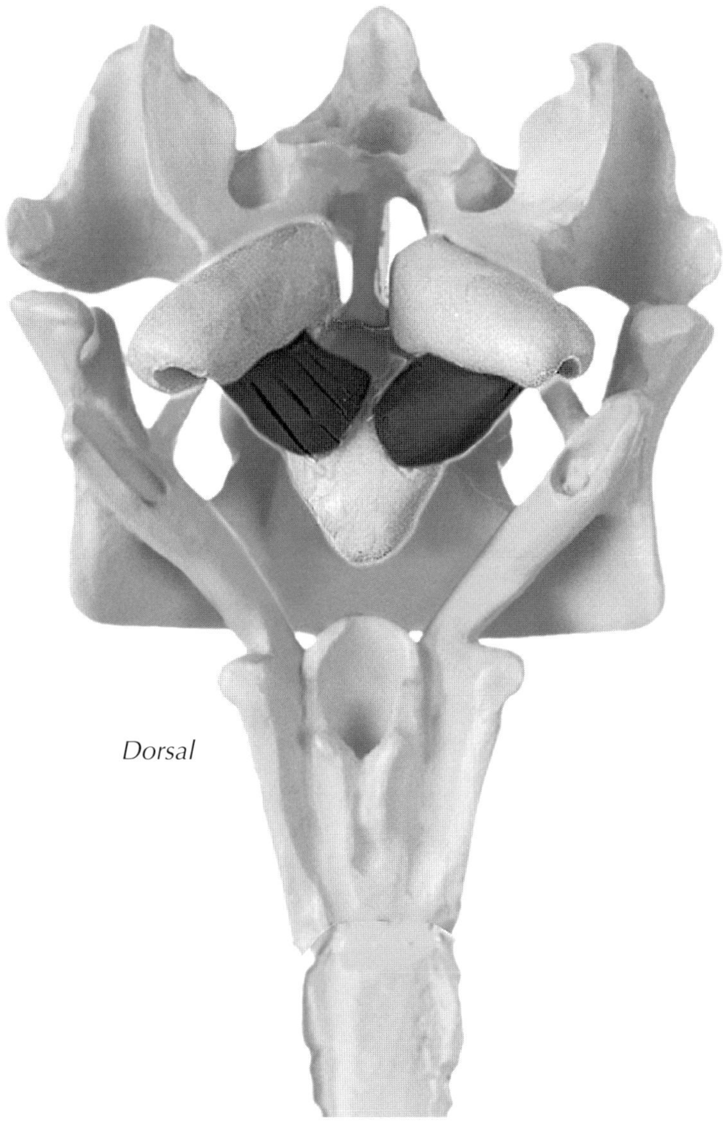

Dorsal

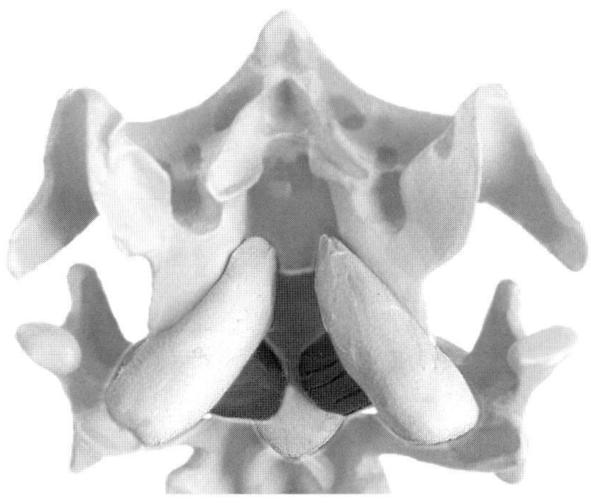

Craniodorsal

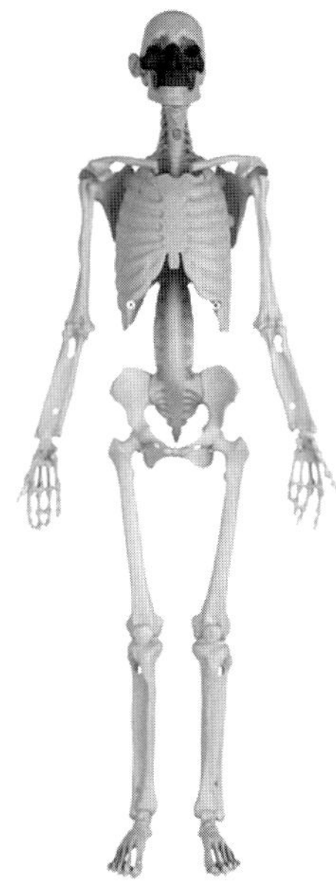

Actions

Elevates soft palate.

Attachments

Cranial

Inferior surface of petrous part of temporal bone, medial side of eustachian (auditory) tube.

Caudal

Upper surface of aponeurosis of the soft palate.

*luh•**vay**•tur*

From the Latin, "levatus"+"or," meaning one that raises.

***vee**•lee*

From the Latin, "veli," meaning belonging to an awning or covering.

*pal•uh•**tee**•nee*

From the Latin, "palatum," meaning belonging to the palate.

Musculus uvulae

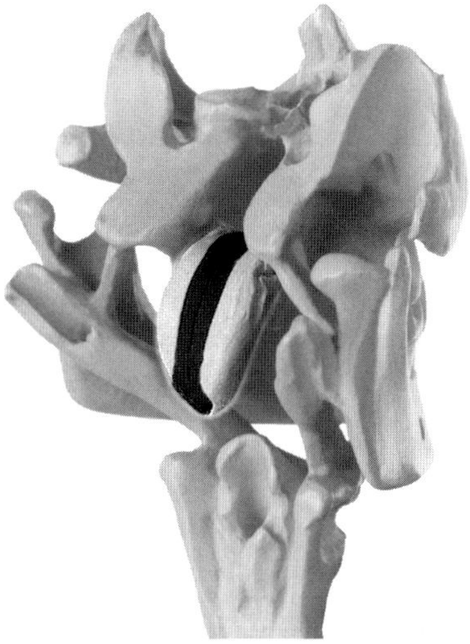

Dorsolateral

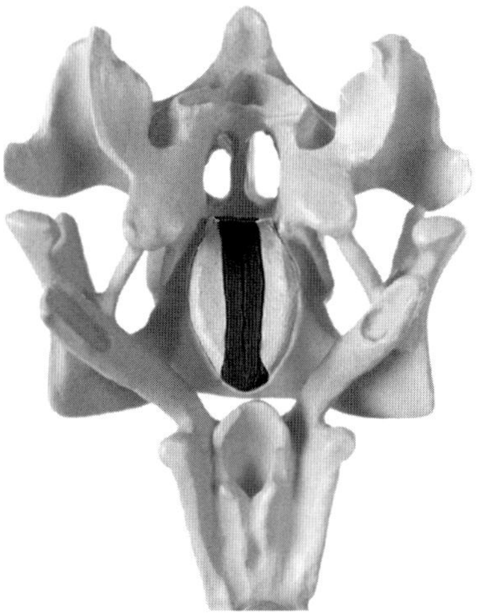

Dorsal

Caudodorsal

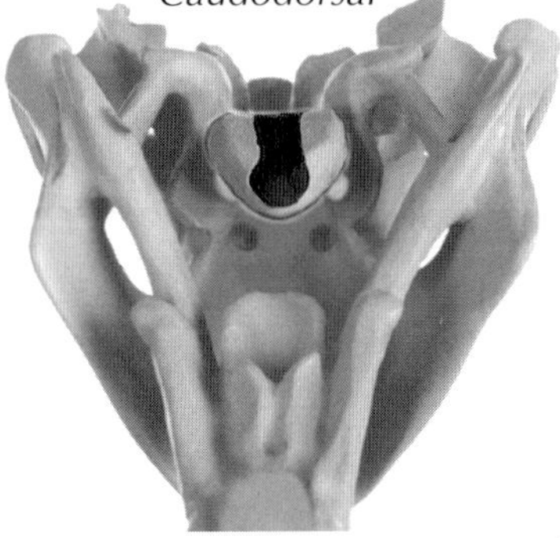

Caudodorsal

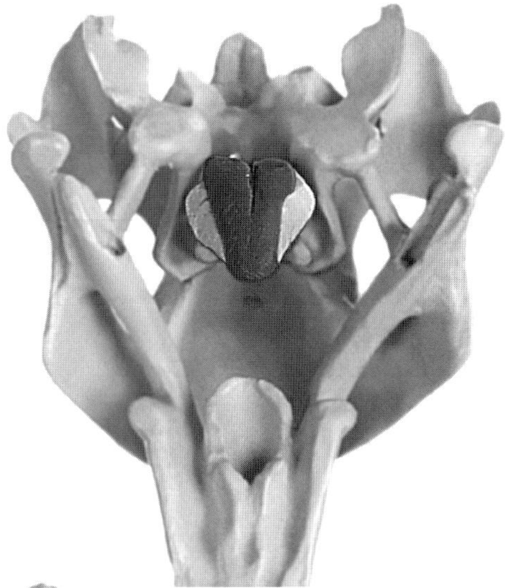

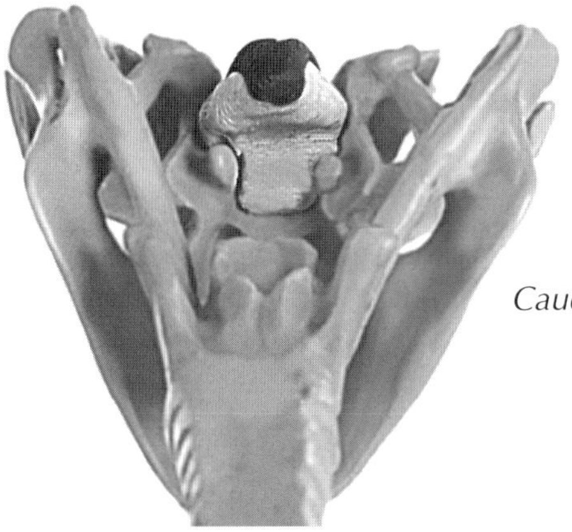

Caudodorsal

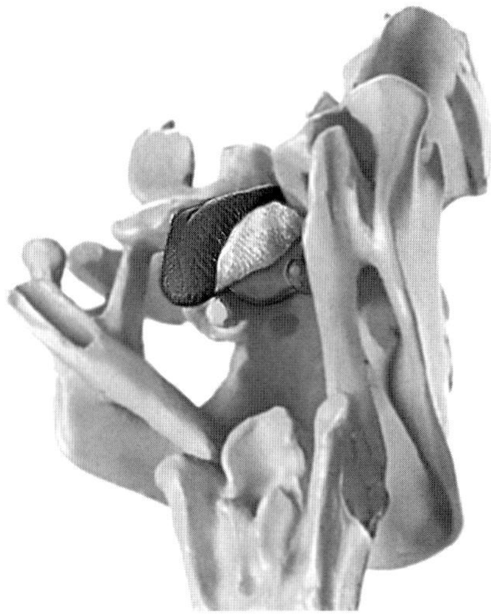

Caudodorsolateral

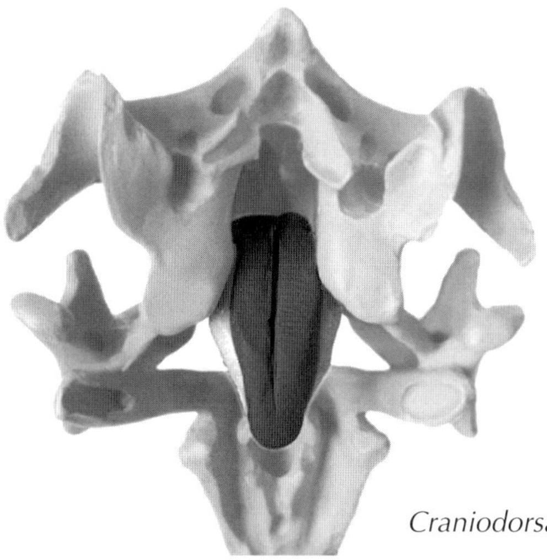

Craniodorsal

Musculus uvulae

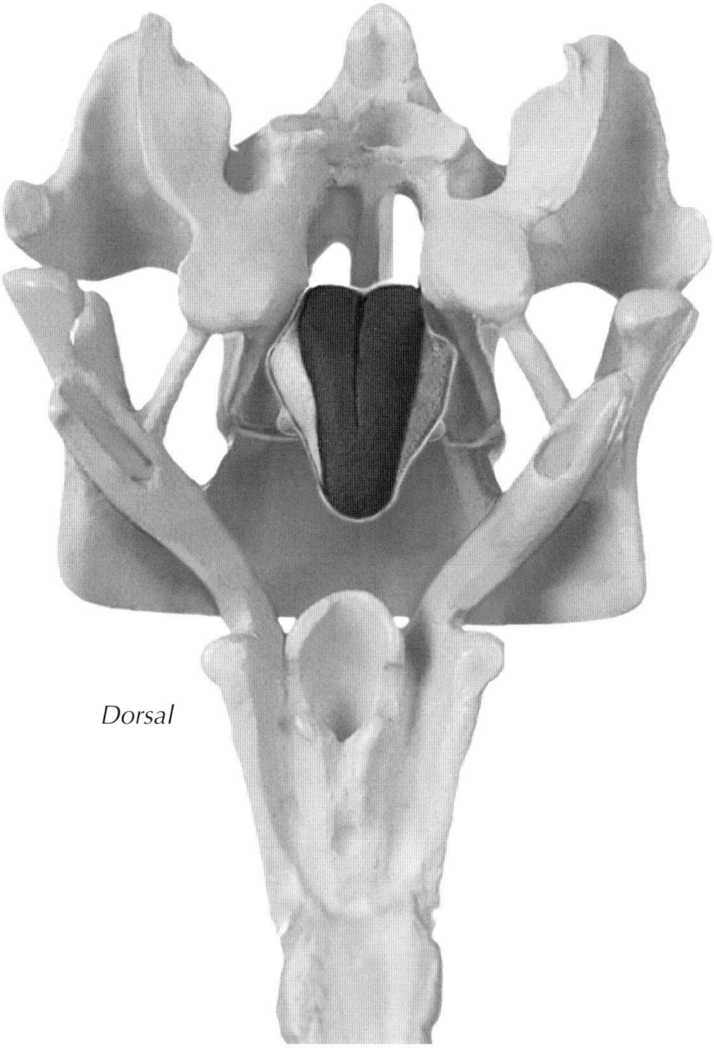

Dorsal

Cranial

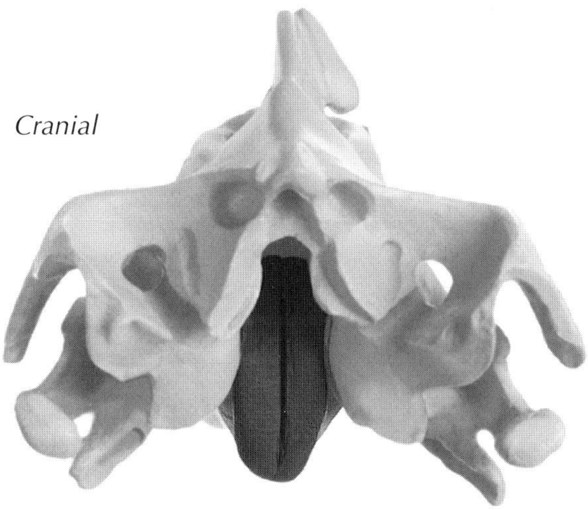

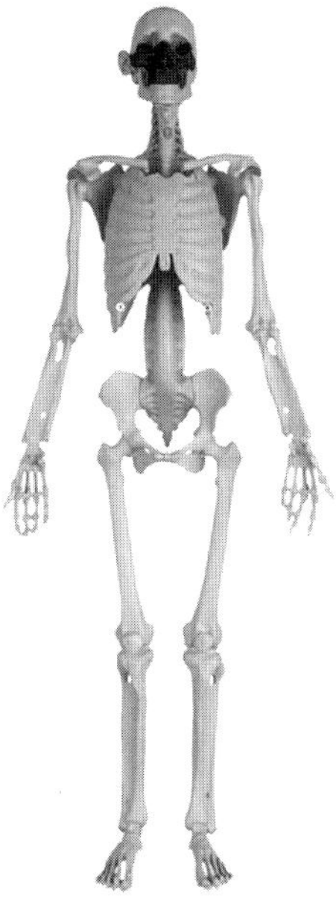

Actions

Elevates uvula.

Attachments

Cranial

Posterior nasal spine of the palatine bone.

Caudal

Mucous membrane of the uvula.

musk • *you* • *lus*

From the Latin, "musculus," meaning muscle.

you • *view* • *lee*

From the Latin, "uvulae," meaning belonging to a small grape (uvula).

Constrictor pharyngis medius

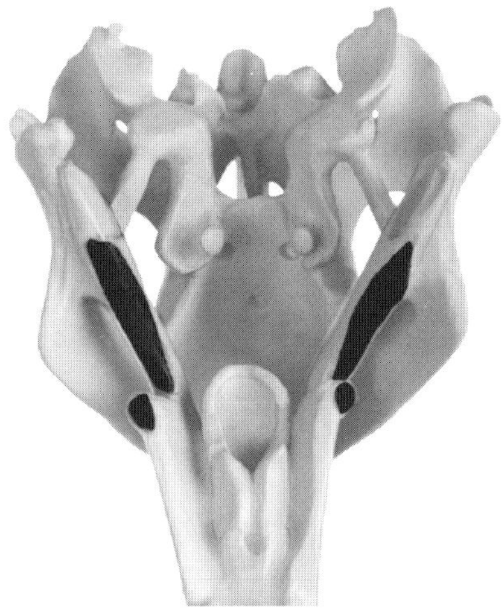

Caudodorsal

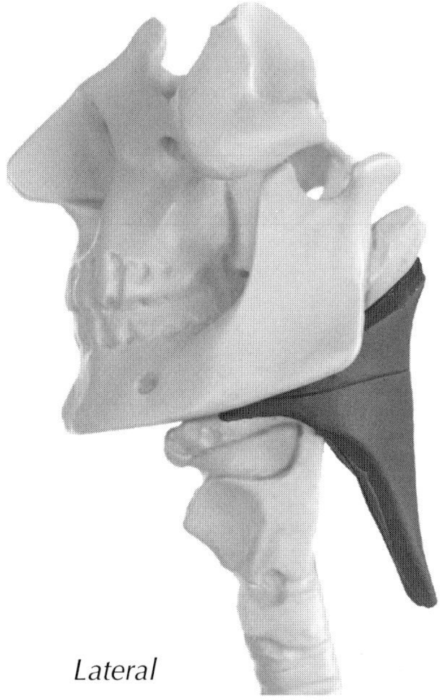

Lateral

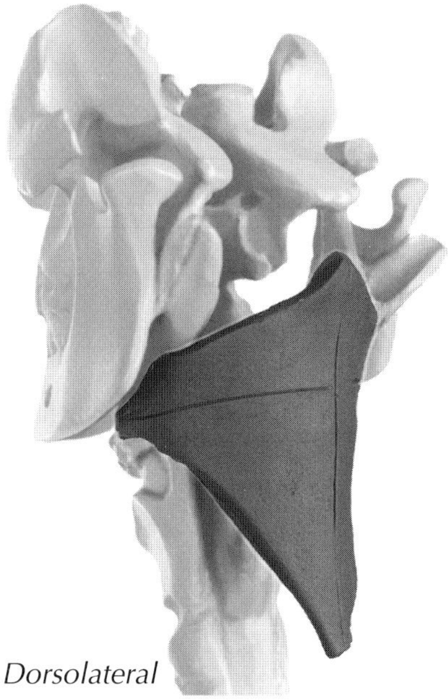

Dorsolateral

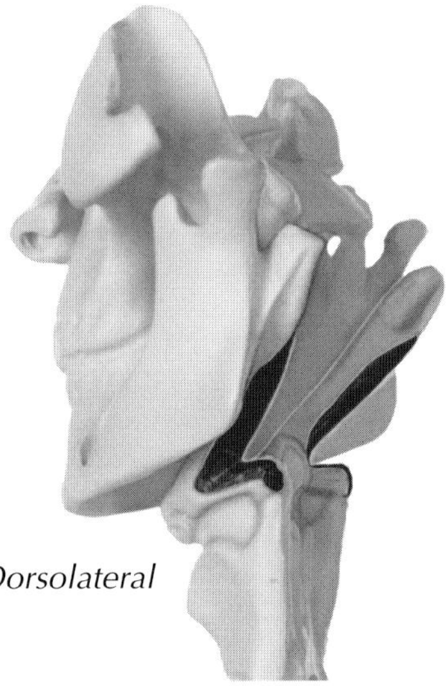

Dorsolateral

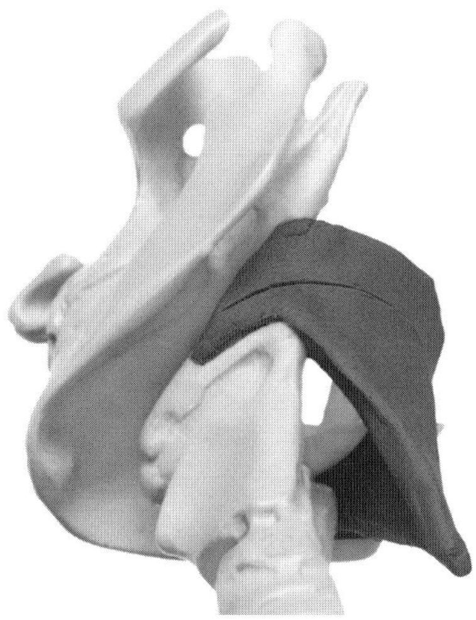

Caudolateral

Craniodorsolateral

Caudoventral

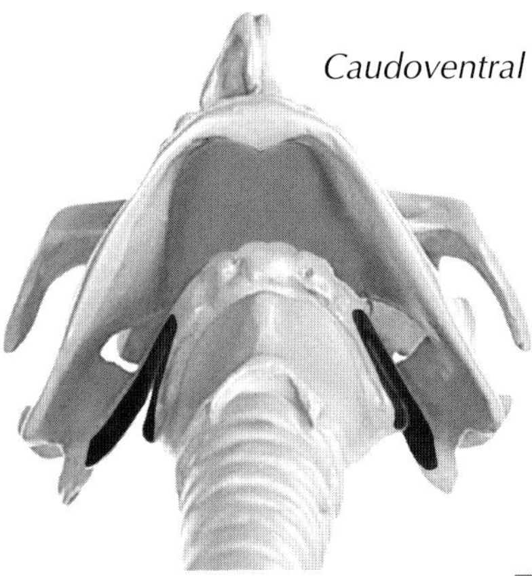

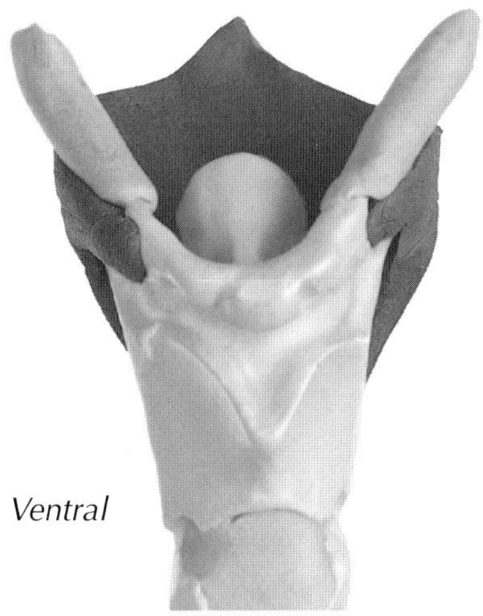

Ventral

Constrictor ***pharyngis*** *medius*

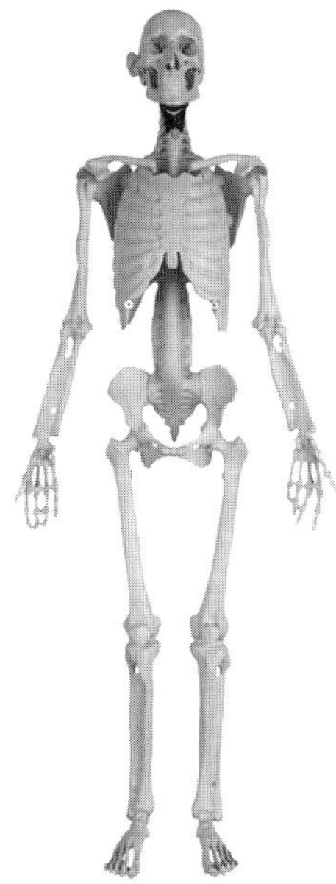

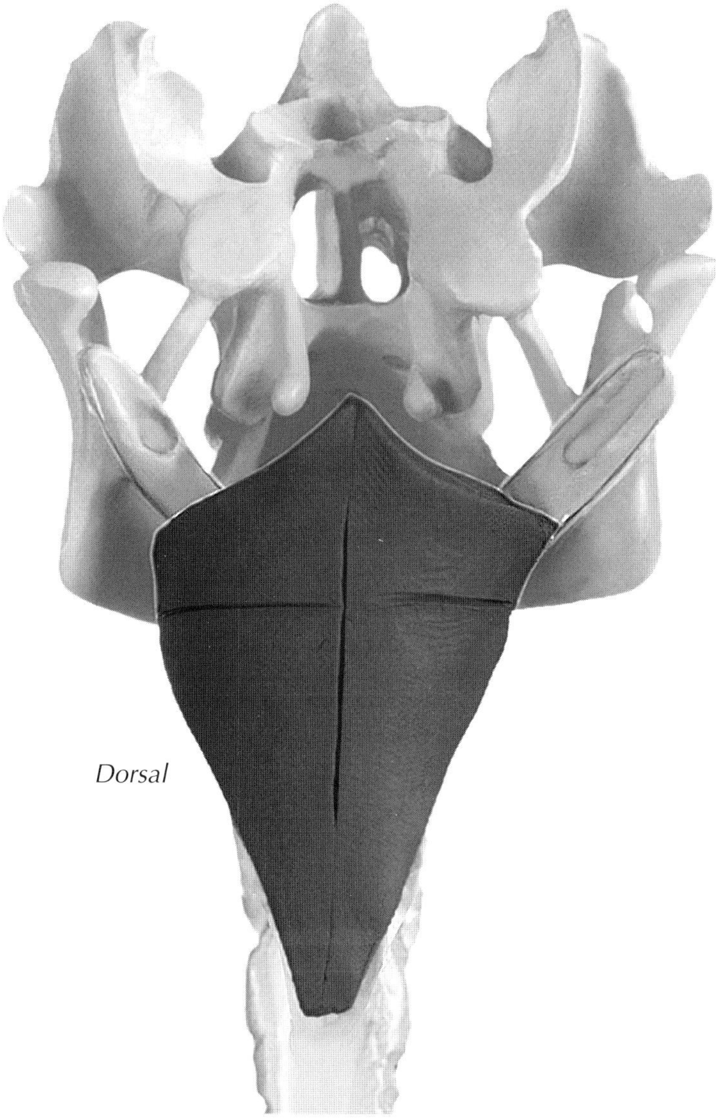

Dorsal

Actions

Constricts pharynx for swallowing.

Attachments

Ventral

Greater and lesser cornus of the hyoid bone, stylohyoid ligament.

Dorsal

Posterior midline raphé of the pharynx; upper part overlaps constrictor superior.

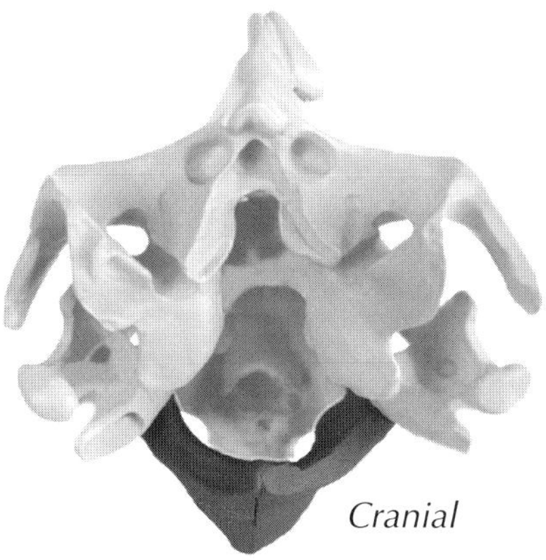

Cranial

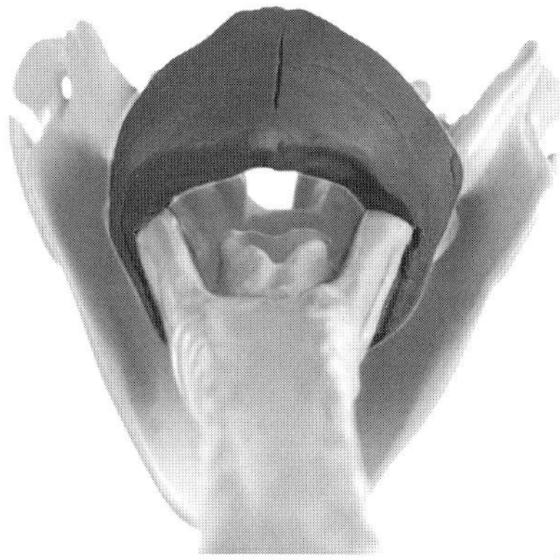

Dorsocaudal

*kon•***strik***•tur*

From the Latin, "constrictus"+"or," meaning one that binds or ties up.

fare•*in*•*jyus*

From the Greek, "pharynx," meaning throat.

mee•*dee*•*us*

From the Latin, "medius," meaning middle.

Constrictor pharyngis inferior

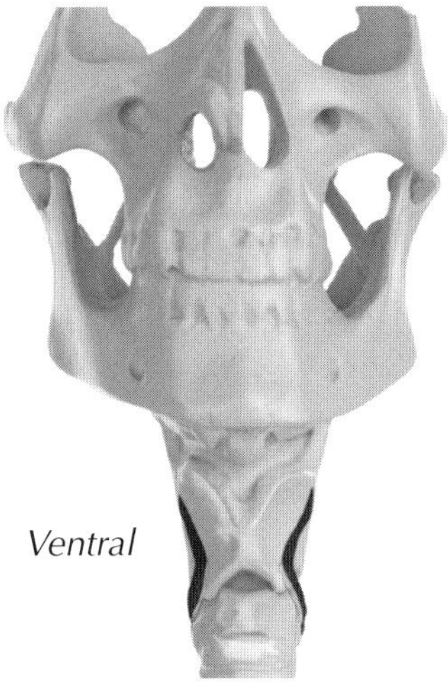

Ventral

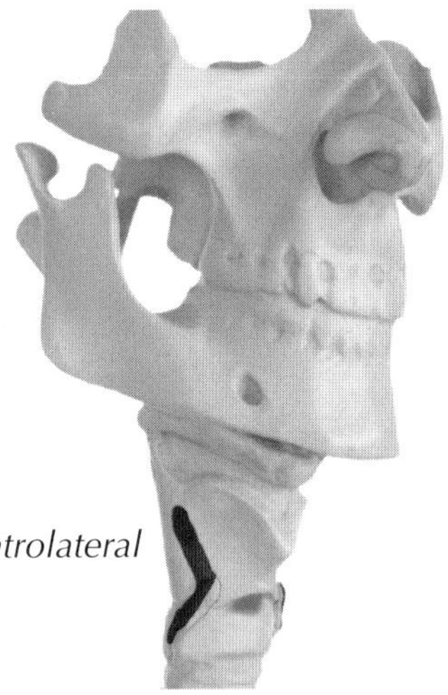

Ventrolateral

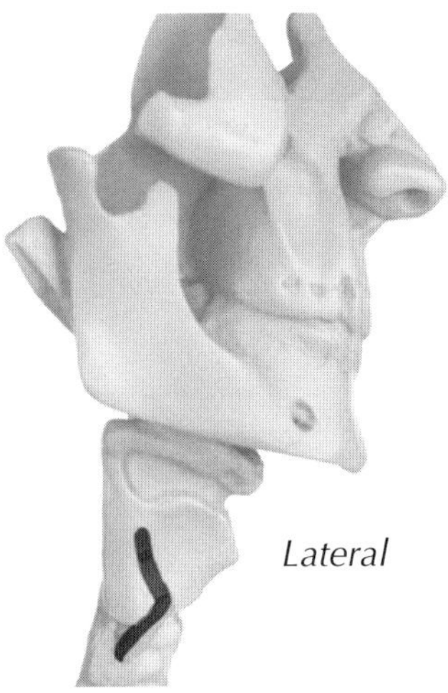

Lateral

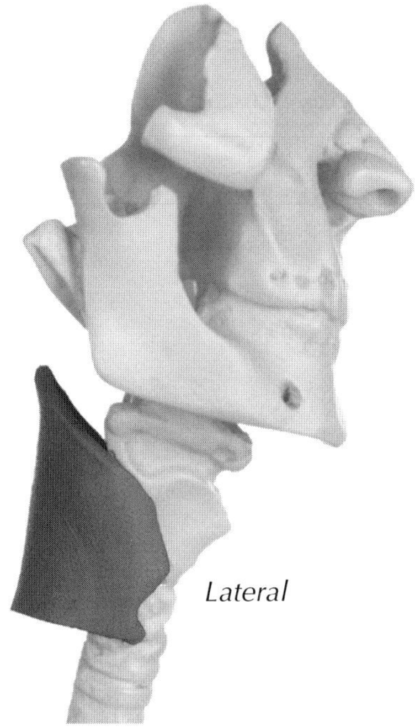

Lateral

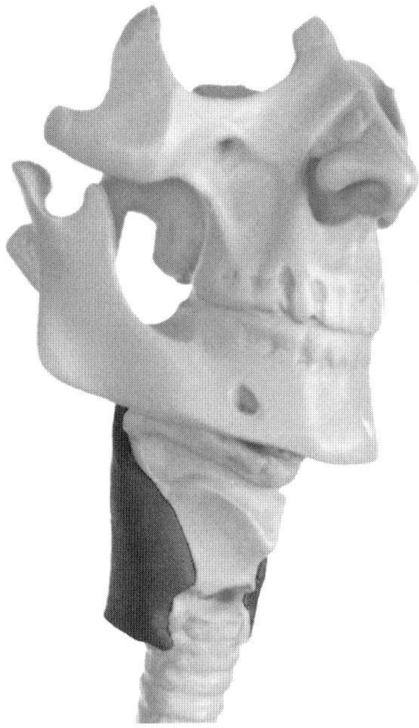

Ventrolateral

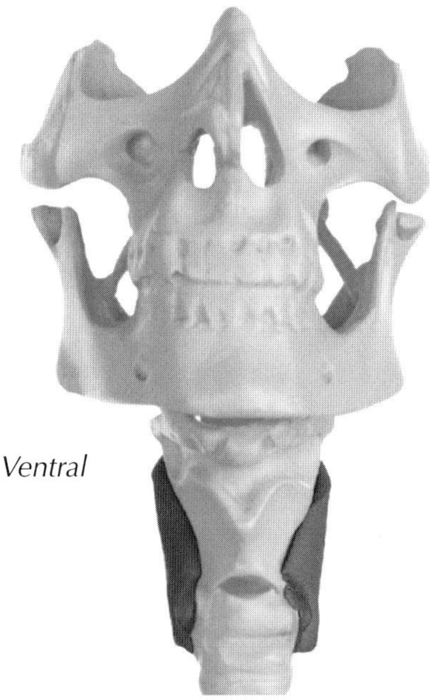

Ventral

Craniodorsolateral

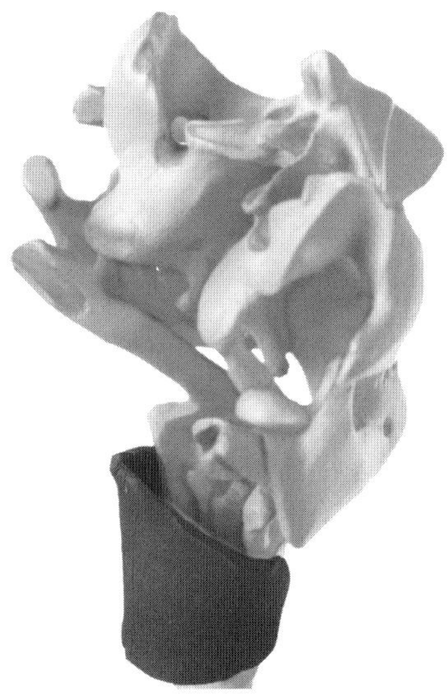

Constrictor pharyngis inferior

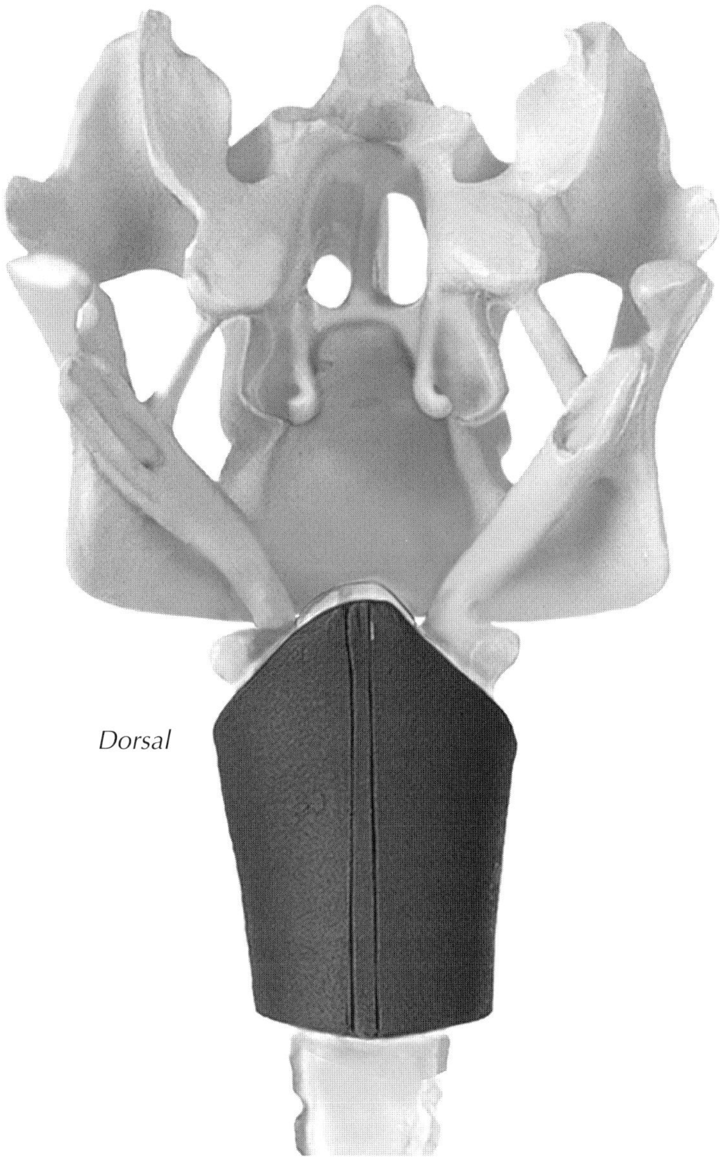

Dorsal

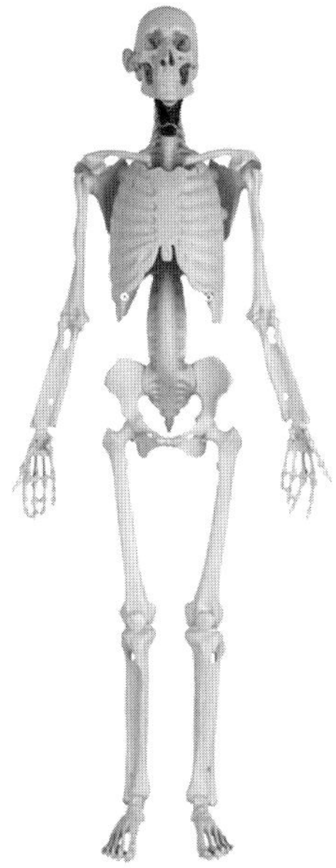

Actions

Constricts lower pharnyx for swallowing.

Attachments

Ventral

Oblique line of thyroid cartilage, lateral cricoid cartilage.

Dorsal

Posterior midline raphé of the pharynx; upper parft overlaps constrictor medius.

Caudodorsal

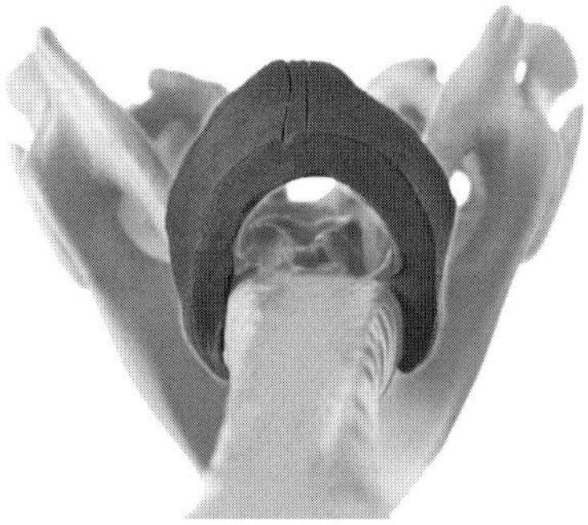

Cranial

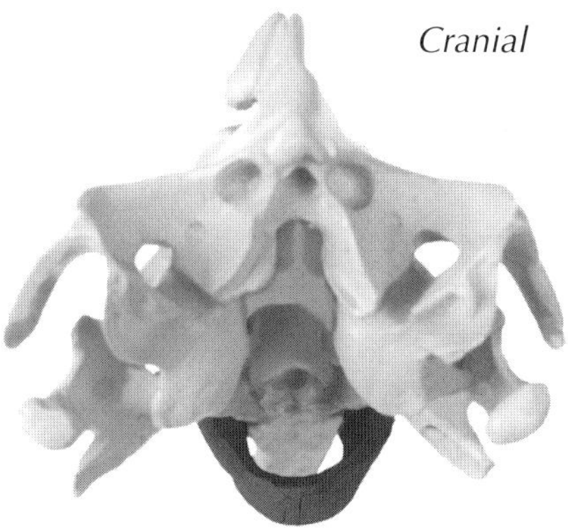

*kon•**strik**•tur*

From the Latin, "constrictus"+"or," meaning one that binds or ties up.

***fare**•in•jyus*

From the Greek, "pharynx," meaning throat.

*in•**fee**•ree•ur*

Latin, meaning lower or beneath.

Stylopharyngeus

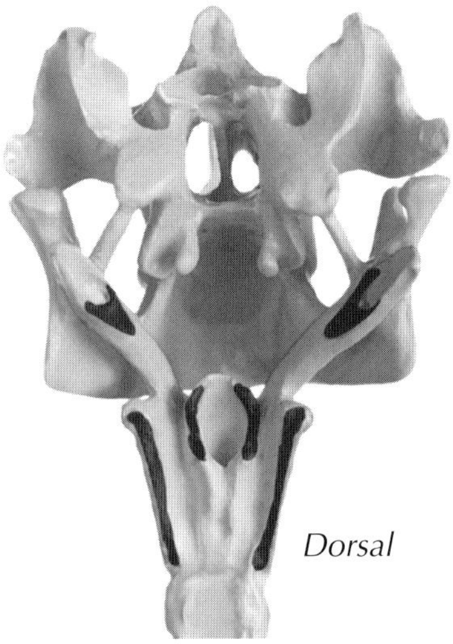

Dorsal

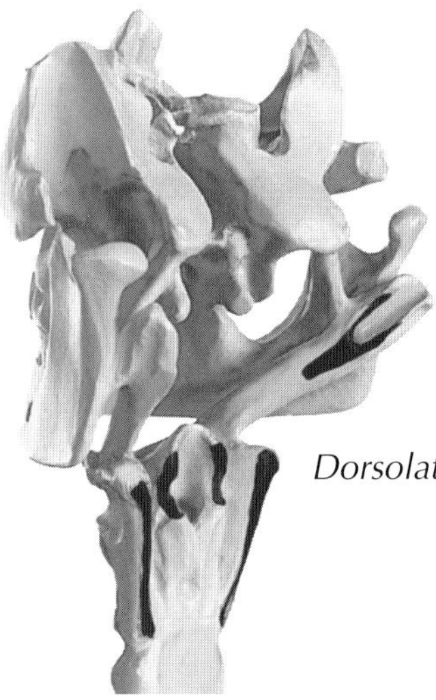

Dorsolateral

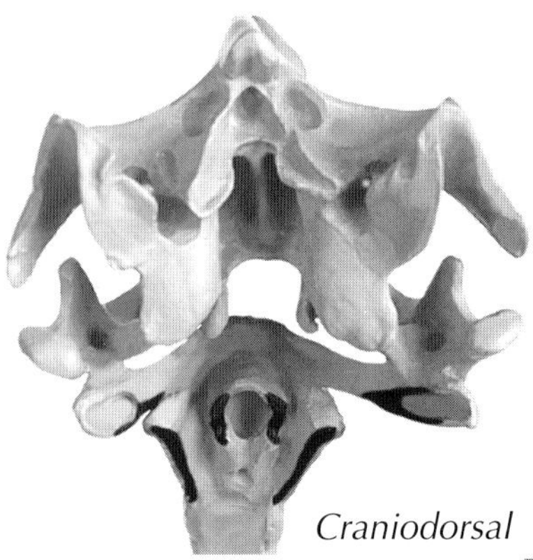

Craniodorsal

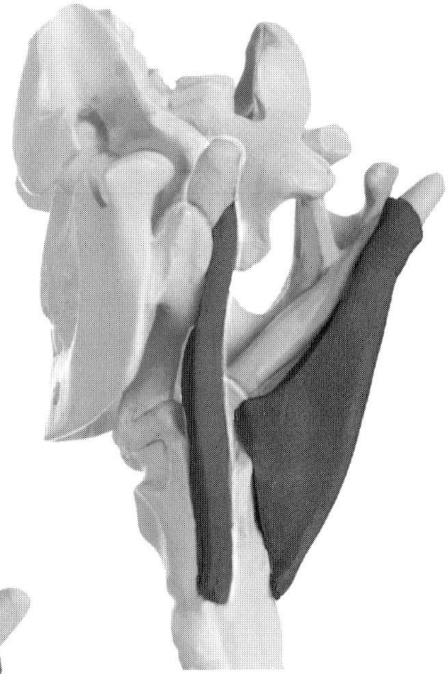

Dorsolateral

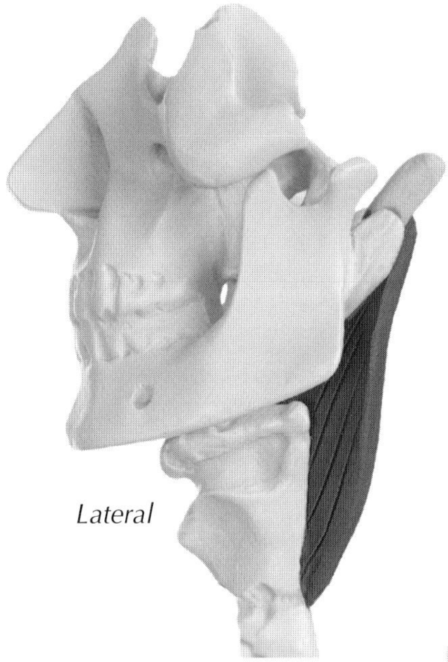

Lateral

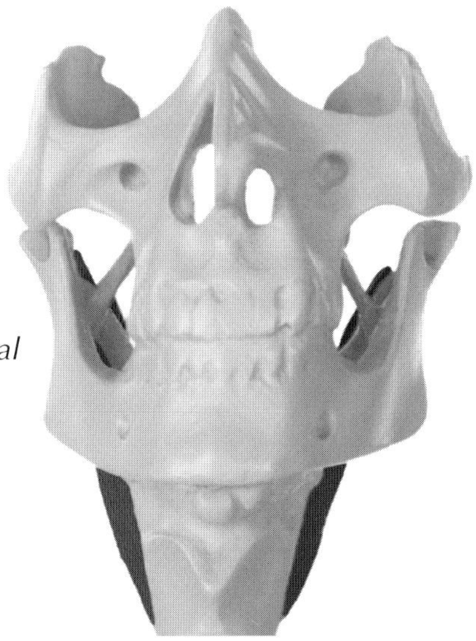

Ventral

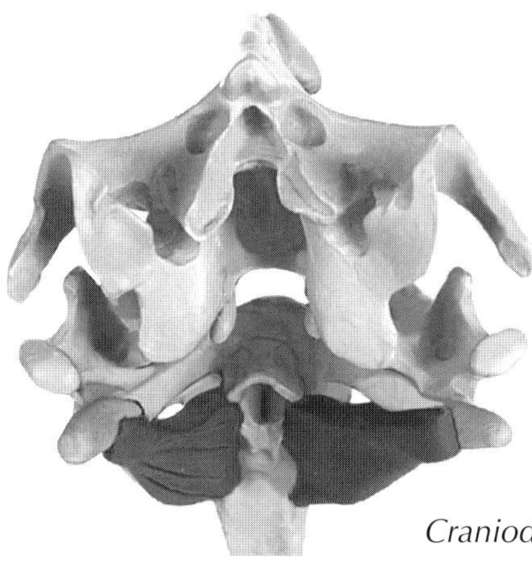

Craniodorsal

Stylopharyngeus

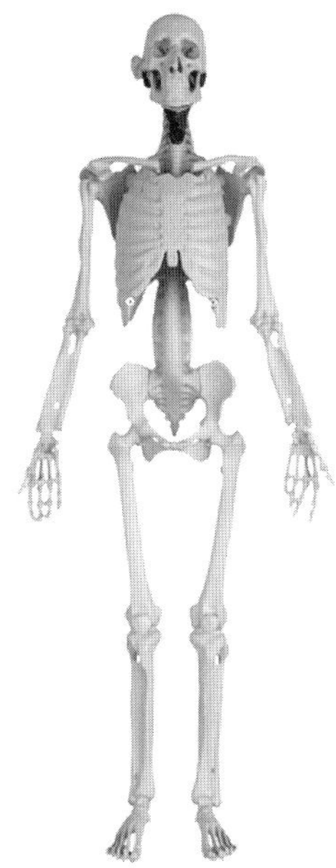

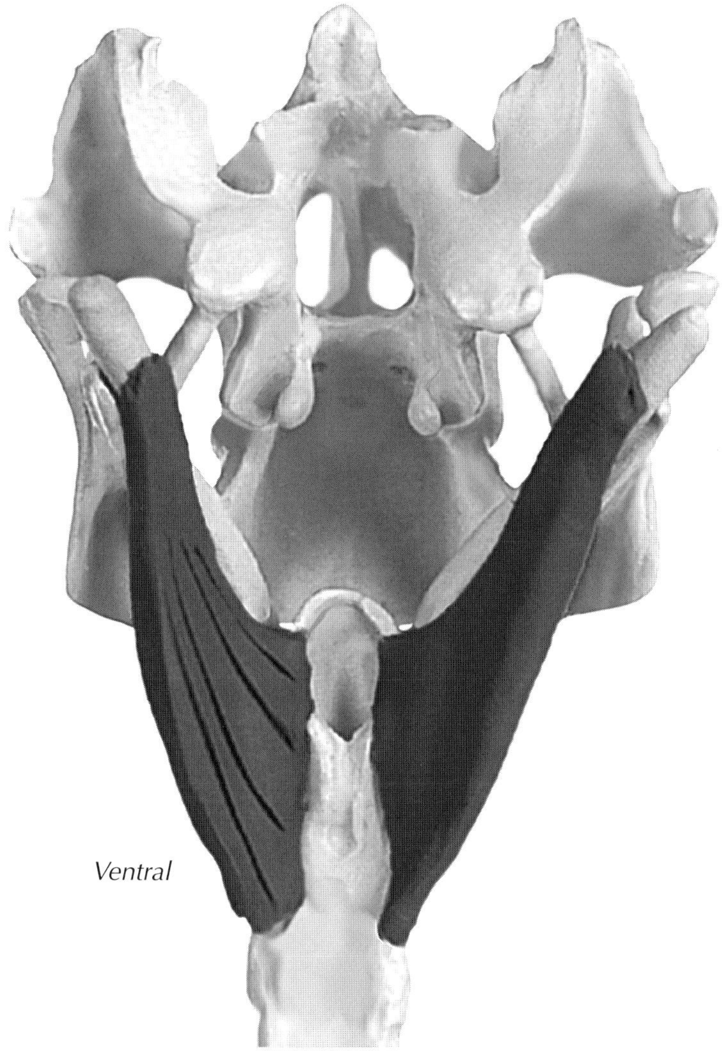

Ventral

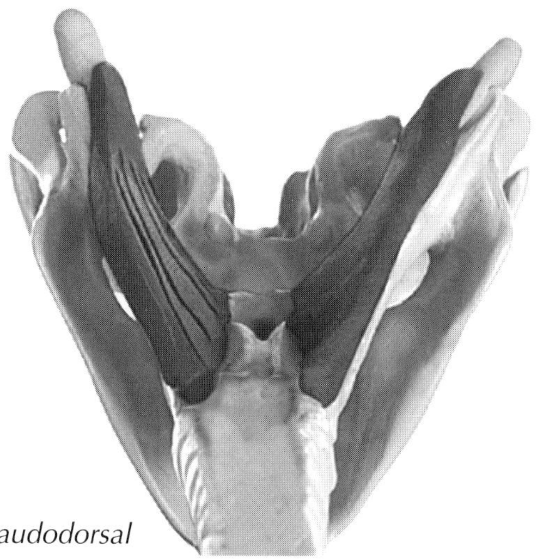

Caudodorsal

Actions

Elevates the pharynx.

Attachments

Cranial

Medial side of styloid process.

Caudal

Dorsal edge of thyroid cartilage and epiglottic fold.

*sty•low•fur•**in**•jyus*

From the Latin, "stylus," meaning pen, and from the Greek, "pharynx," meaning throat.

Buccinator

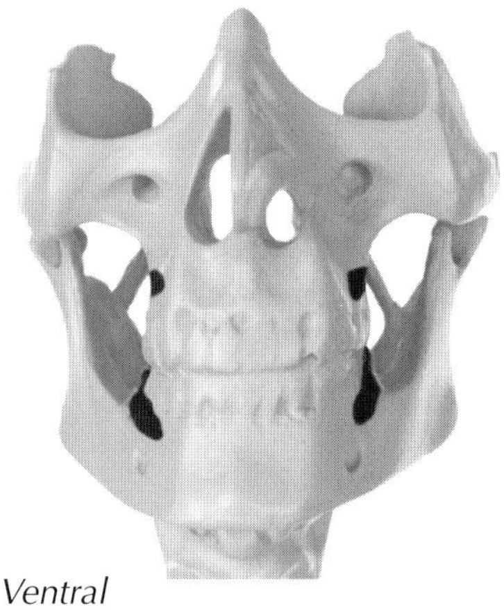

Ventral

Ventrolateral

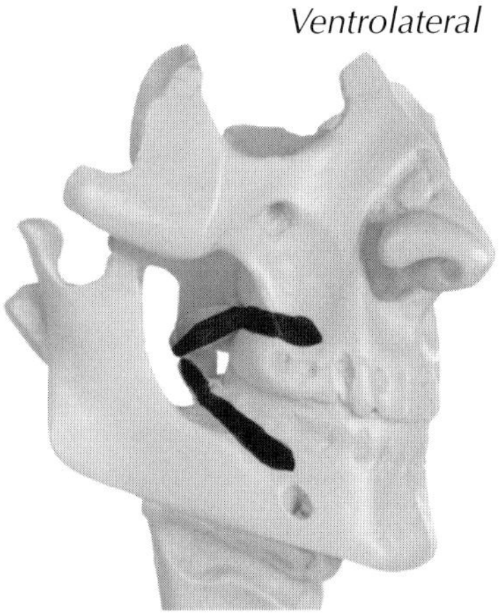

Lateral

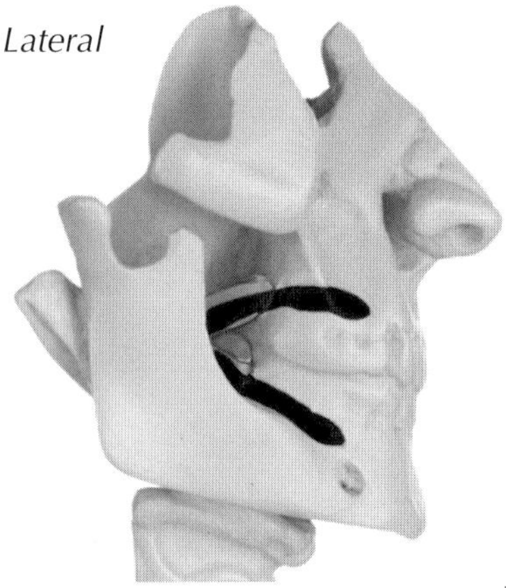

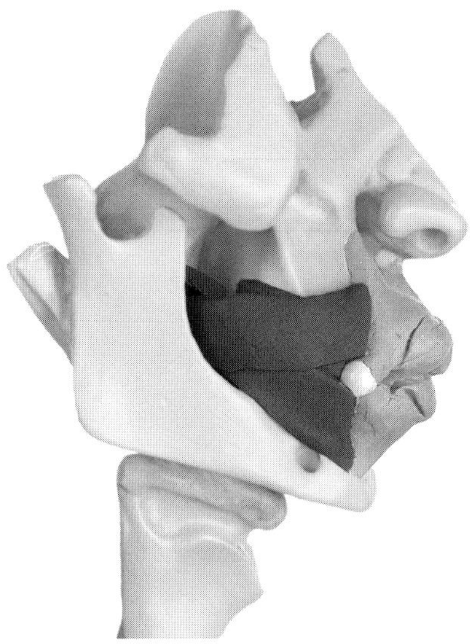

Lateral

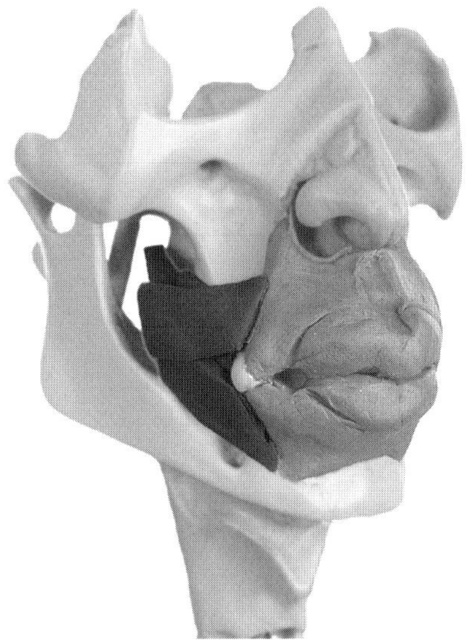

Ventrolateral

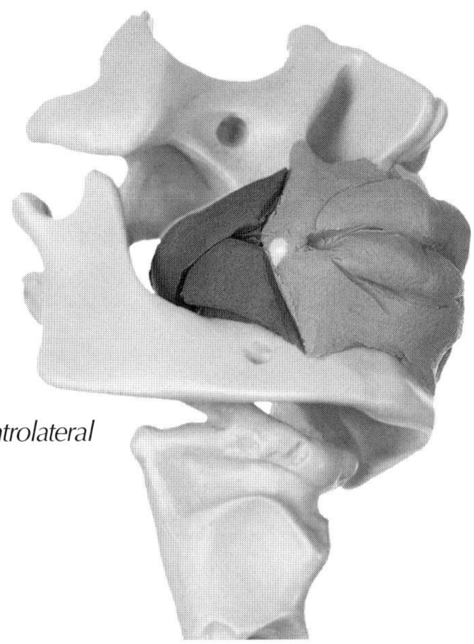

Caudoventrolateral

Dorsal

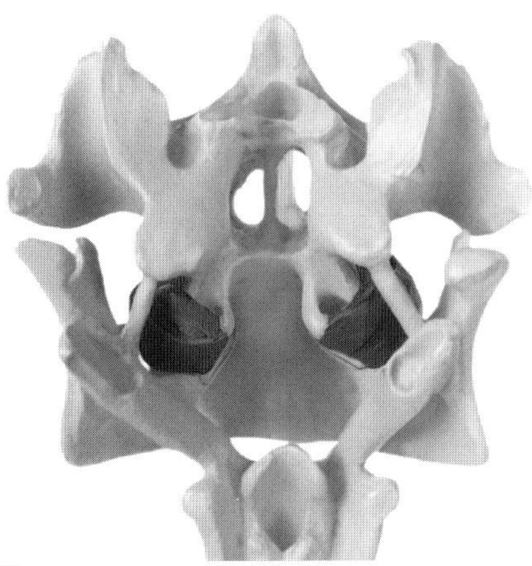

Dorsolateral

Buccinator

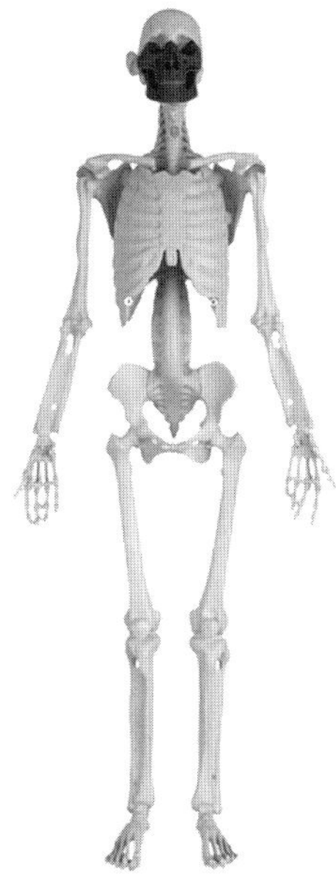

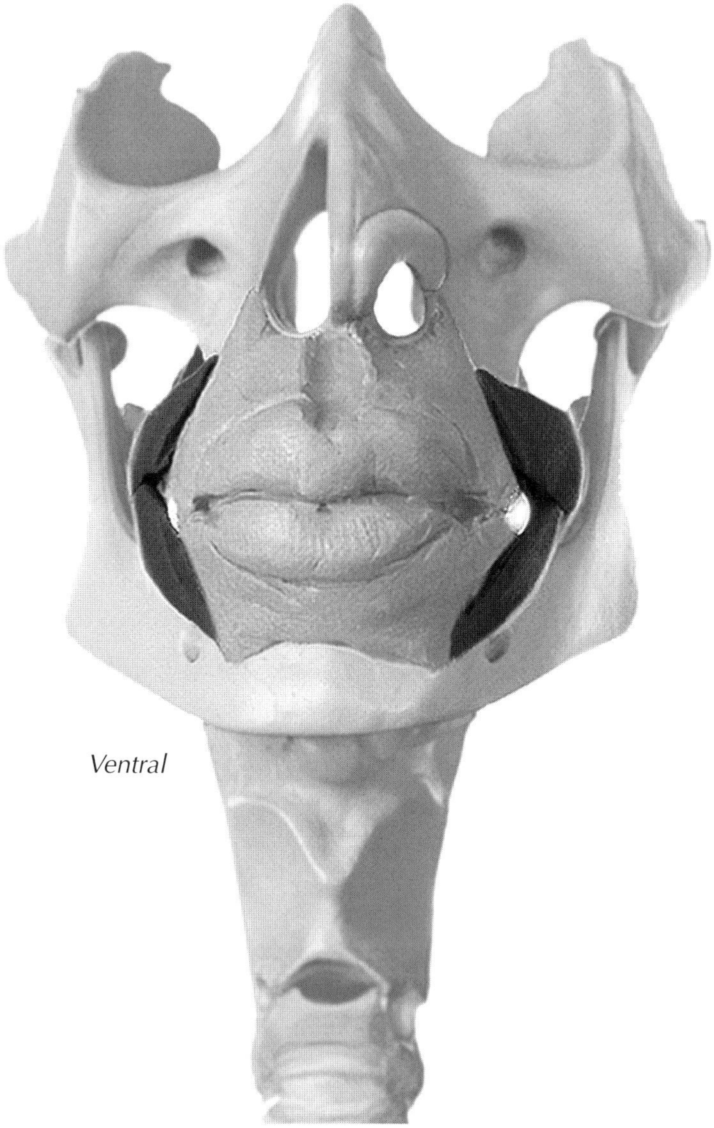

Ventral

Cranial

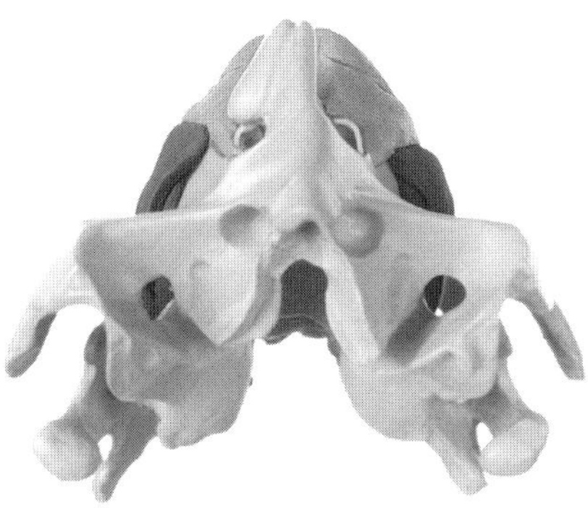

Caudal

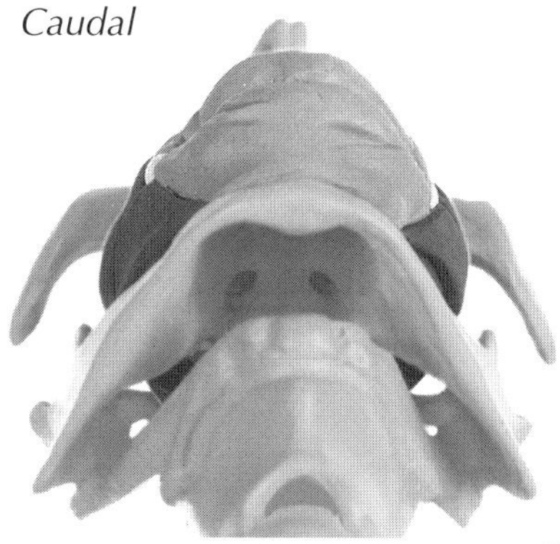

Actions

Compresses the cheeks and retracts the corners of the mouth.

Attachments

Dorsal

Pterygomandibular ligament, alveolar processes of maxilla and mandible.

Ventral

Modiolus at the corner of the mouth (see the note for orbicularis oris).

***buks**•ih•nay•tur*

From the Latin, "buccinatus"+"or," meaning one who plays the trumpet.

Masseter
pars profunda

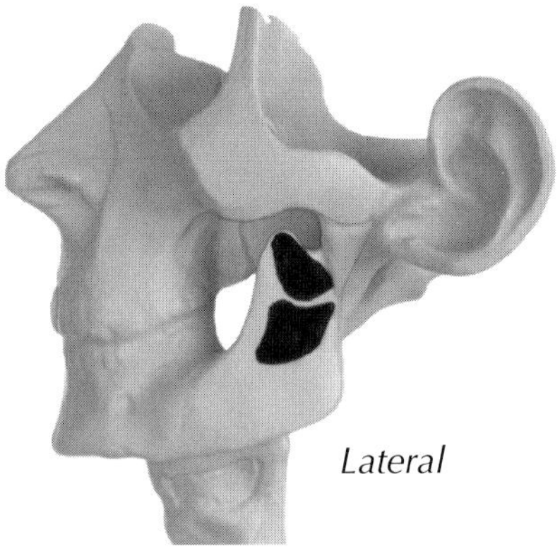

Lateral

Dorsolateral

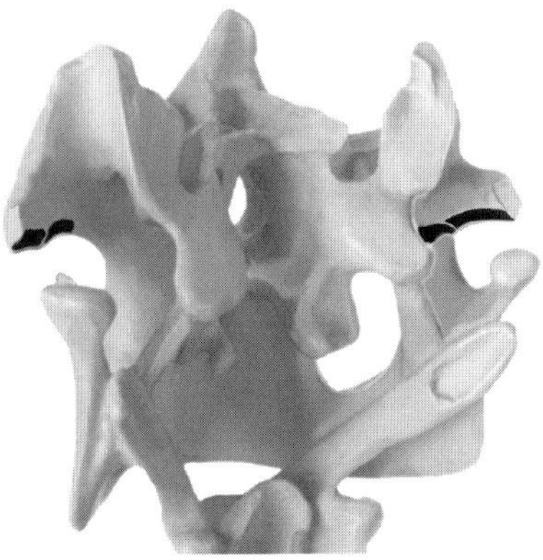

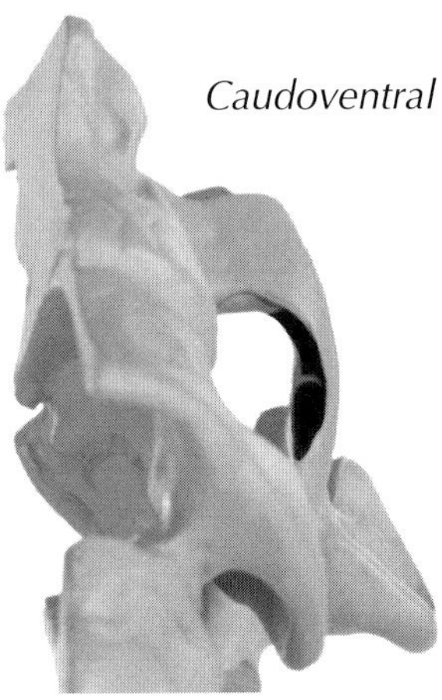

Caudoventral

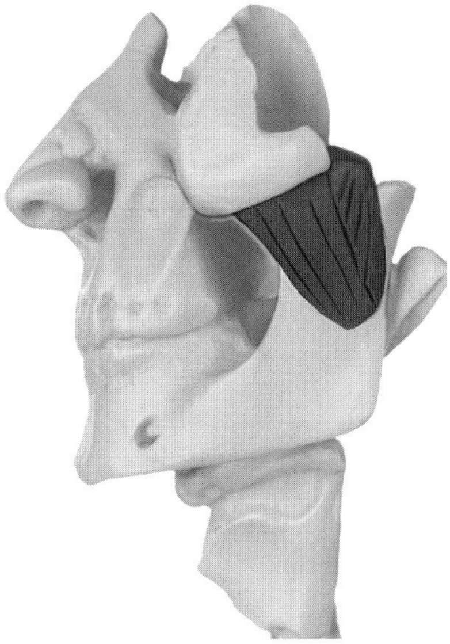

Lateral

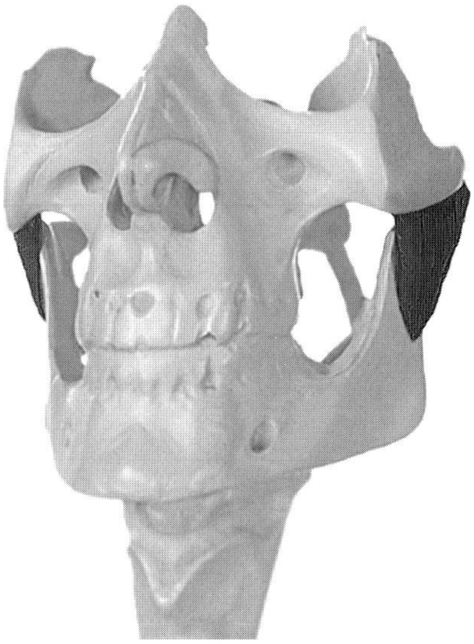

Ventrolateral

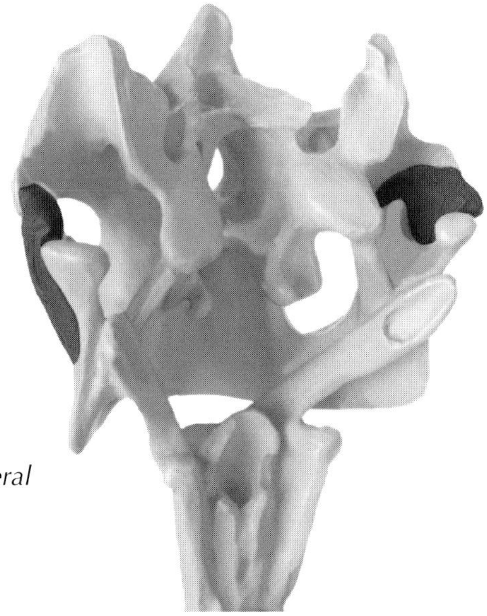

Dorsolateral

Lateral

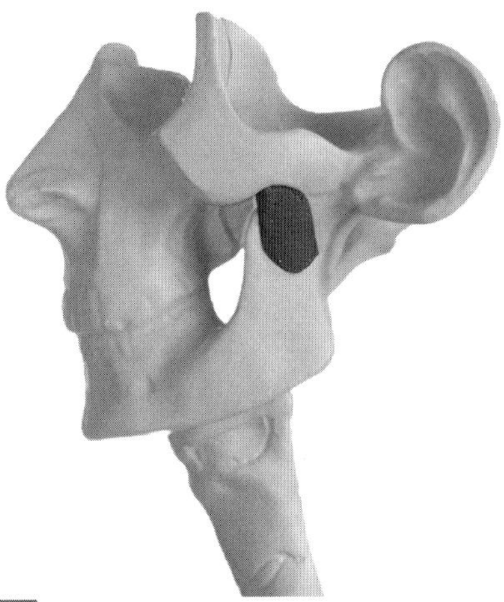

Ventral

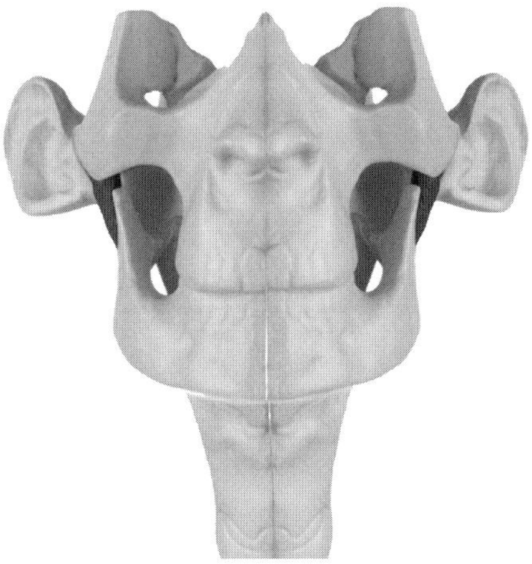

Masseter
pars profunda

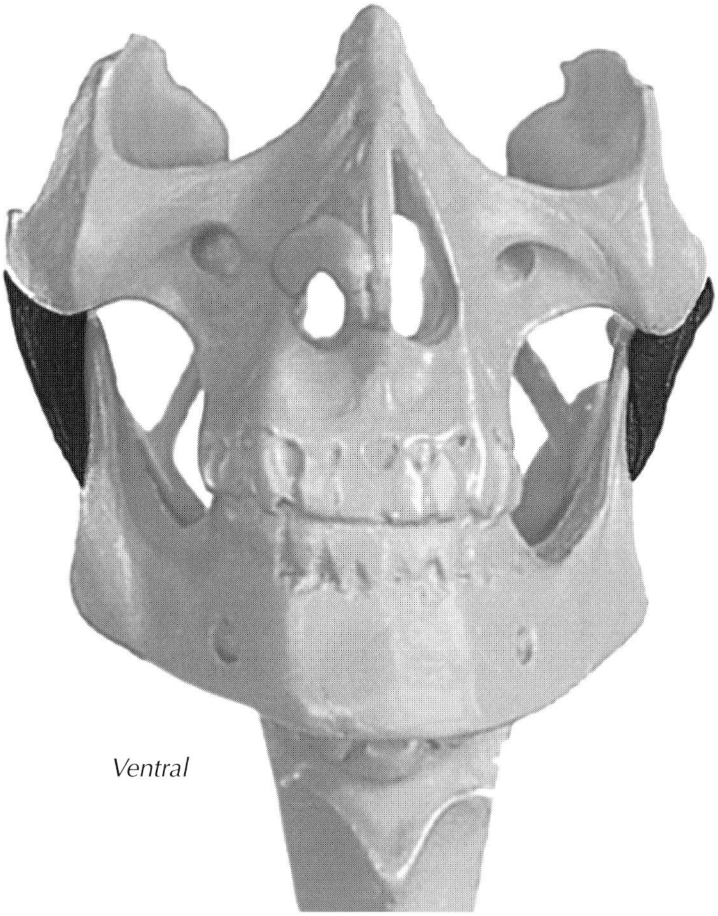

Ventral

Caudal

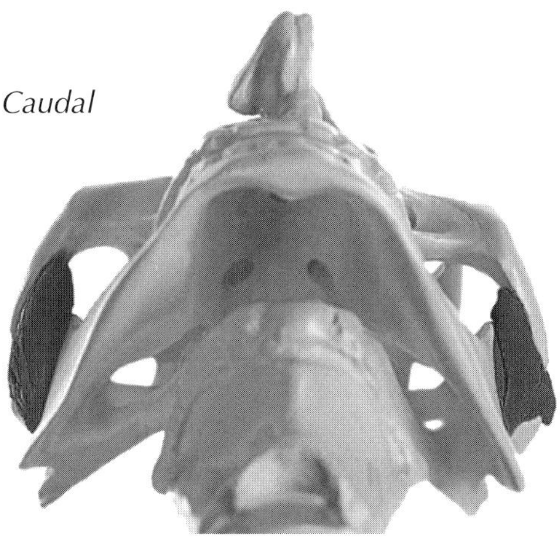

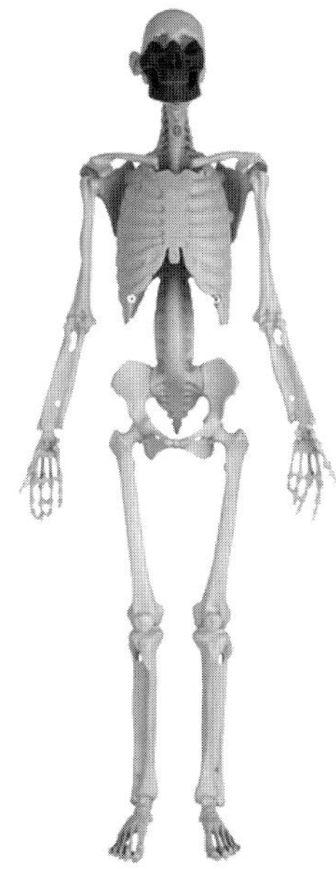

Actions
Elevates mandible, occludes teeth.

Attachments
Cranial

Medial surface of the zygomatic arch.

Caudal

Lateral surfaces of coronoid process and upper half of the mandible.

Note
Masseter has three layers: deep, middle and superficial. Nomina Anatomica describes only two: deep, or *profunda*, and superficial. The superficial layer is shown in MYOLOGIK™5, the Extrinsic volume.

*muh•***see***•tur*

From the Greek, "maseter," meaning one who chews.

parz

Latin, meaning part.

*pro•***fun***•da*

Latin, meaning deep.

Nasalis
pars alaris

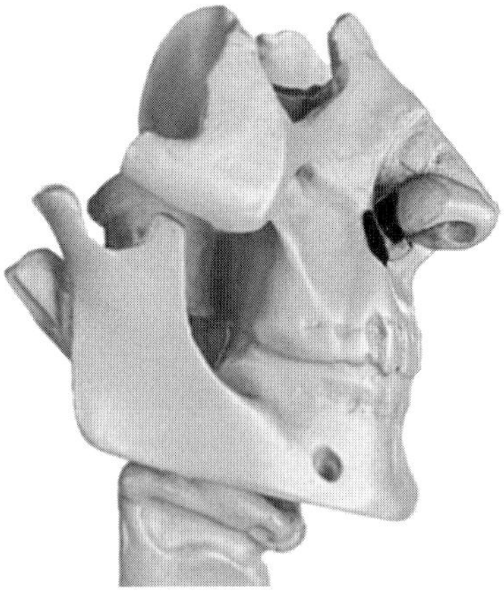

Lateral

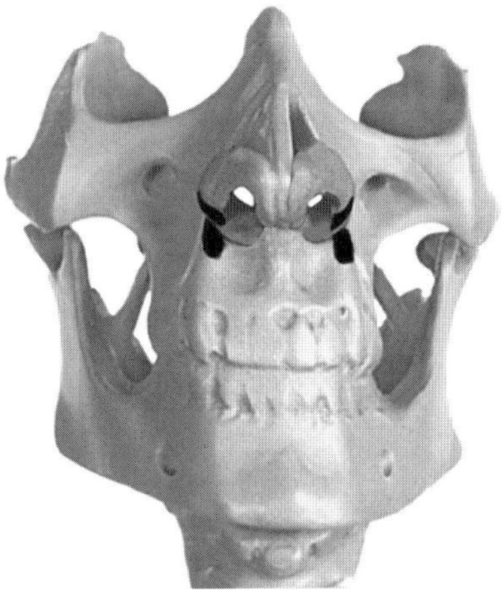

Ventral

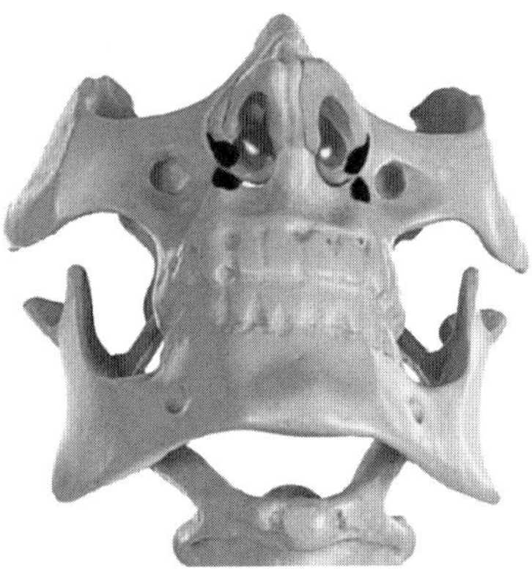

Caudoventral

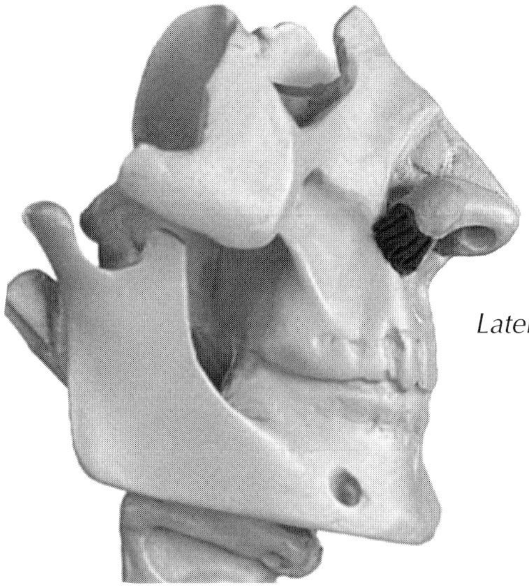

Lateral

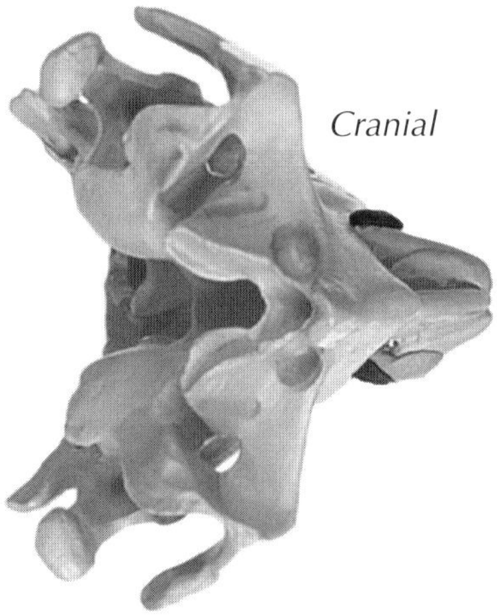

Cranial

Ventrolateral

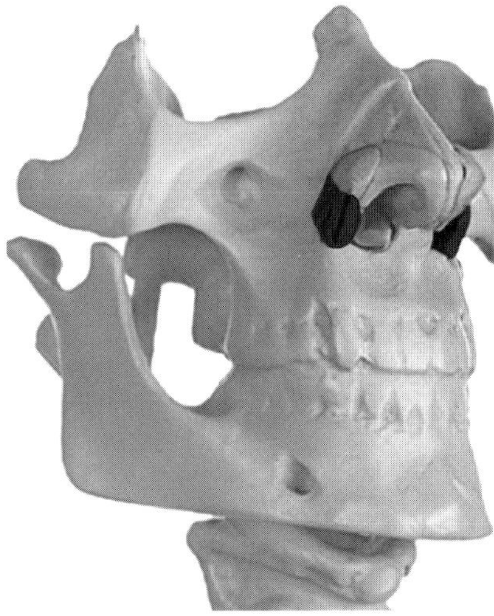

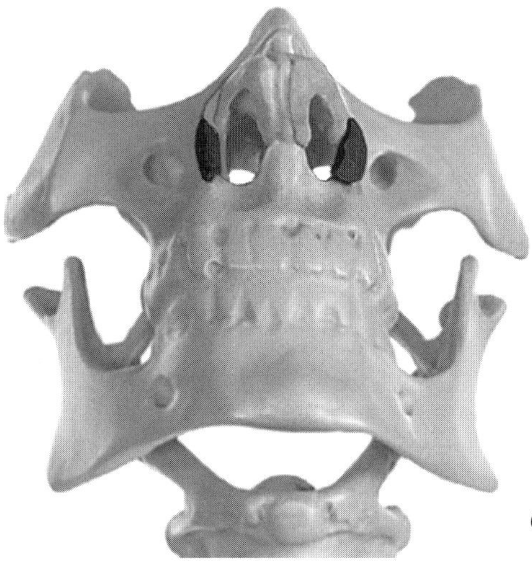

Caudoventral

Nasalis
pars alaris

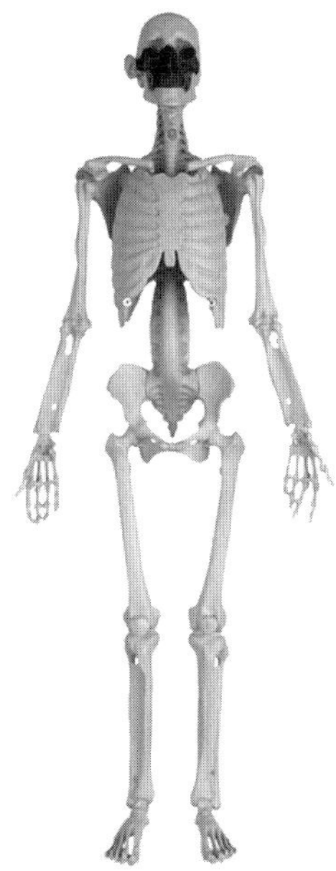

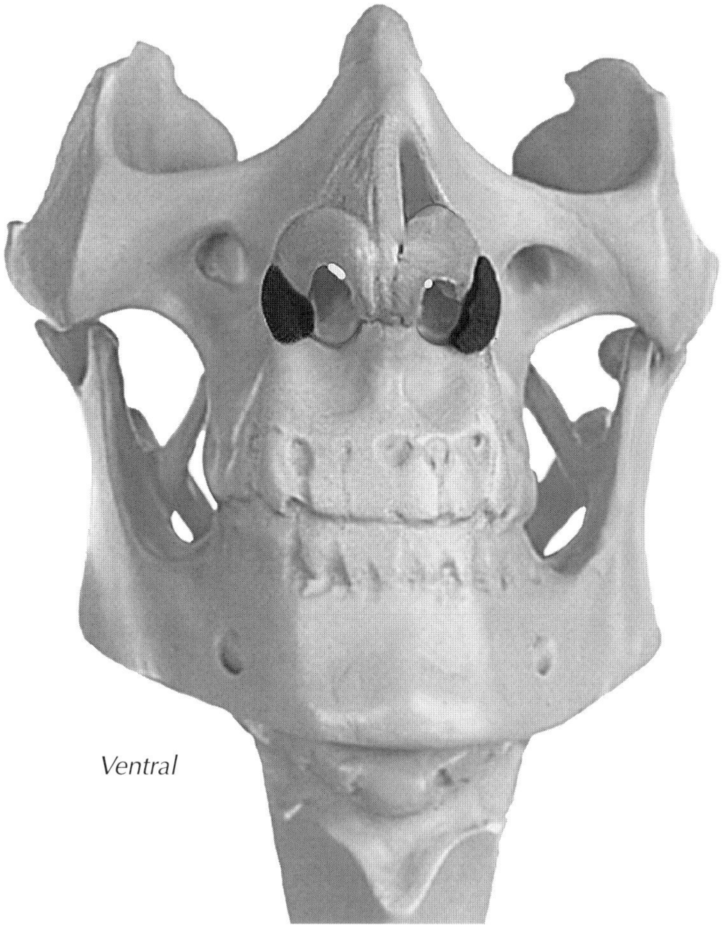

Ventral

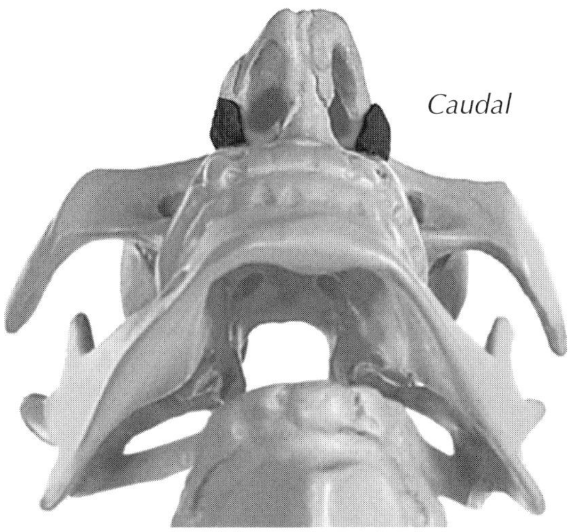

Caudal

Actions

Dilates nostril.

Attachments

Cranial

Edge of the nasal notch of the maxilla, lateral edge of external bony nare.

Caudal

Skin at margin of nostril.

*nay***•zal•***us*

From the Latin, "nasus"+"alis," meaning related to the nose.

parz

Latin, meaning part.

*a***•lar•***is*

Latin, meaning belonging to the wing or ala.

Nasalis
pars transversa

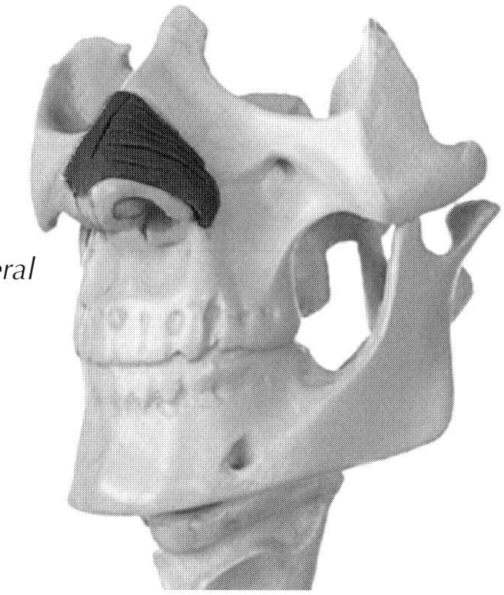

Ventrolateral

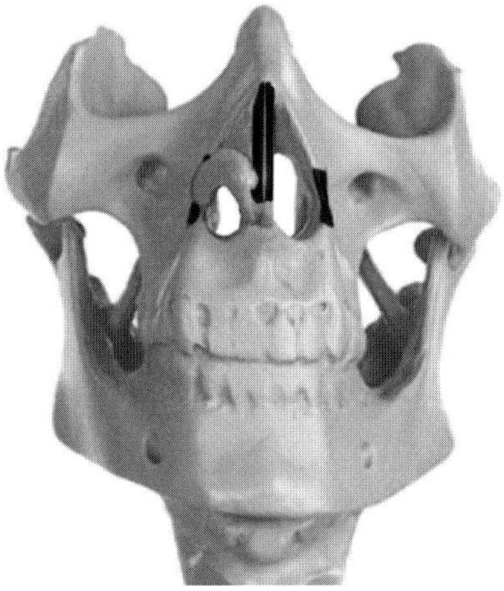

Ventral

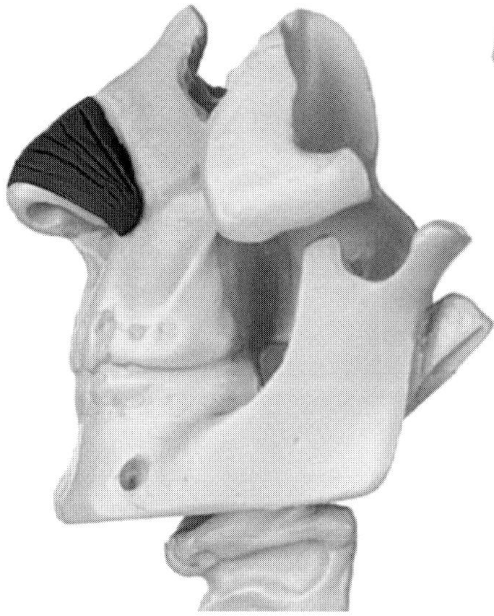

Lateral

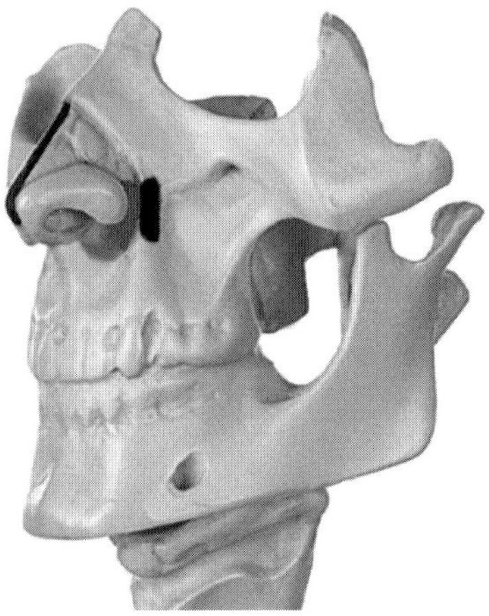

Ventrolateral

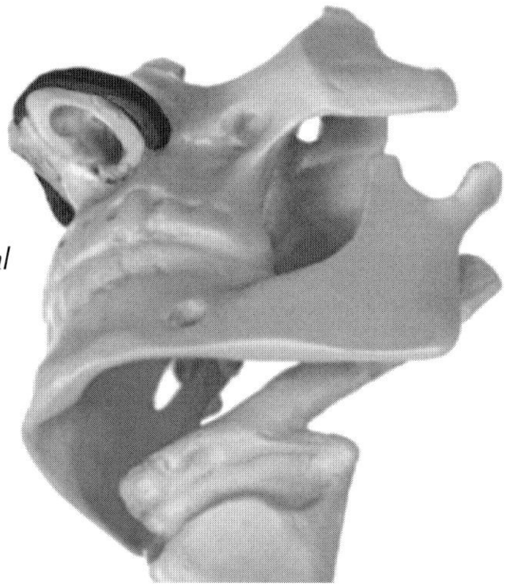

Caudoventrolateral

Cranial

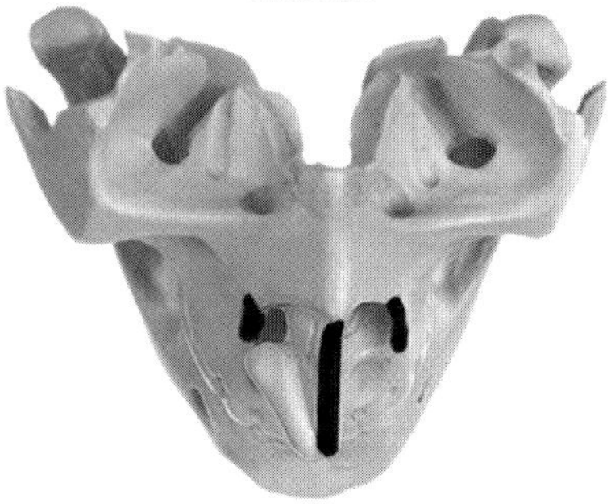

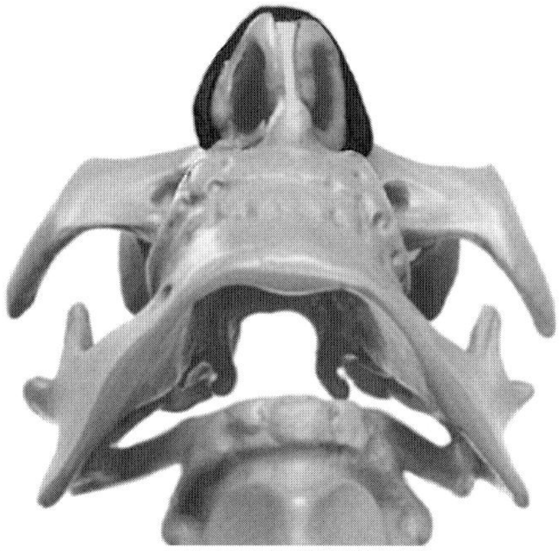

Caudoventral

Nasalis
pars transversa

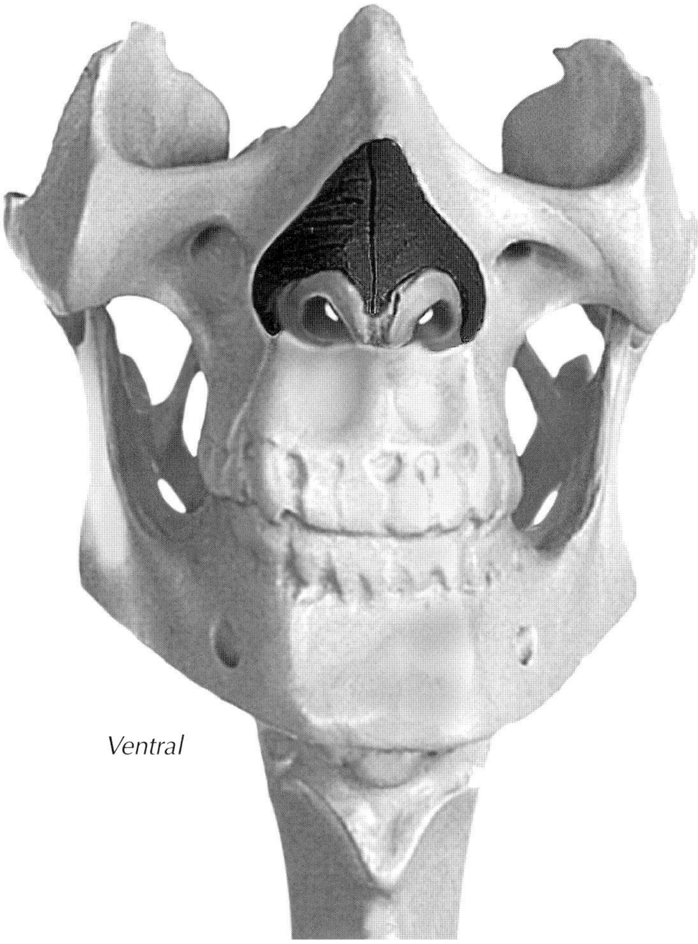

Ventral

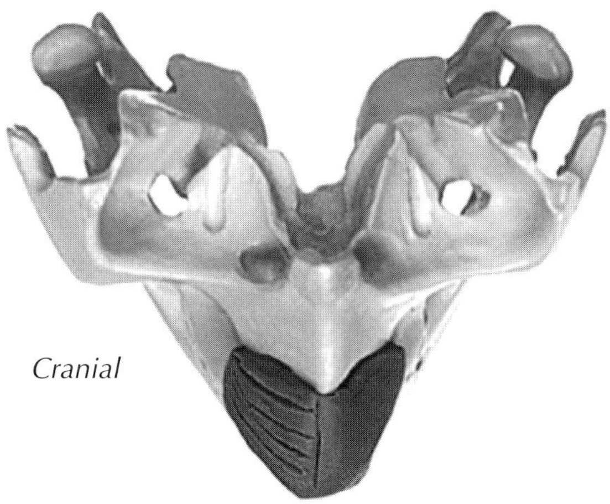

Cranial

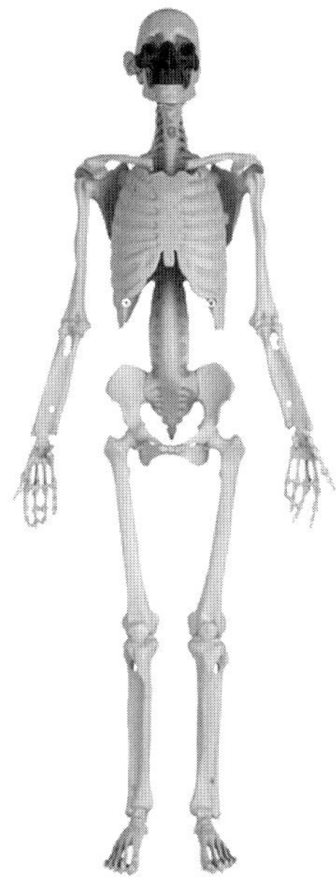

Actions

Compresses nostril.

Attachments

Lateral

Canine eminence of maxilla.

Medial

Nasal cartilages.

*nay•**zal**•us*

From the Latin, "nasus"+"alis", meaning related to the nose.

parz

Latin, meaning part.

*trans•**ver**•sa*

Latin, meaning crossing.

Depressor septi

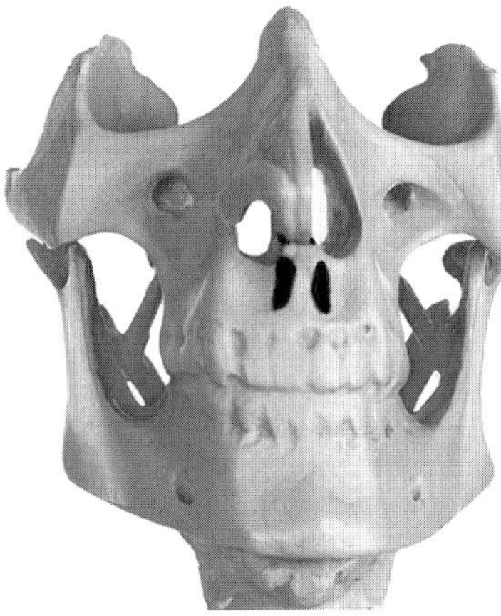

Ventral

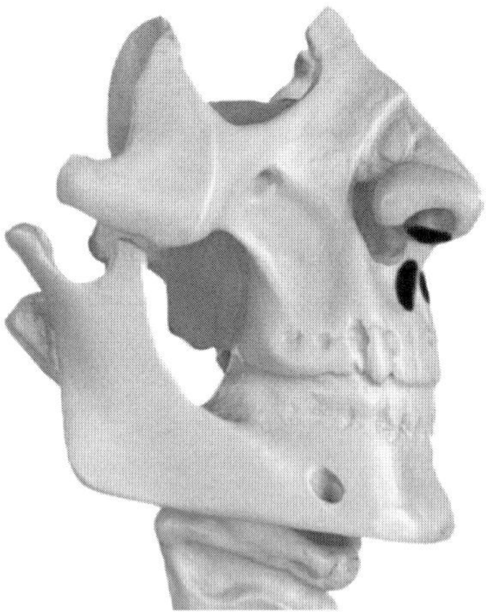

Ventrolateral

Ventrolateral

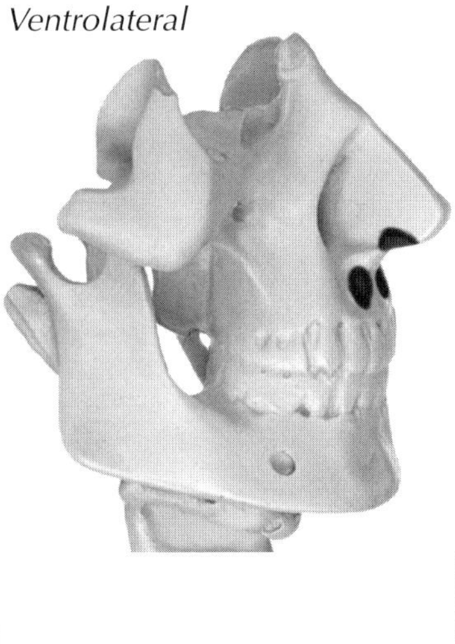

Ventrolateral

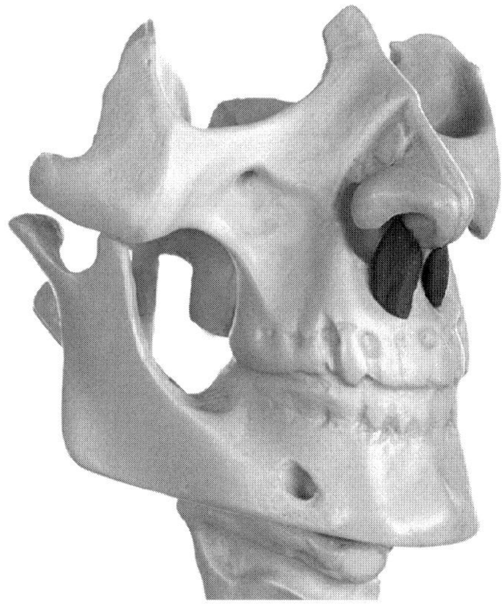

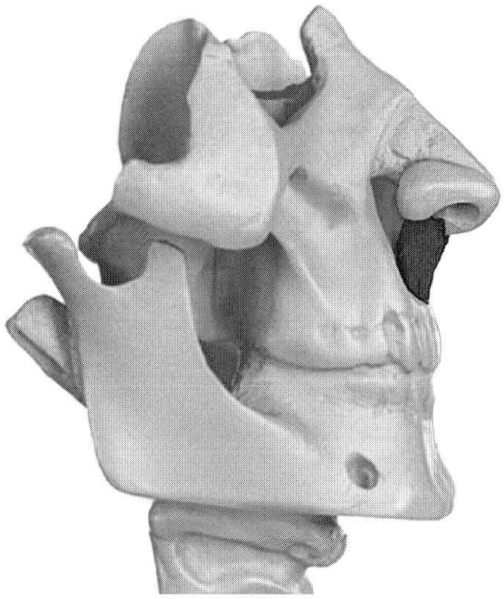

Lateral

Lateral

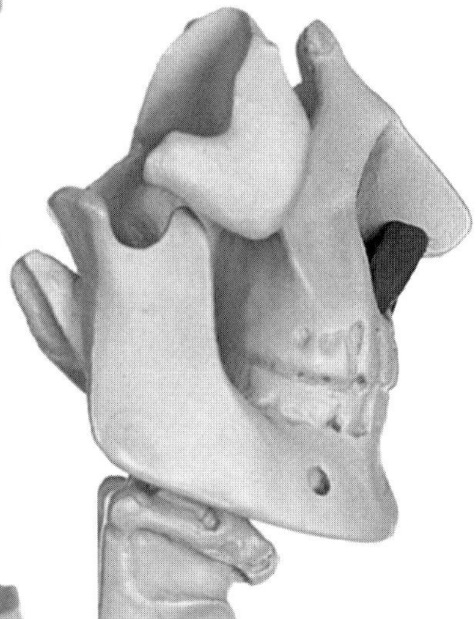

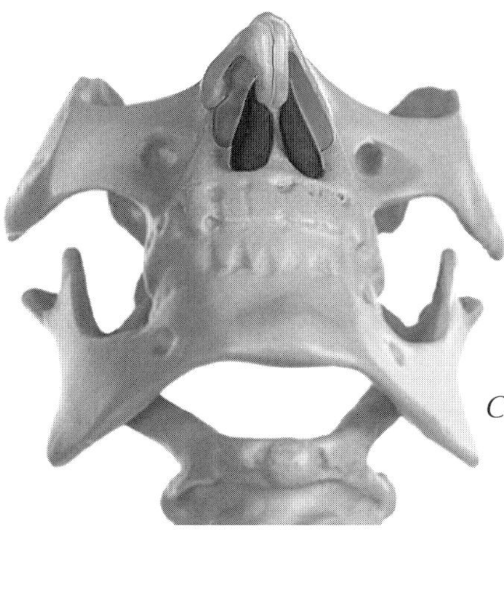

Caudoventral

Depressor septi

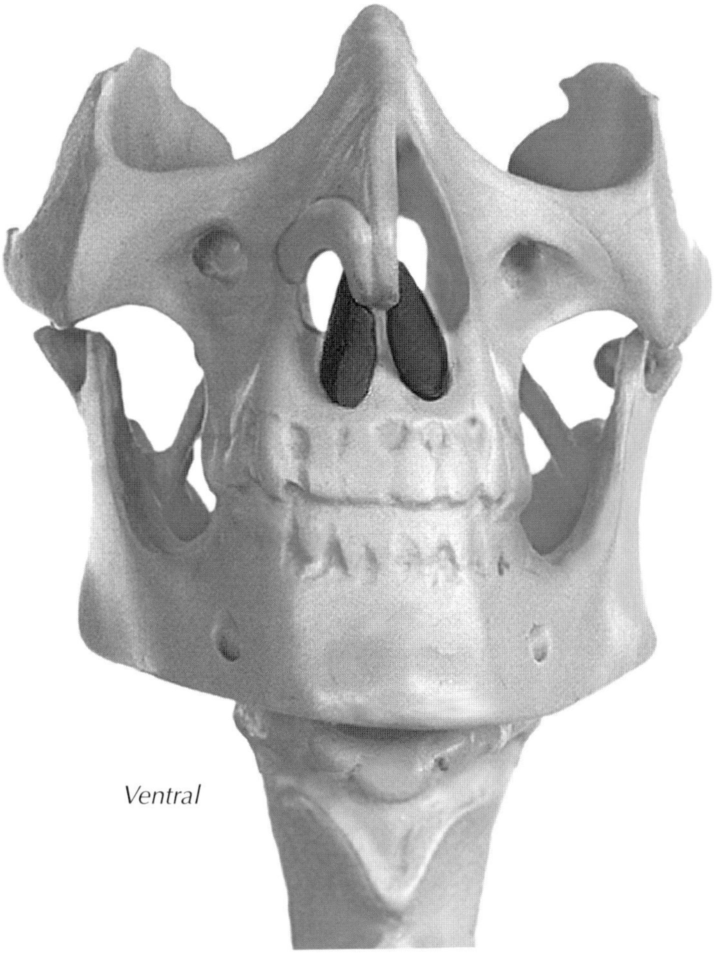

Ventral

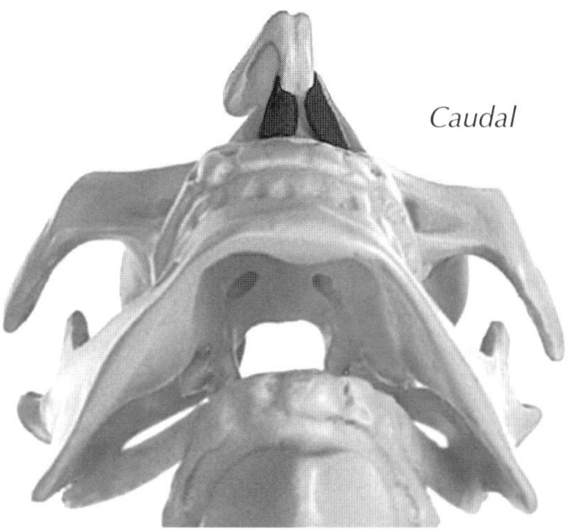

Caudal

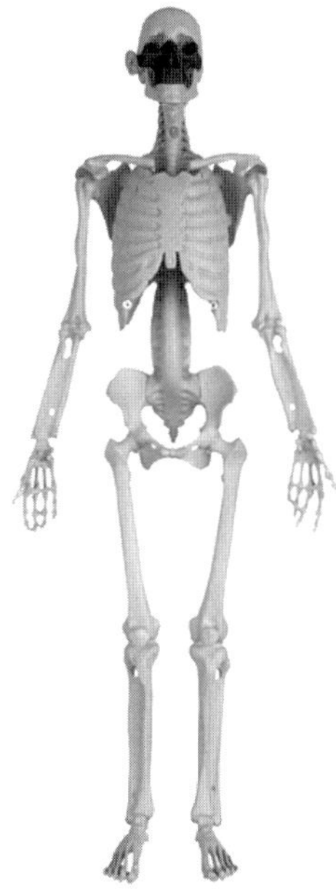

Actions
Pulls nasal septum downward or caudally.

Attachments
Caudal
Incisive fossa of maxilla.
Cranial
Nasal septum.

*duh•**pres**•ur*

From the Latin, "depressus"+"or", meaning to one that presses down.

***sep**•tee*

From the Latin, "septum," meaning fence or wall.

Mentalis

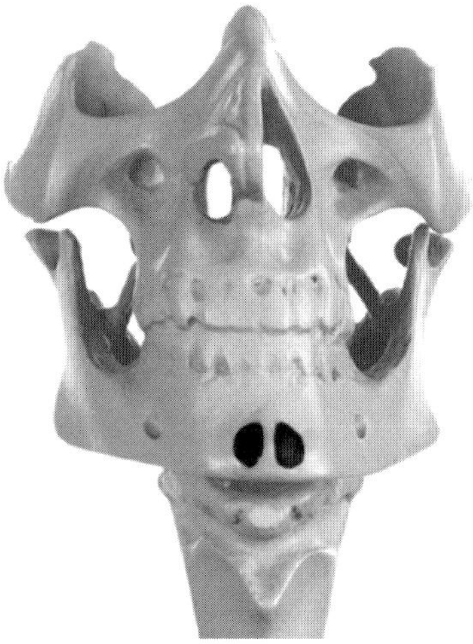

Ventral

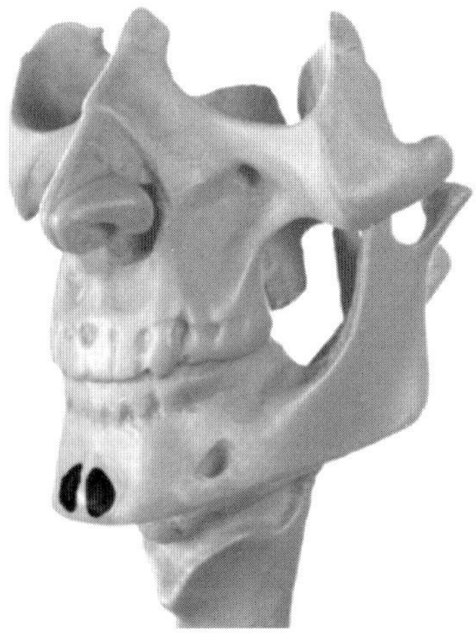

Ventrolateral

Cranial

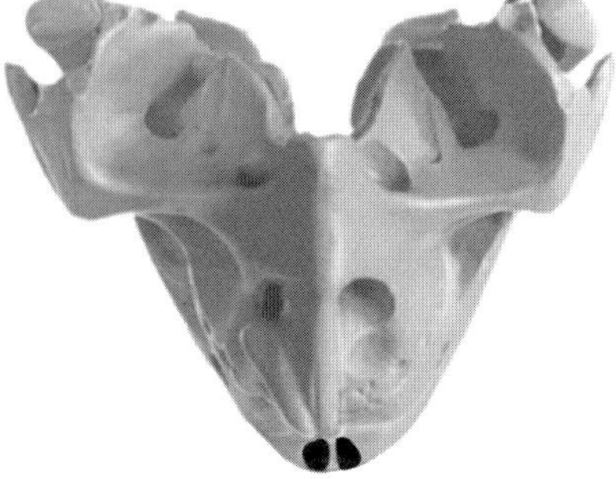

Ventrolateral

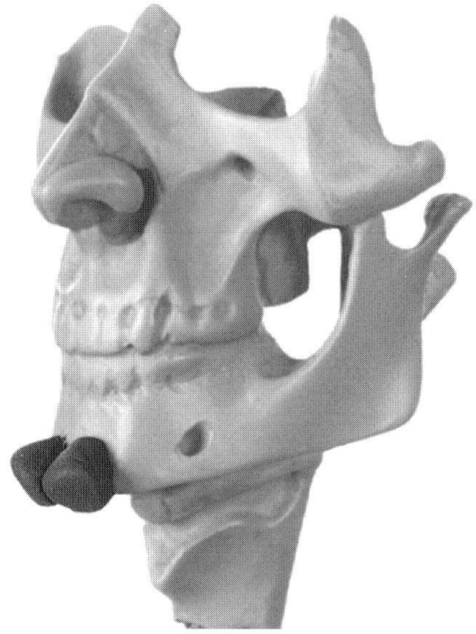

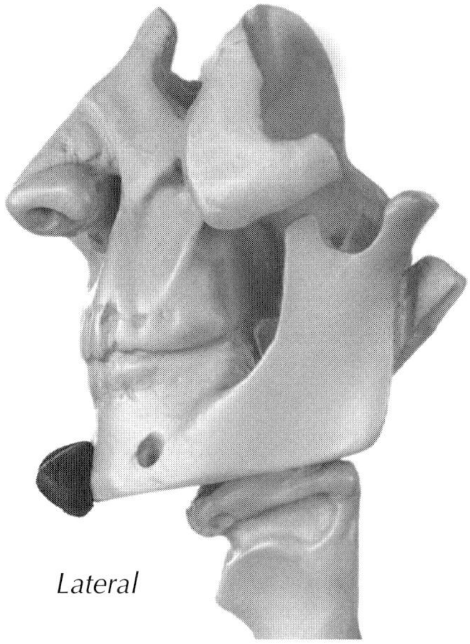

Lateral

Caudoventrolateral

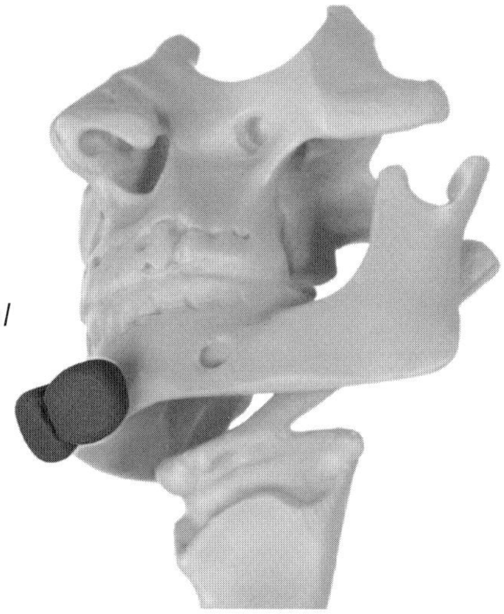

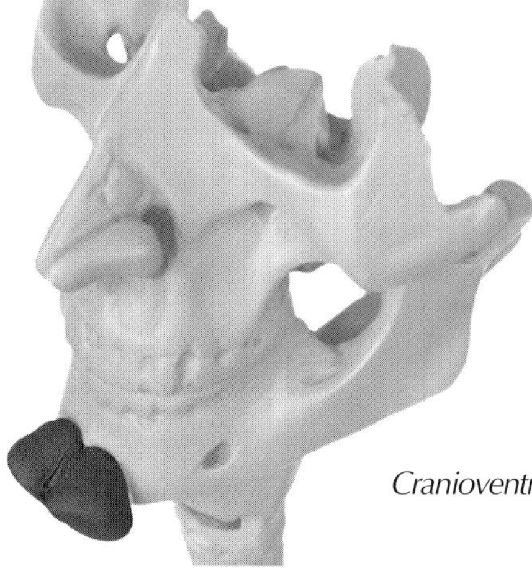

Cranioventrolateral

Mentalis

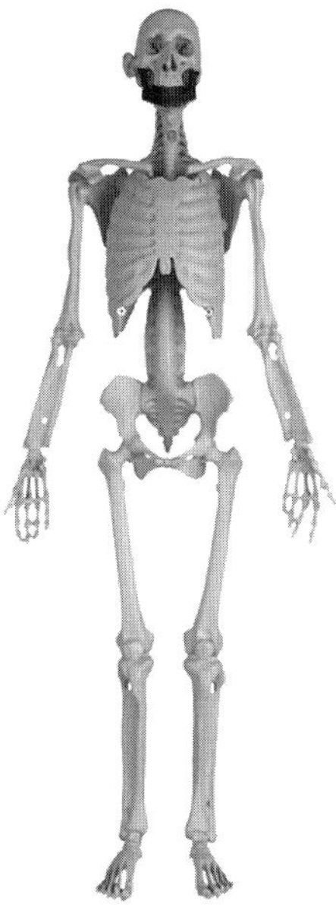

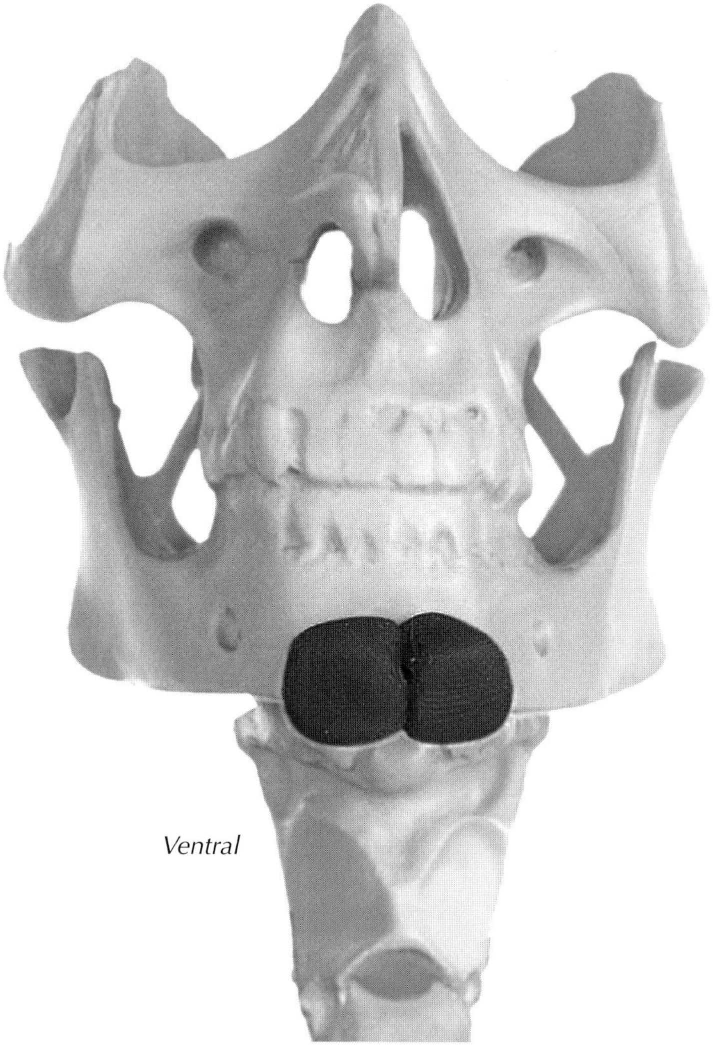

Ventral

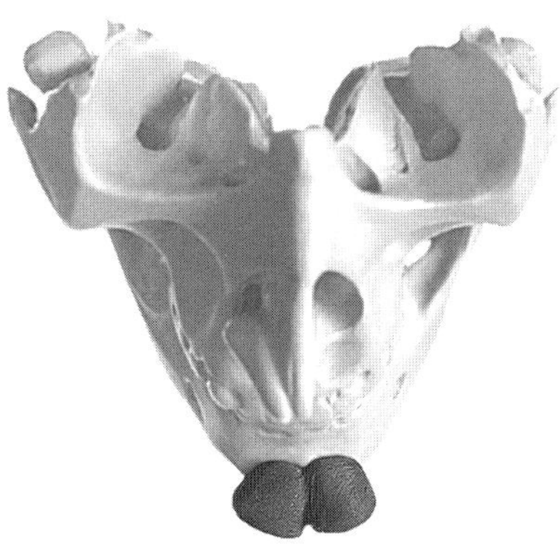

Cranioventral

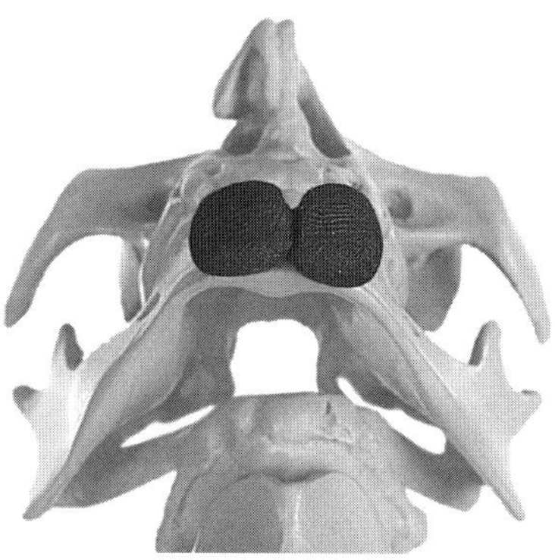

Caudal

Actions

Raises and protrudes lower lip, puckers the skin of the chin.

Attachments

Cranial

Incisive fossa of mandible.

Caudal

Skin of the chin.

men•***tal***•*us*

From the Latin, "mentum"+"alis," meaning related to the chin.

Incisivis superioris

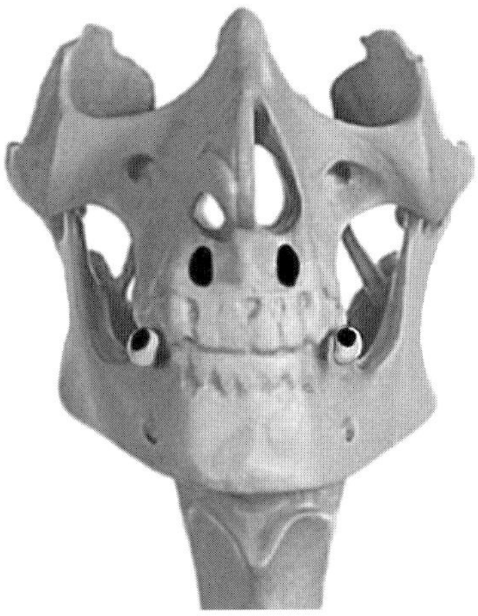

Ventral

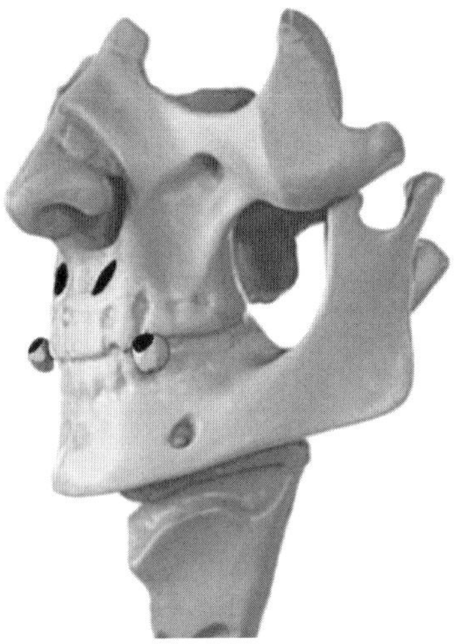

Ventrolateral

Cranioventral

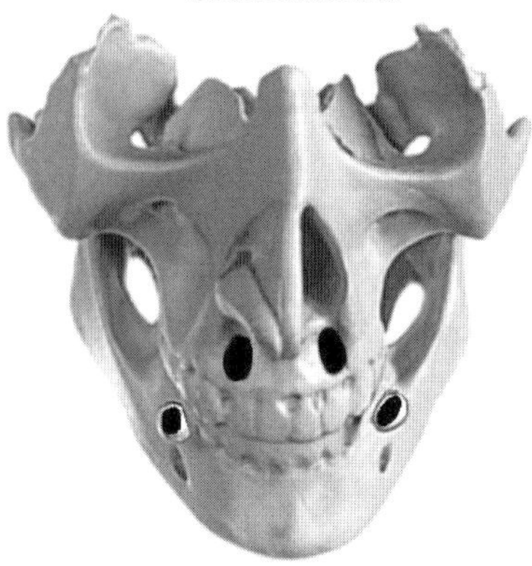

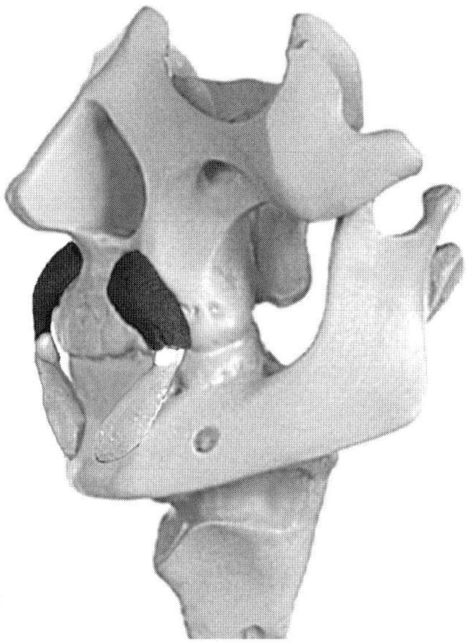

Ventrolateral

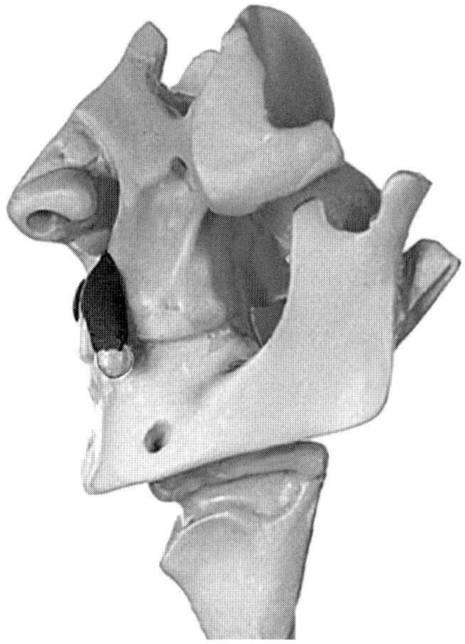

Lateral

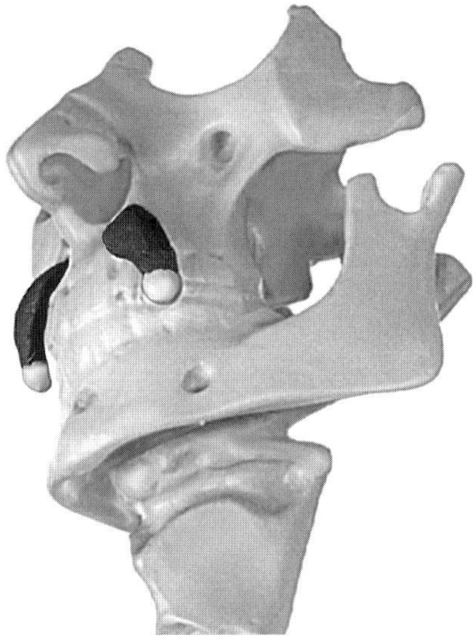

Caudolateroventral

Caudoventral

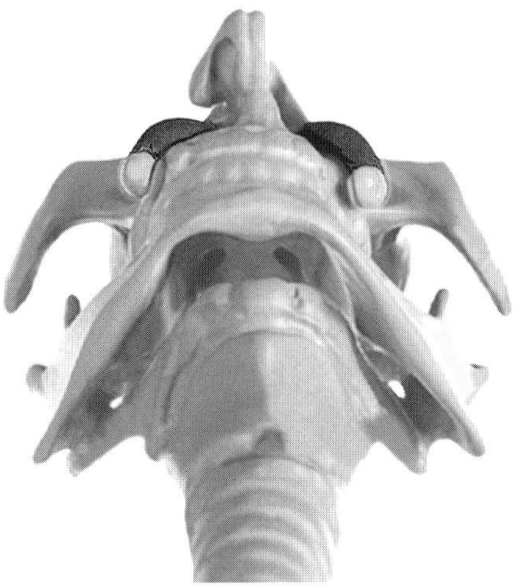

Incisivis superioris

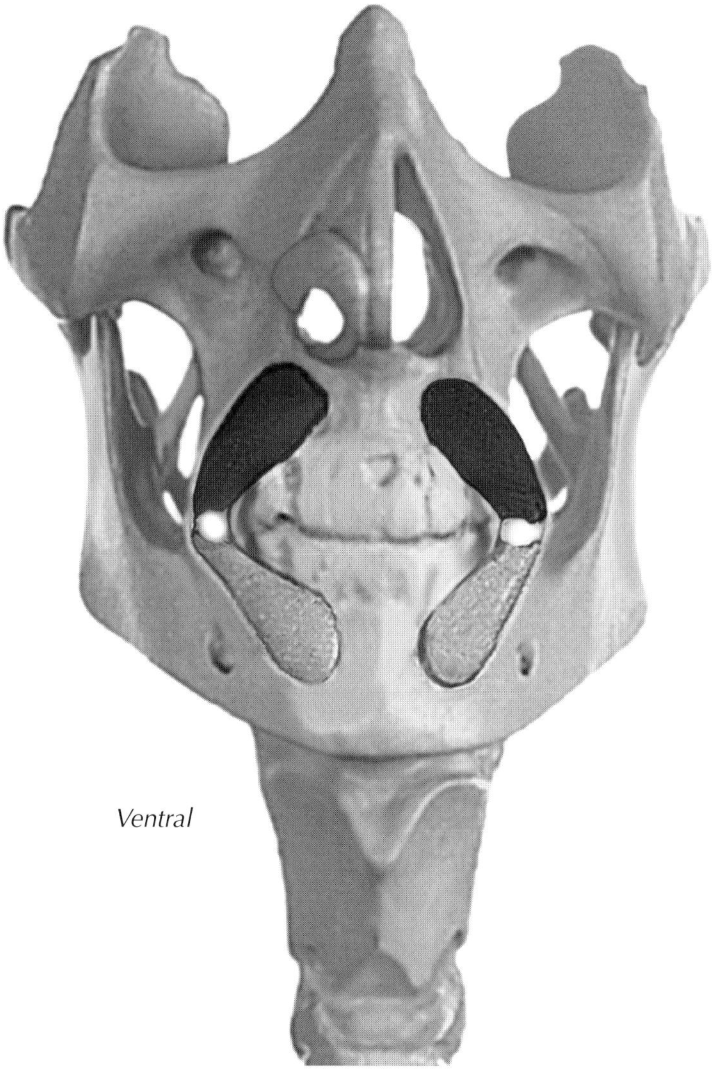

Ventral

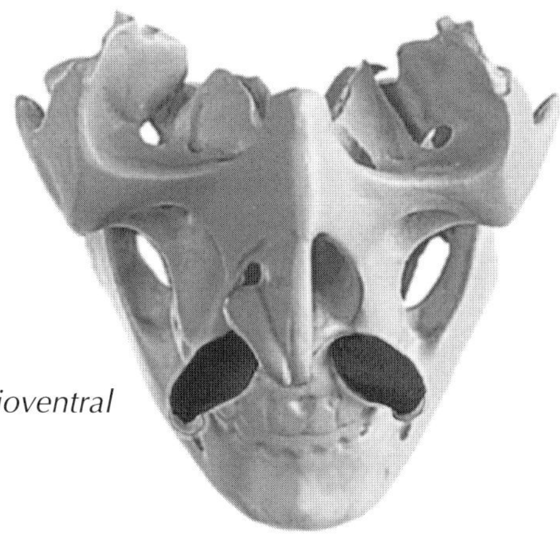

Cranioventral

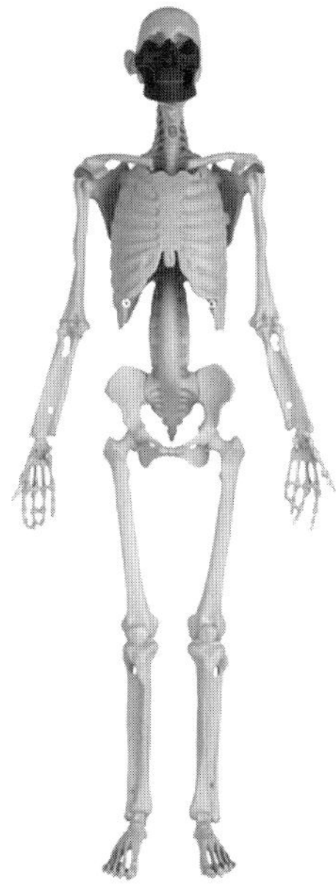

Actions

Draws modiolus upward and medially.

Attachments

Medial

Incisive fossa of maxilla superior to lateral incisor.

Lateral

Superficial slip: blends with levator anguli oris and attaches to body and apex of modiolus.

Deep slip: superior cornu and base of modiolus.

Note

The modiolus is a fibromuscular mass at the corner of the mouth. The slips of incisivis are not shown separately.

*in•**sy**•suh•vus*

Latin, meaning sharp.

*soo•**pee**•ree•or•is*

From the Latin, "superior," meaning above or upper.

Incisivis inferior

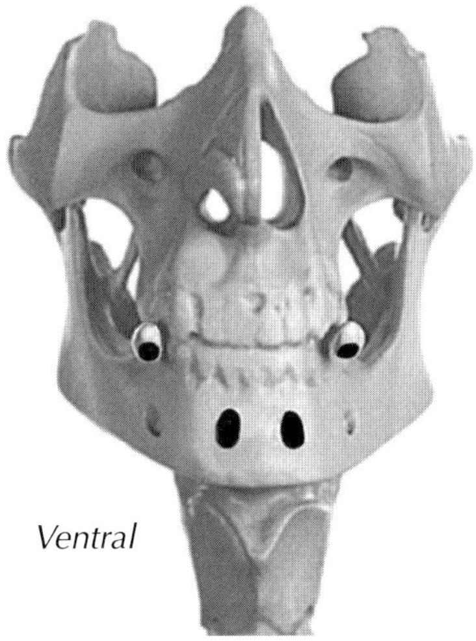

Ventral

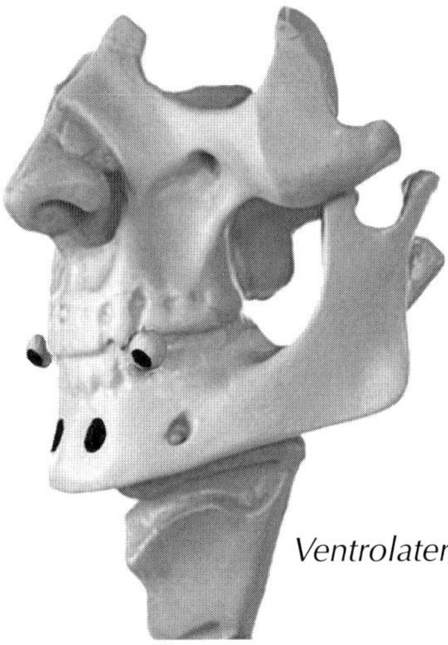

Ventrolateral

Caudoventral

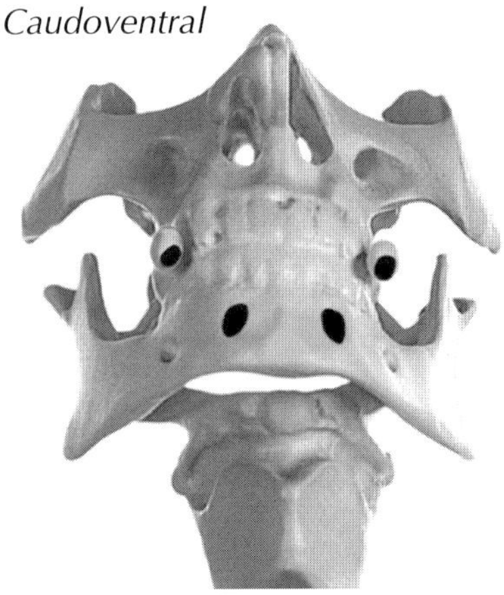

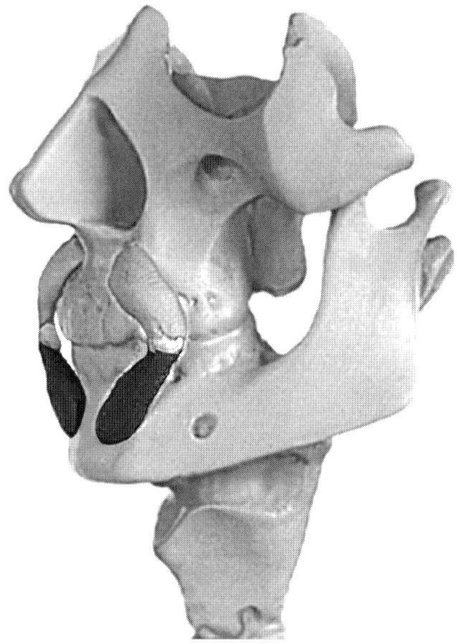

Ventrolateral

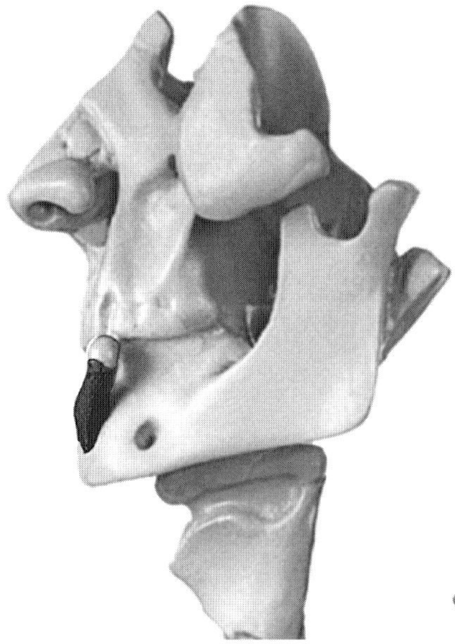

Lateral

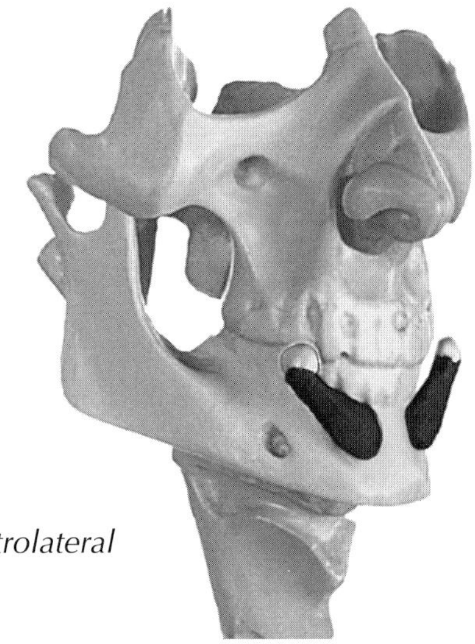

Ventrolateral

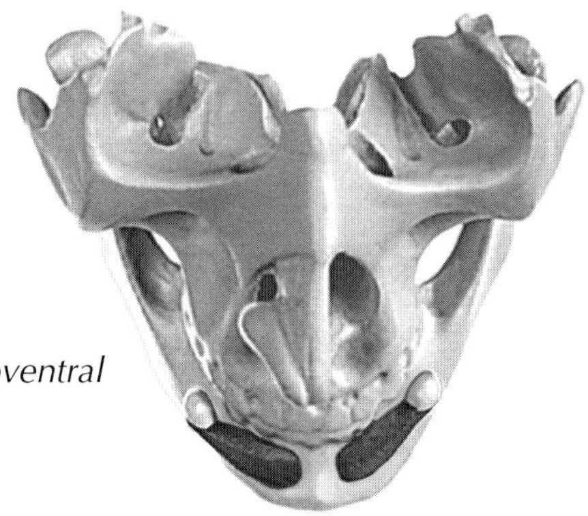

Cranioventral

Incisivis inferior

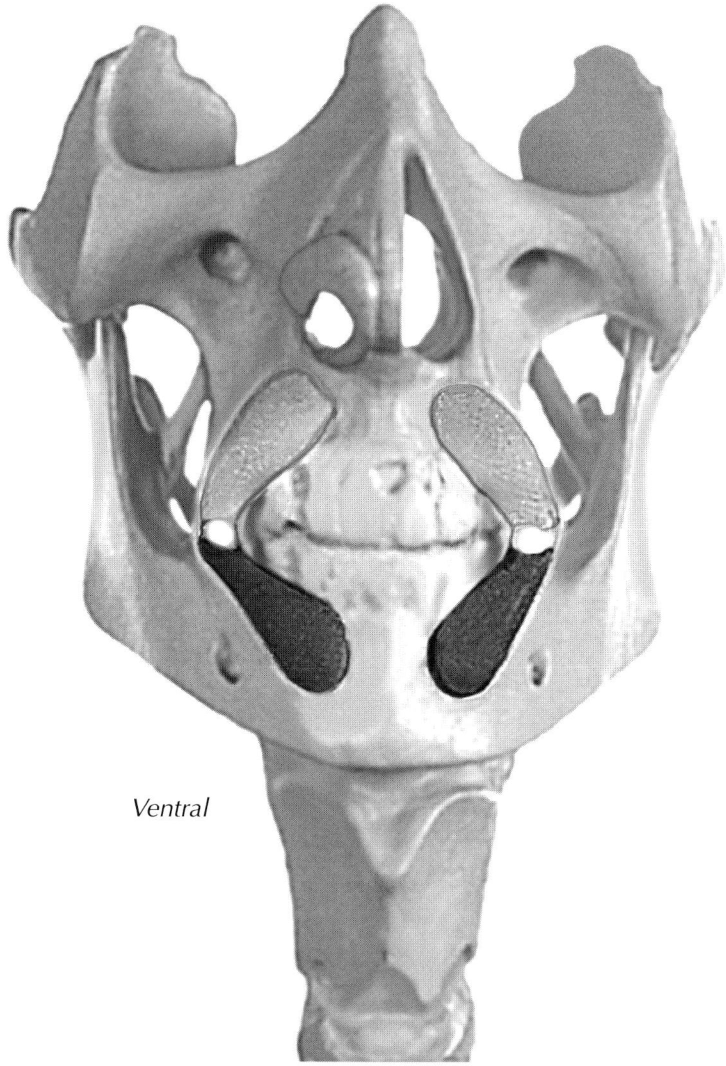

Ventral

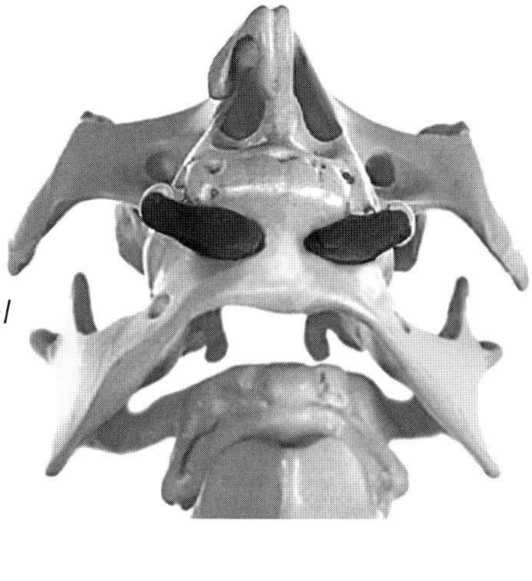

Caudoventral

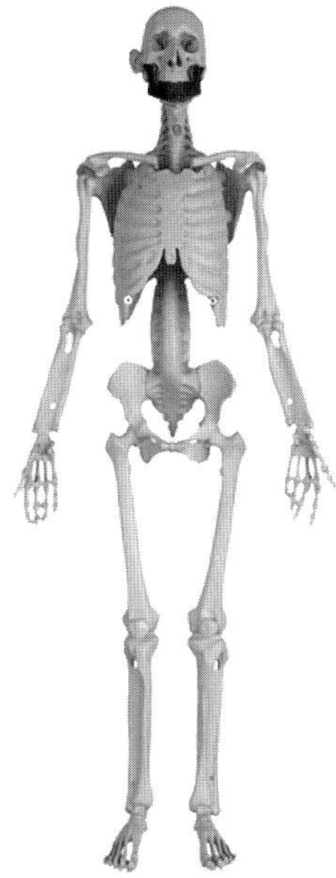

Actions

Draws modiolus downward and medially.

Attachments

Medial

Incisive fossa of mandible inferior to lateral incisor.

Lateral

<u>Superfical slip</u>: blends with pars peripheralis orbicularis oris and attaches to body and apex of modiolus
<u>Deep slip</u>: inferior cornu and base of modiolus.

Note

Separate slips are shown here as one.

*in•**sy**•suh•vus*

Latin, meaning sharp.

*in•**fee**•ree•ur*

From the Latin, "infra," meaning lower or beneath.

Orbicularis oris
pars peripheralis

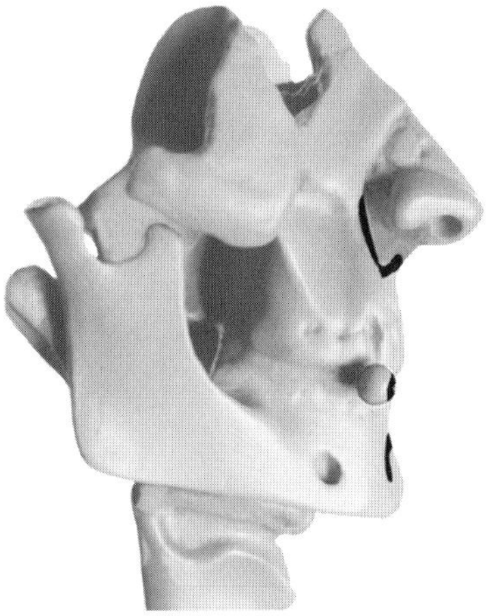

Lateral

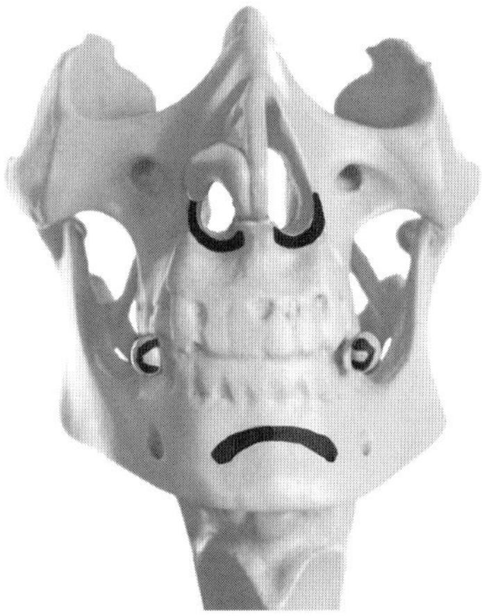

Ventral

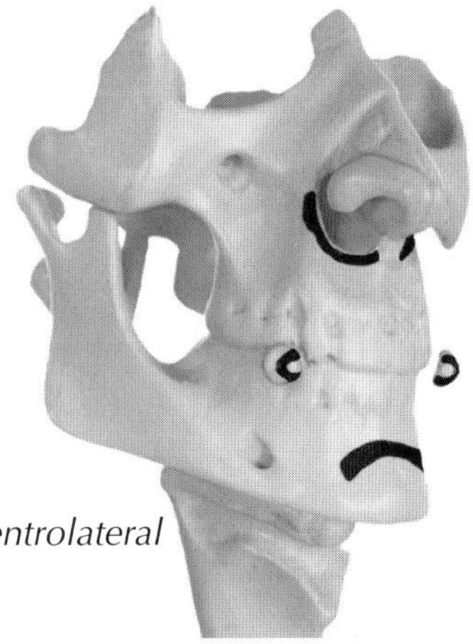

Ventrolateral

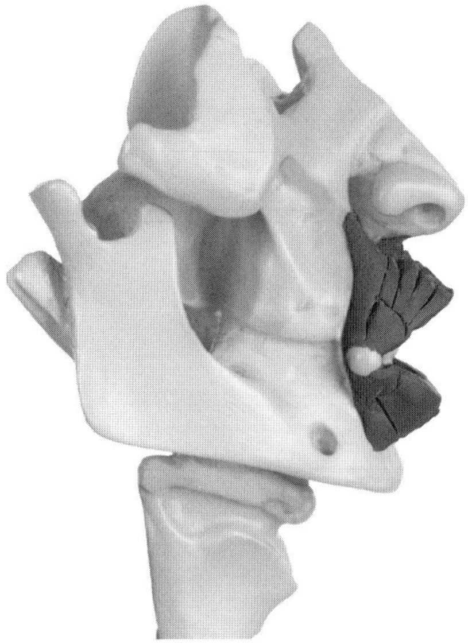

Lateral

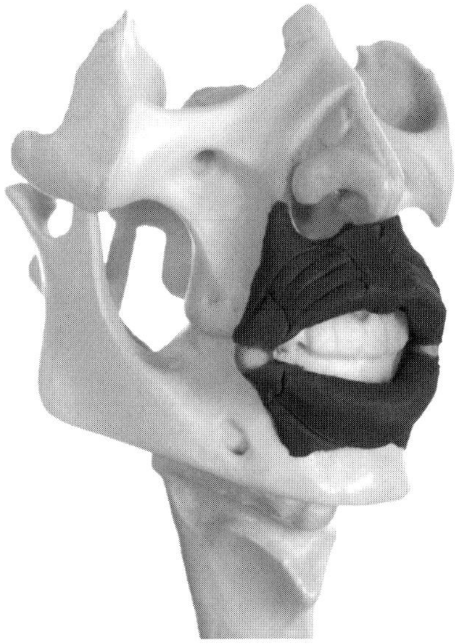

Ventrolateral

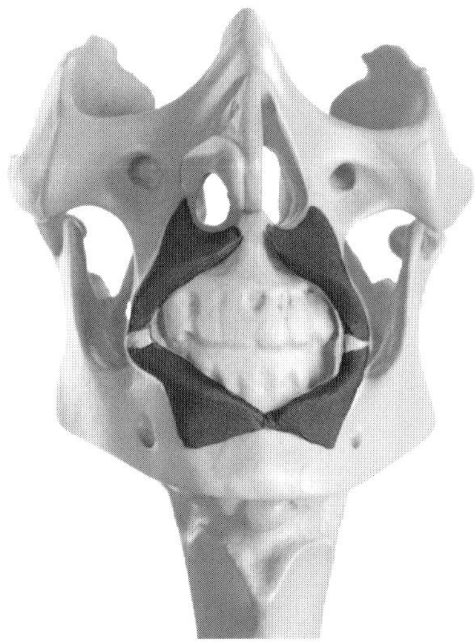

Ventral

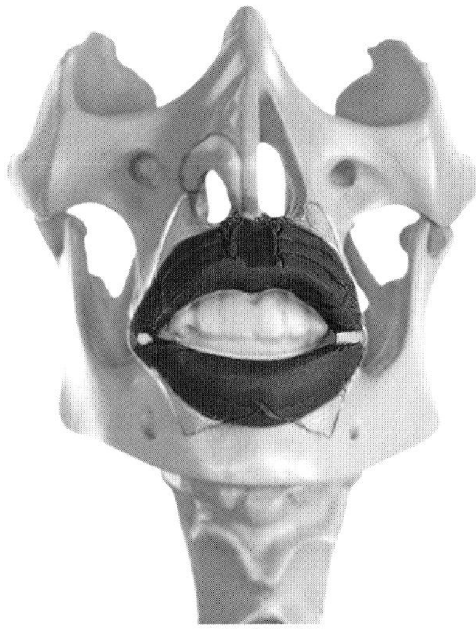

Ventral

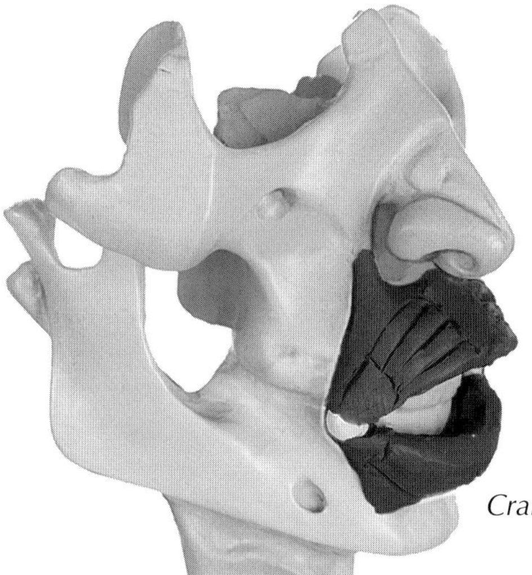

Cranioventrolateral

Orbicularis oris
pars peripheralis

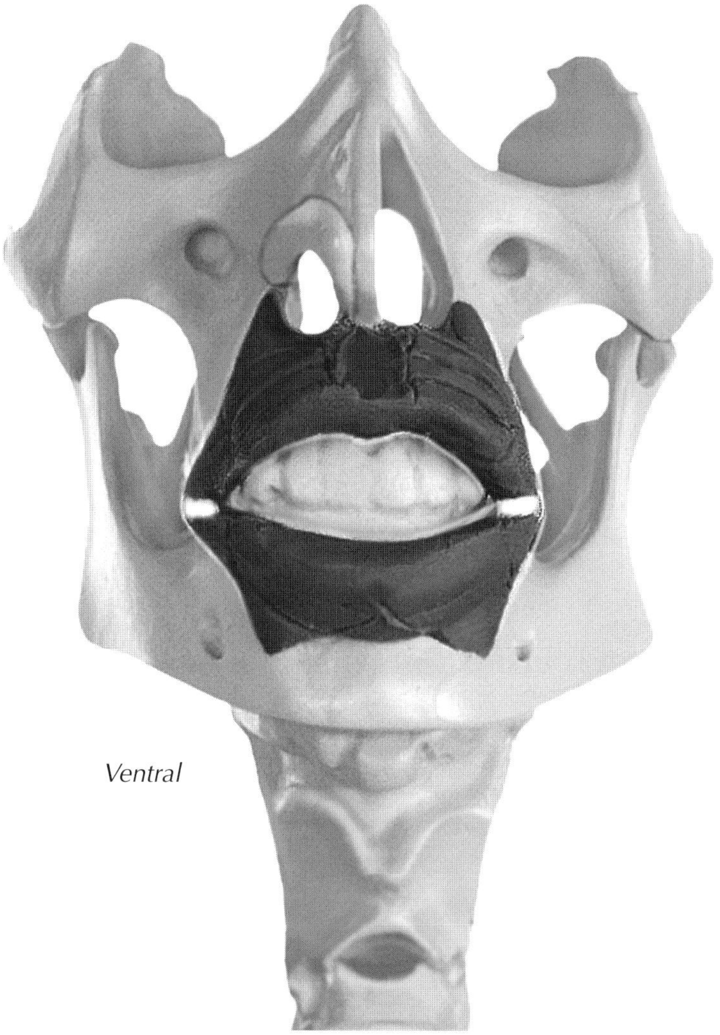

Ventral

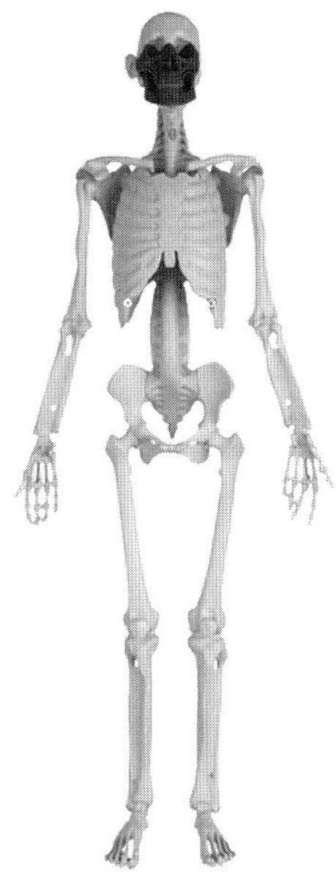

Actions

Draw lips over teeth, move modioli side to side, pucker lips.

Attachments

Lateral

Entire thickness of modiolus.

Medial

Deep fans of upper lip to edge of nasal apperture at nasolabial sulcus. Lower lip fans attach to mandible at mentolabial sulcus. Superficial fans cross midline, interlacing with opposites.

Note

The superficial fans of the upper lip cross one other to attach to the skin. The tension between them creates a furrow known as the "philtrum."

*or•bik•you•**lar**•us*

Latin, meaning a small disc.

***or**•us*

Latin, meaning belonging to the mouth.

*per•ih•fur•**ah**•lis*

From the Greek, "periphereia," meaning moving around.

Cranioventral

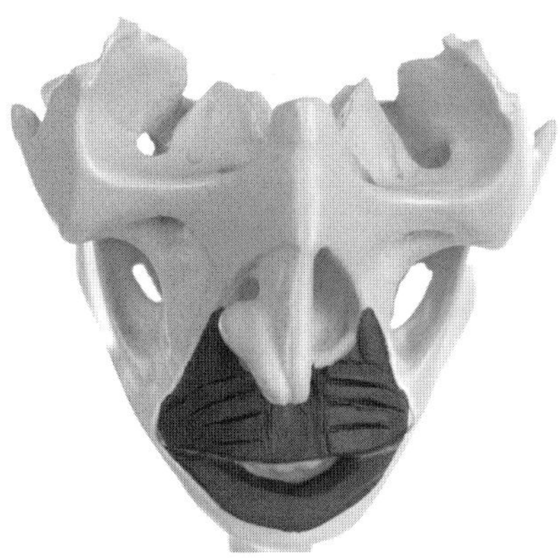

Caudoventral

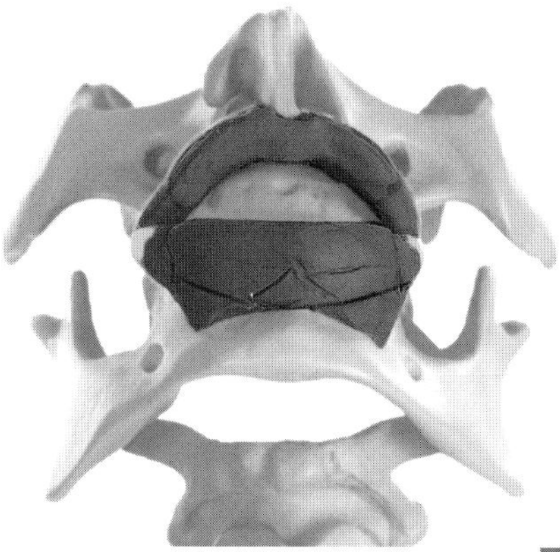

Orbicularis oris
pars marginalis

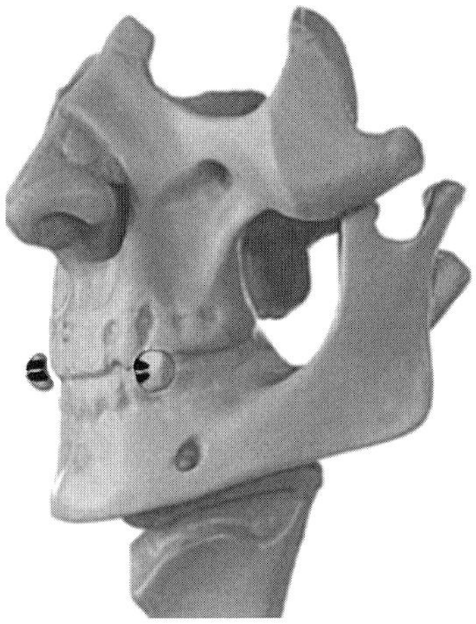

Ventrolateral

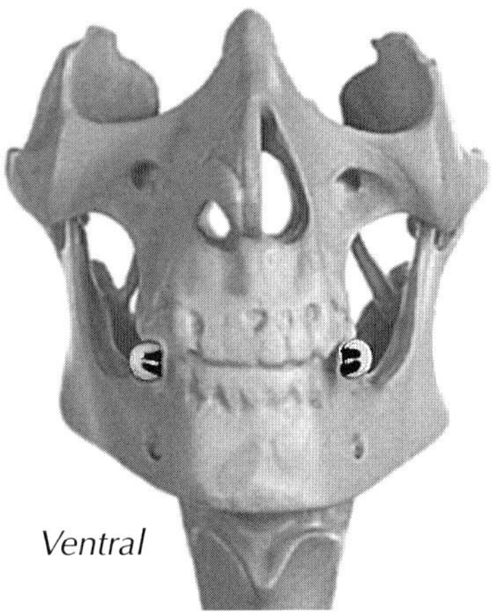

Ventral

Cranioventral

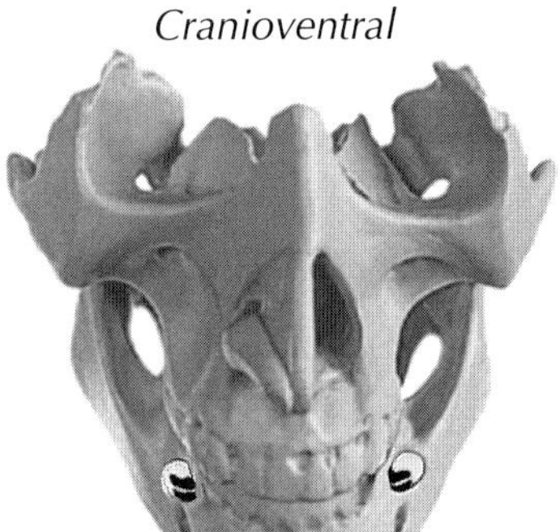

Ventrolateral

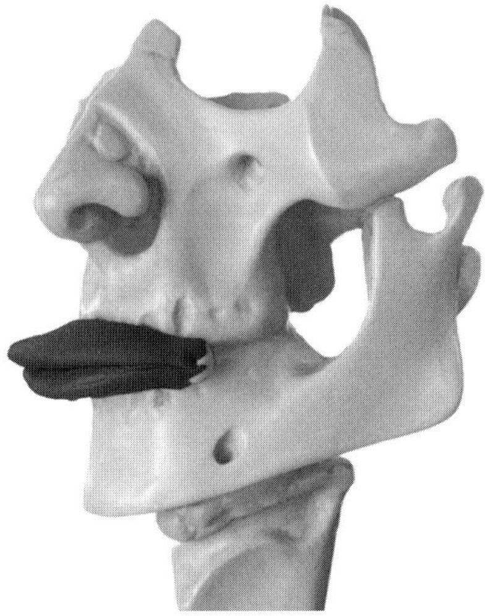

Ventrolateral

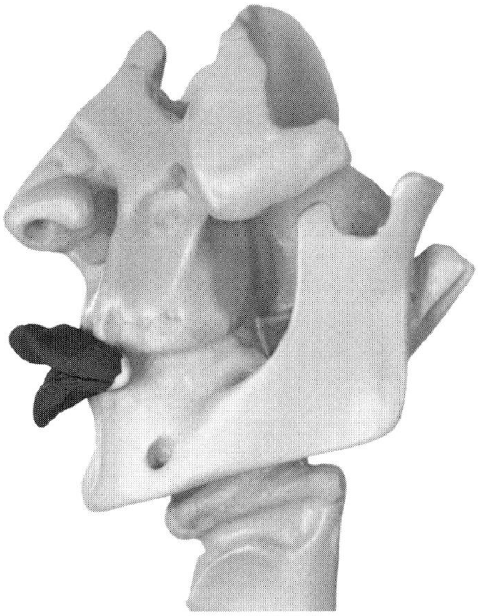

Lateral

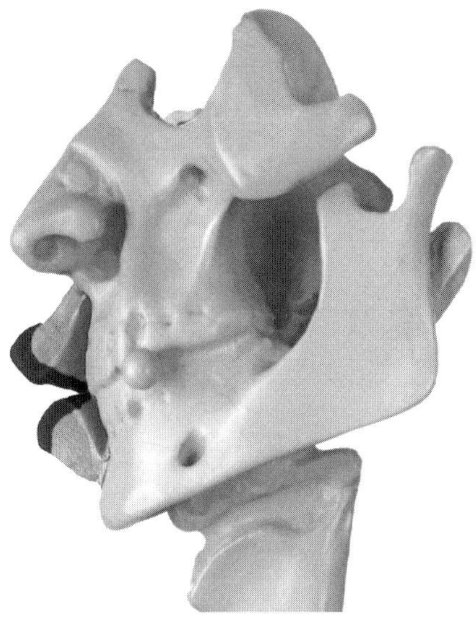

Lateral

Orbicularis oris
pars marginalis

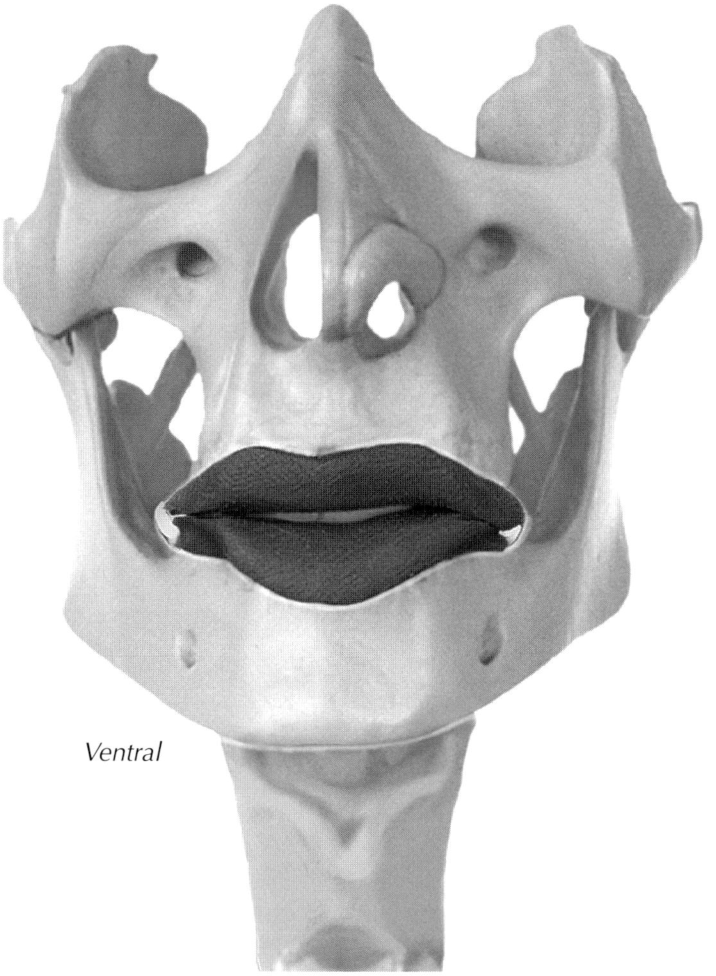

Ventral

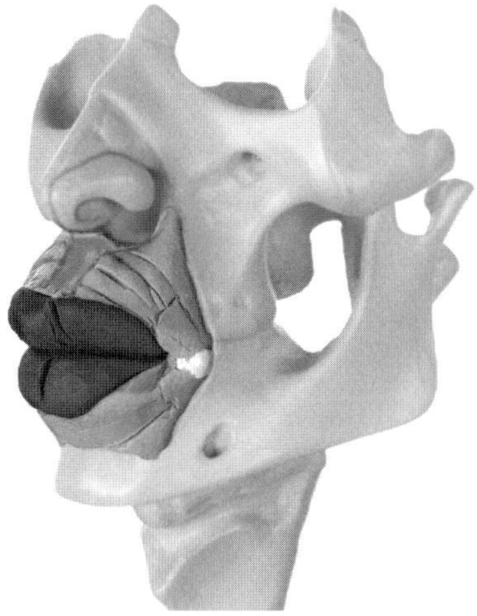

Ventrolateral

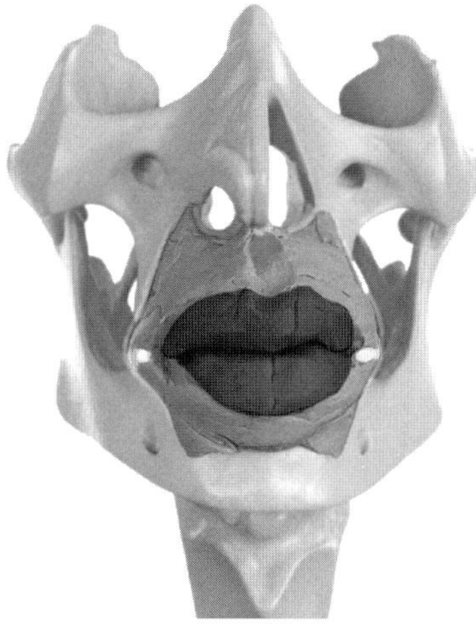

Ventral

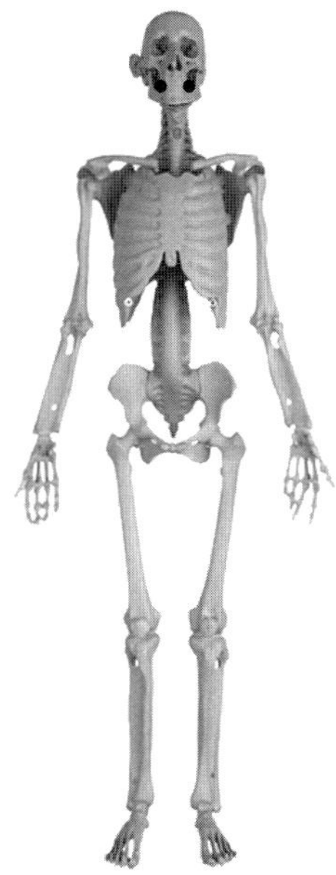

Actions
Purse lips, draw in red lip area, close oral fissure.

Attachments
Lateral

Base of modiolus.

Medial

Fibers run within the red lip margin and attach to the red lip dermis just across the median plane.

Note
The modiolus is a small nodule of fibromuscular tissue at corner of mouth.

*or•bik•you•**lar**•us*

Latin, meaning a small disc.

or•*us*

Latin, meaning belonging to the mouth.

*mar•jin•**al**•lis*

Latin, "margin"+"alis," meaning related to a border.

Levator anguli oris

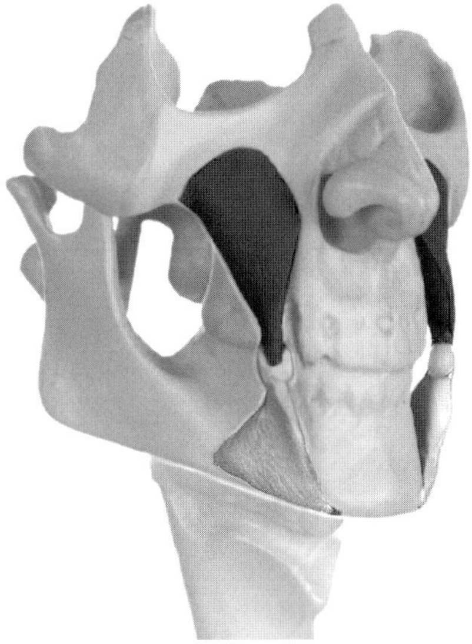

Ventrolateral

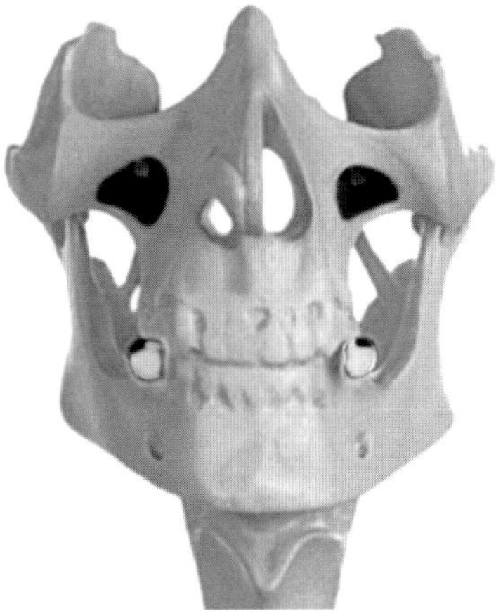

Ventral

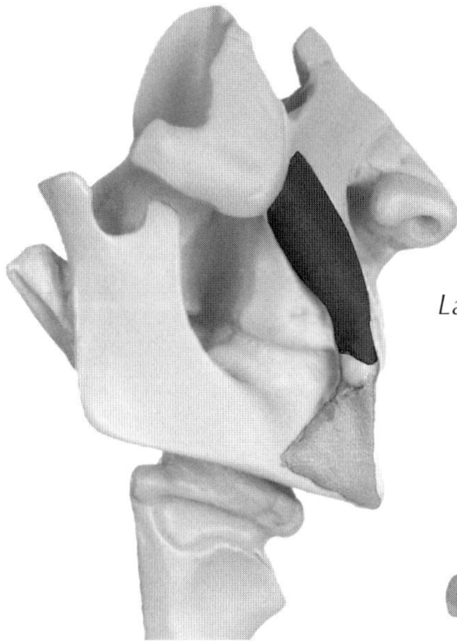

Lateral

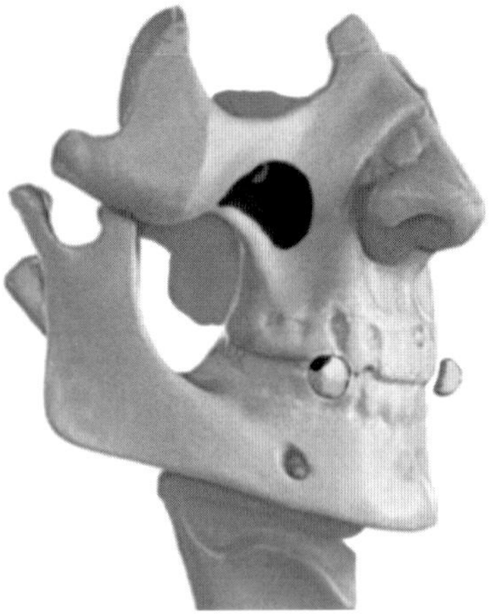

Ventrolateral

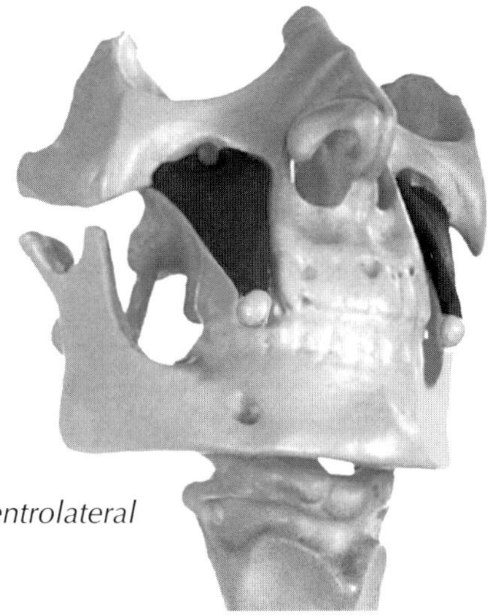

Ventrolateral

Cranioventral

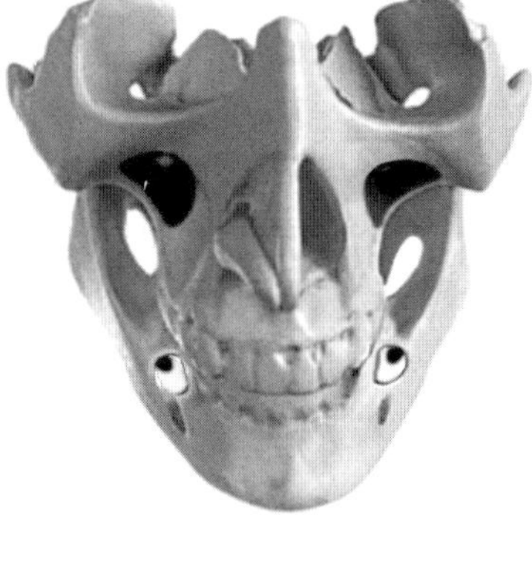

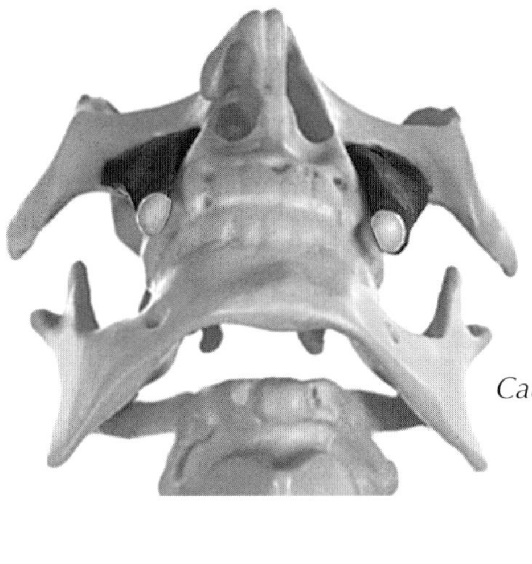

Caudoventral

Levator anguli oris

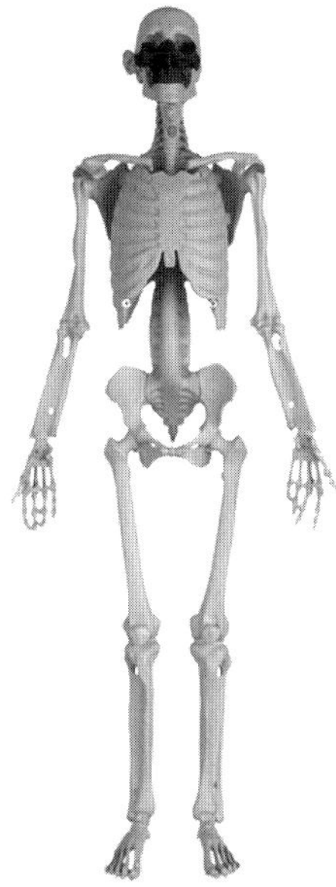

Ventral

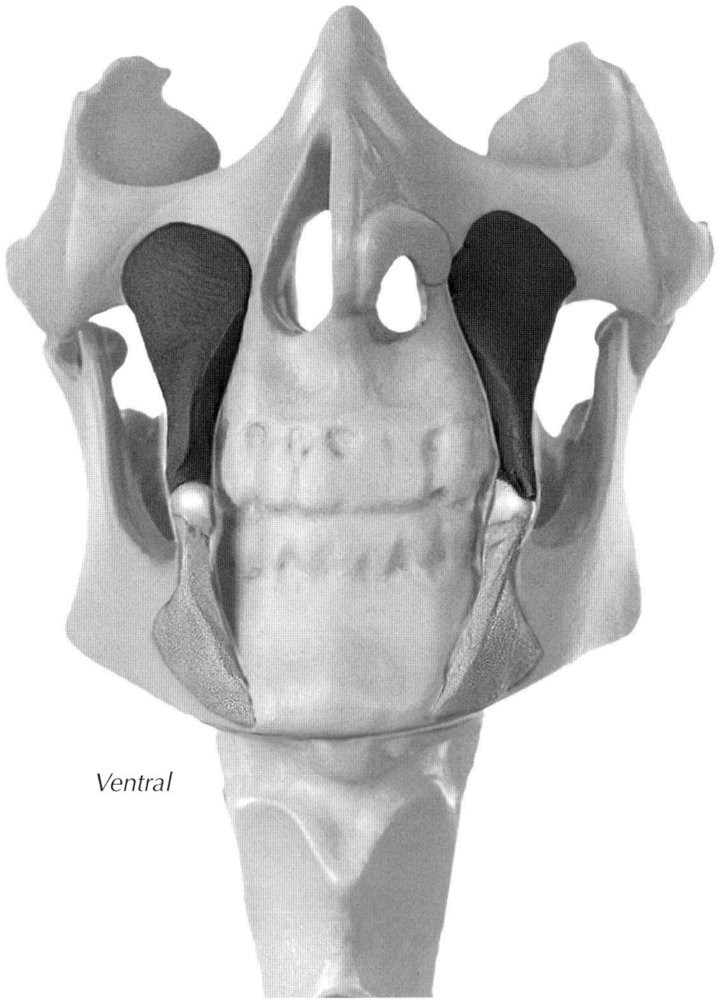

Cranioventral

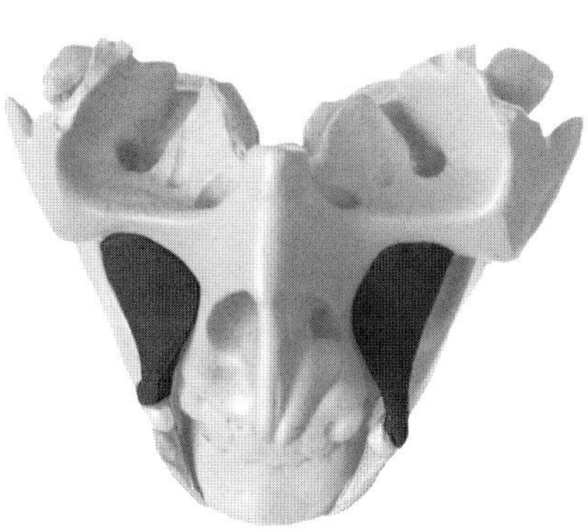

Cranioventral

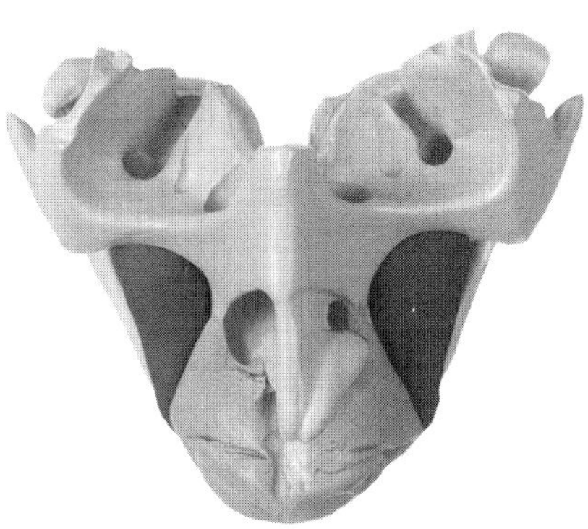

Actions

Elevates corner of the mouth.

Attachments

Cranial

Canine fossa of the maxilla.

Caudal

Modiolus at the corner of the mouth.

Note

The modiolus is a small nodule of fibromuscular tissue at the corner of the mouth.

*luh•***vay***•tur*

From the Latin, "levatus"+"or," meaning one that raises.

ang•*you*•*lee*

Latin, meaning belonging to a corner or angle.

or•*us*

Latin, meaning belonging to the mouth.

Levator labii
superioris

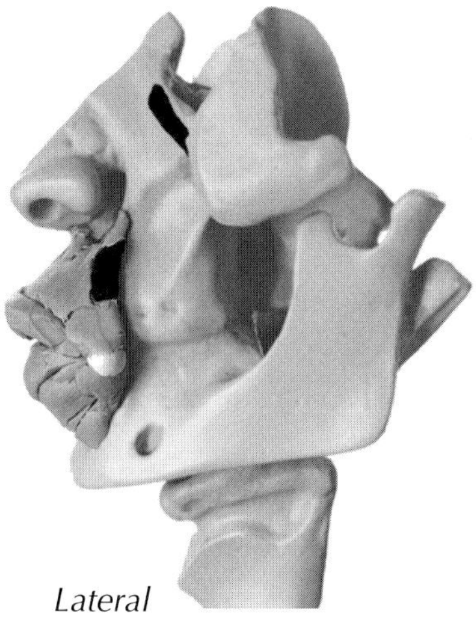

Lateral

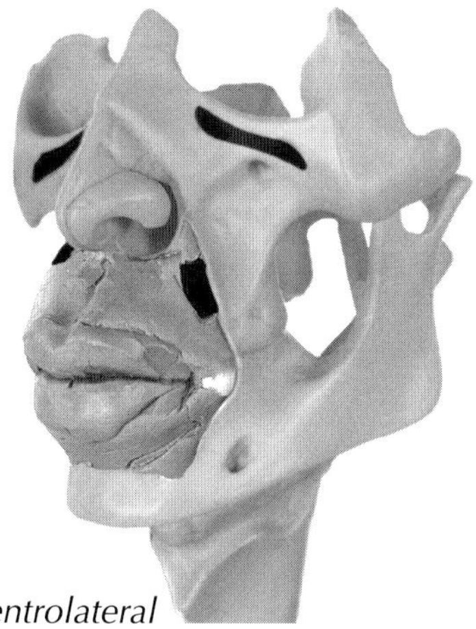

Ventrolateral

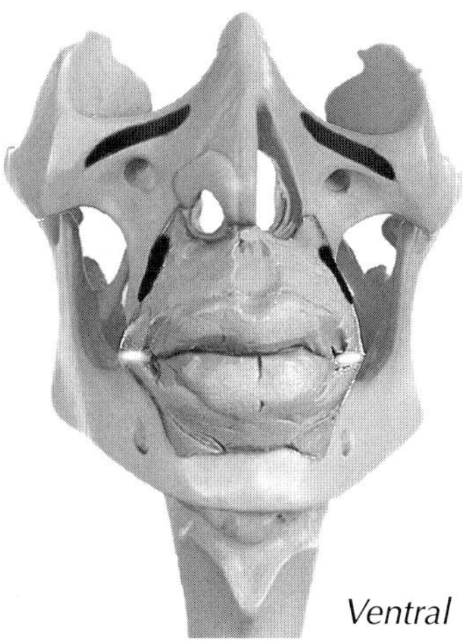

Ventral

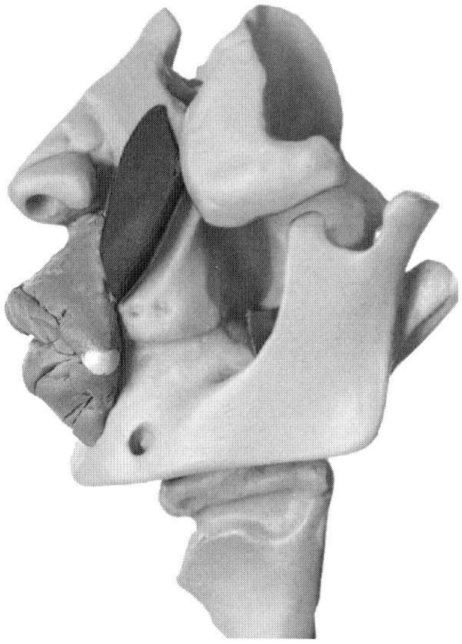

Lateral

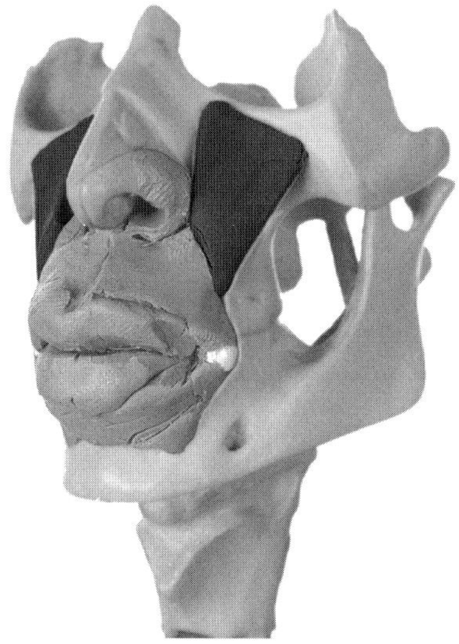

Ventrlateral

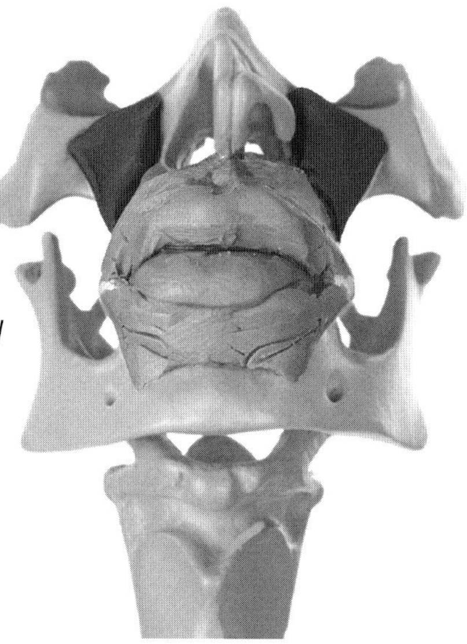

Caudoventral

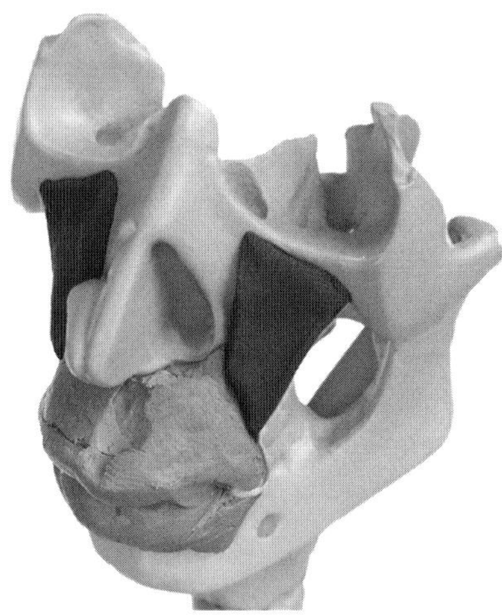

Cranioventrolateral

Levator labii superioris

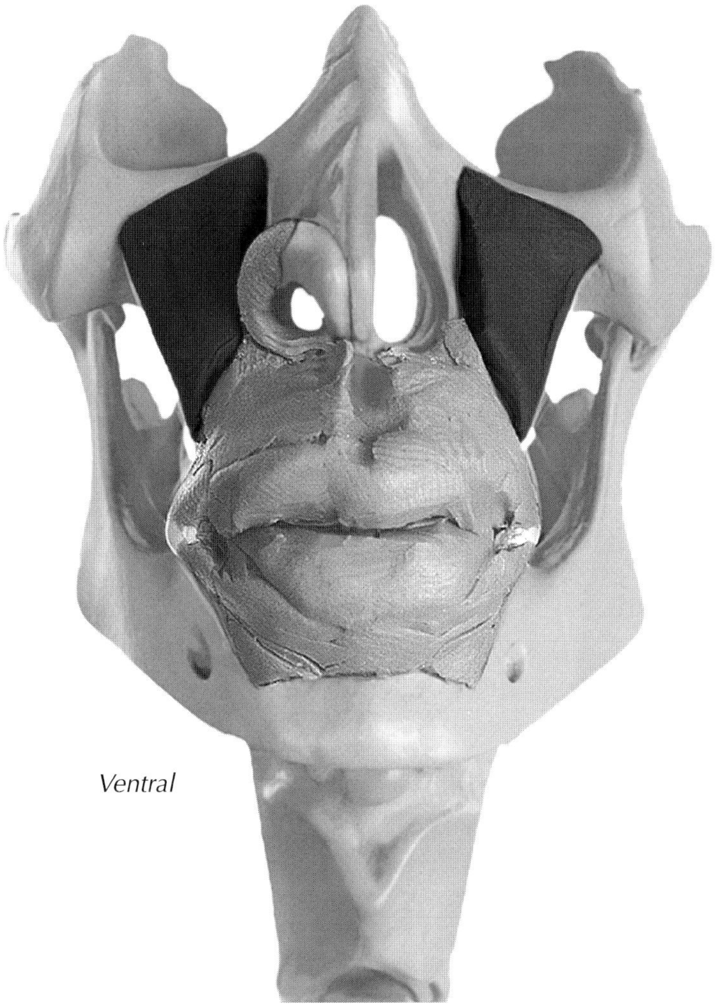

Ventral

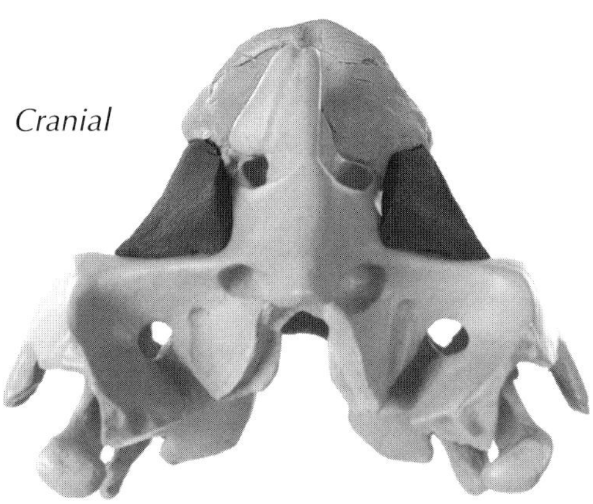

Cranial

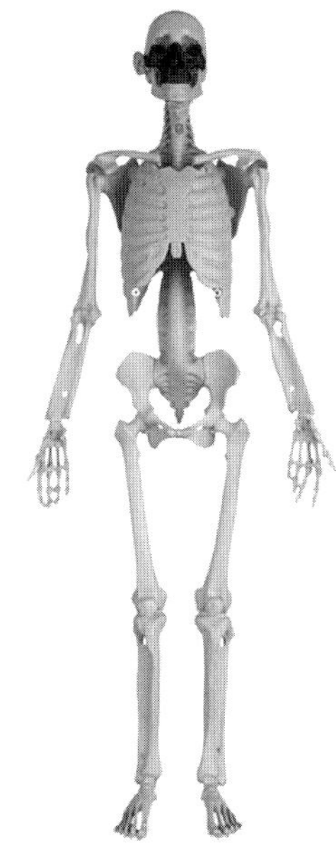

Actions

Elevates upper lip.

Attachments

Cranial

Inferior margin of orbit.

Caudal

Skin and muscle of upper lip.

*luh•**vay**•tur*

From the Latin, "levatus"+"or," meaning one that raises.

lay•bee•eye

Latin, meaning belonging to the lip.

*soo•**pee**•ree•or•is*

From the Latin, "superior,"meaning above or upper.

Levator labii superioris alaeque nasi

Lateral

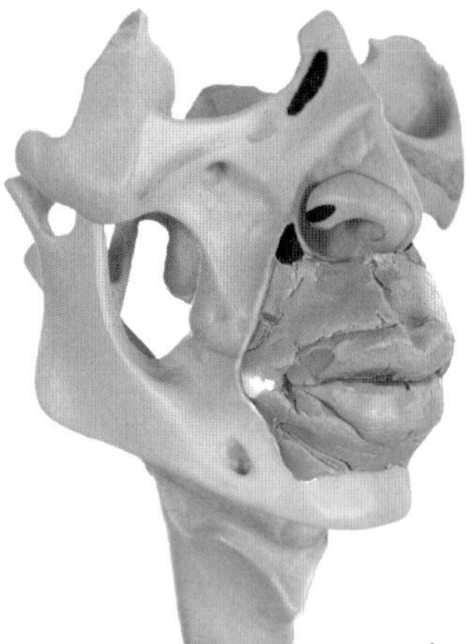

Ventrolateral

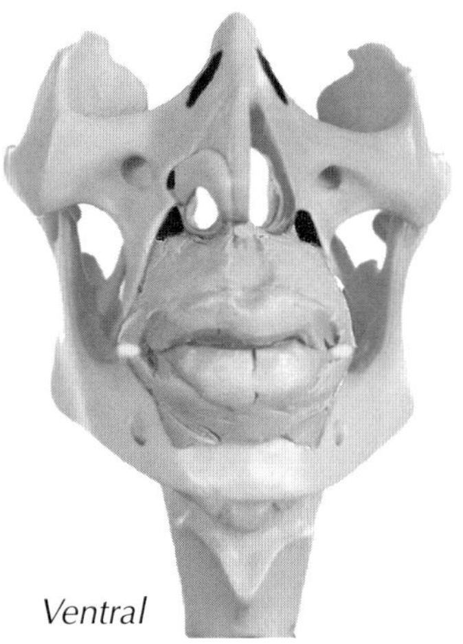

Ventral

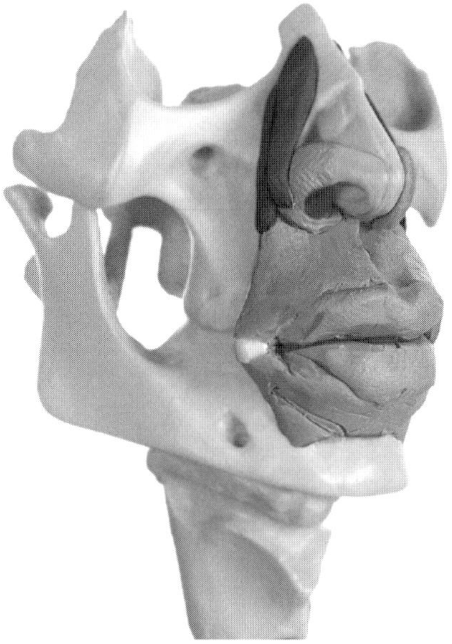

Ventrolateral

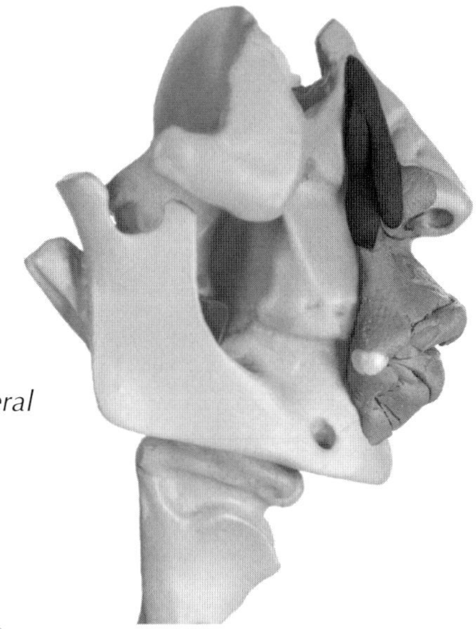

Lateral

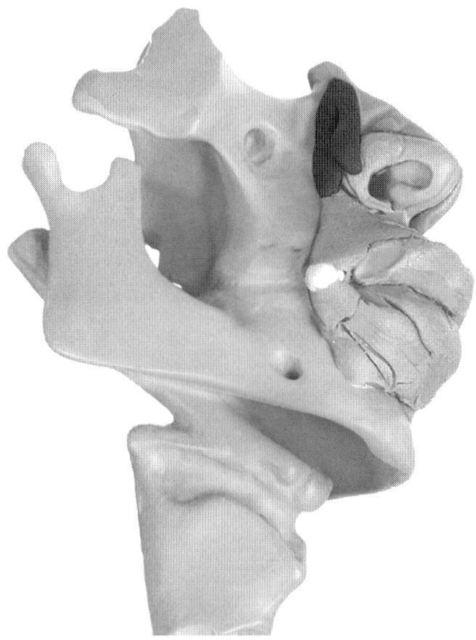

Caudolateral

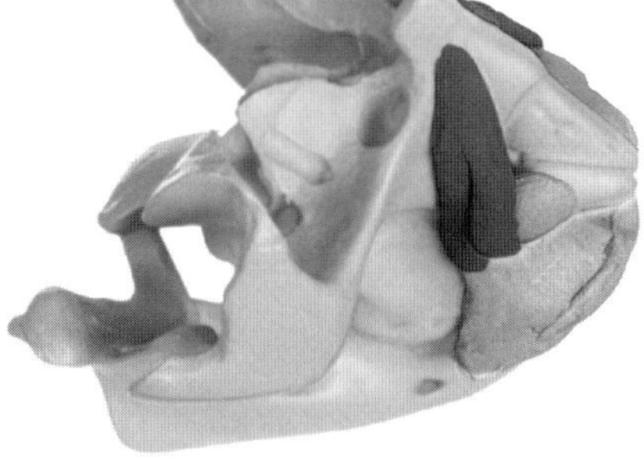

Craniolateral

Levator labii superioris alaeque nasi

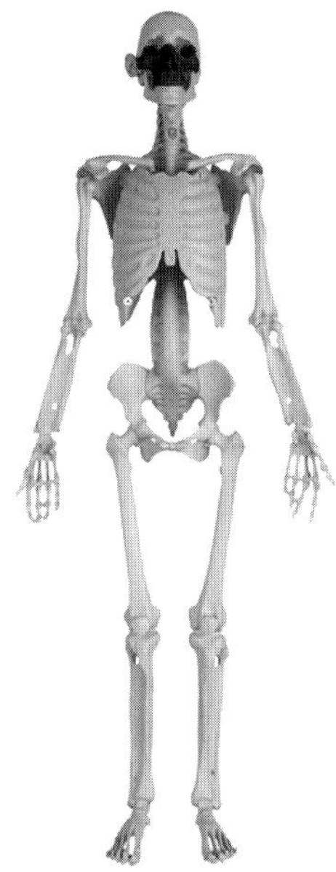

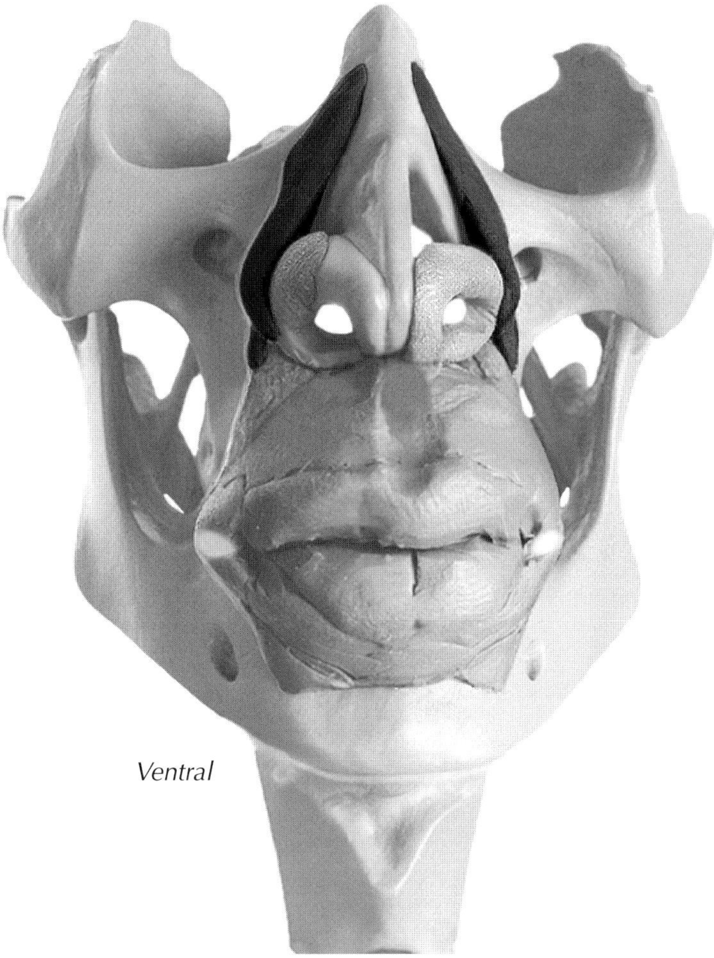

Ventral

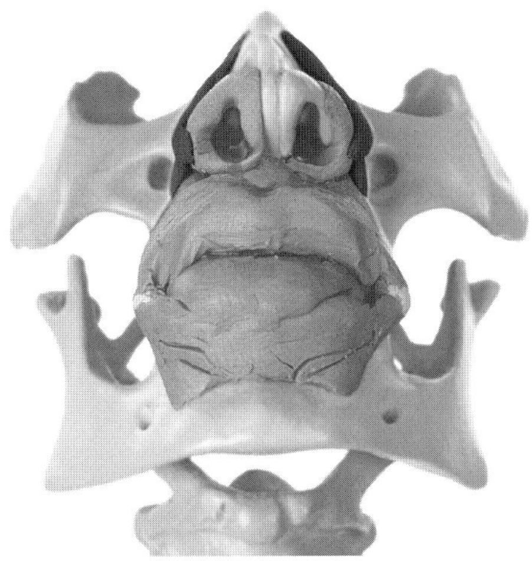

Caudoventral

Actions

Elevates upper lip and dilates nostril.

Attachments

Cranial

Frontal process of maxilla.

Caudal

Alar cartilage, skin of nostril and upper lip.

*luh•**vay**•tur*

From the Latin, "levatus"+"or," meaning one that raises.

***lay**•bee•eye*

Latin, meaning belonging to the lip.

*soo•**pee**•ree•or•is*

Latin, meaning above or upper.

***al**•lih•kwai*

From the Latin, "alae"+"que,"meaning "and the wing."

***nay**•zee*

Latin, belonging to the nose.

Zygomaticus minor

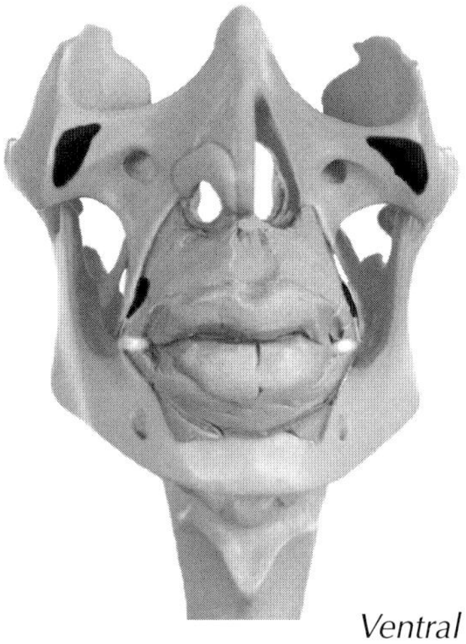

Ventral

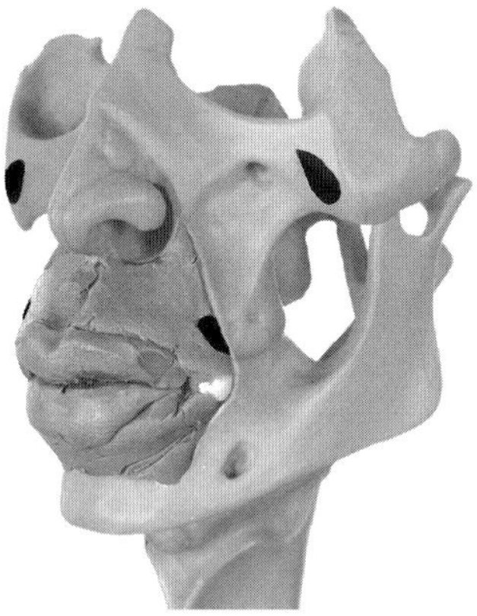

Ventrolateral

Cranioventral

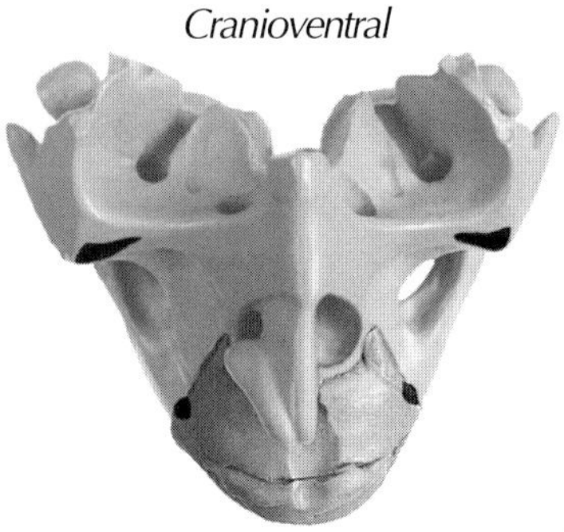

Ventrolateral

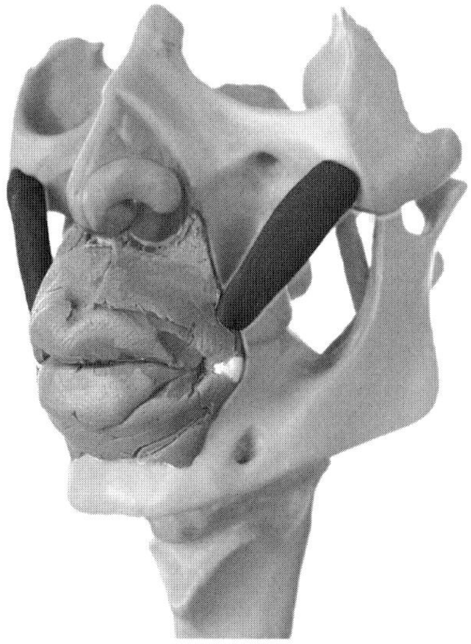

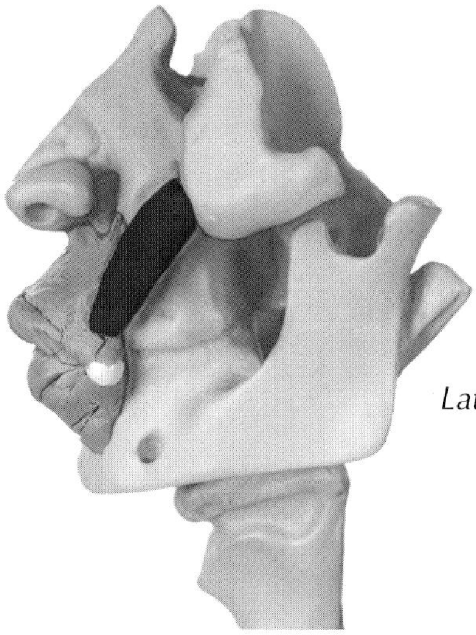

Lateral

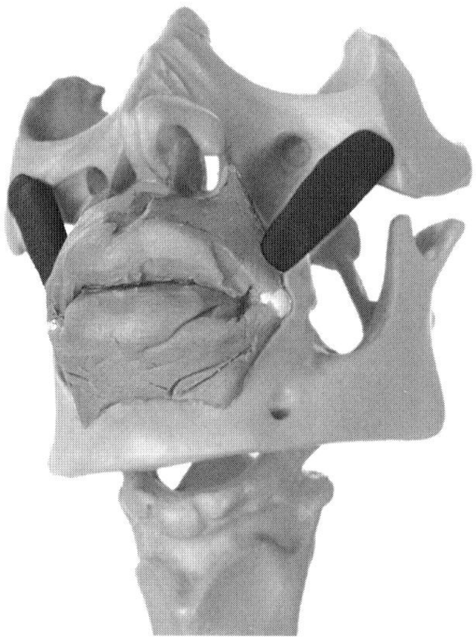

Caudoventrolate

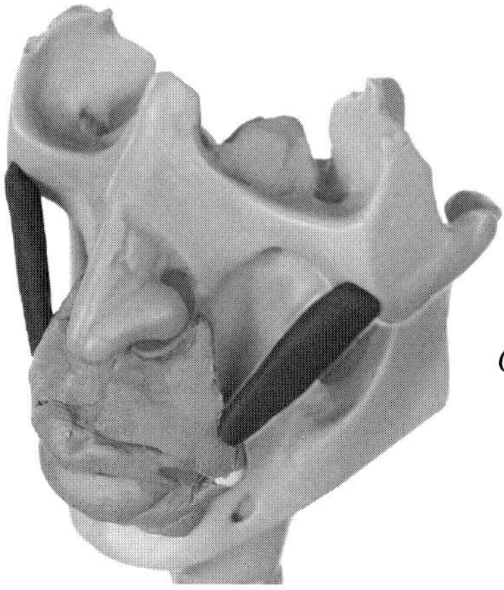

Cranioventrolateral

Zygomaticus minor

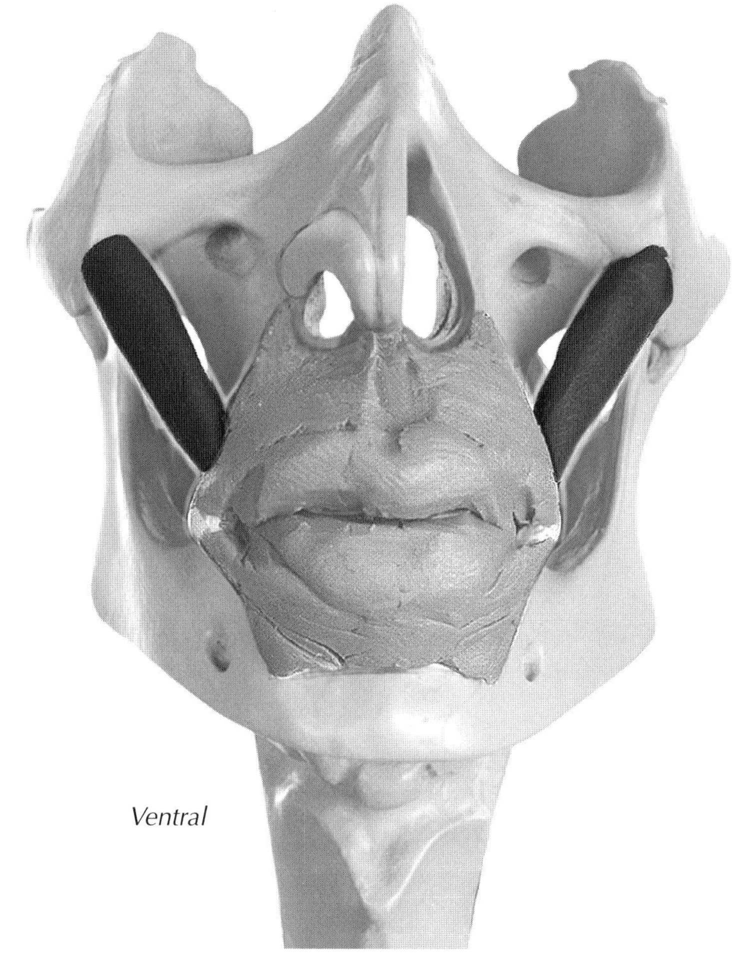

Ventral

Cranioventral

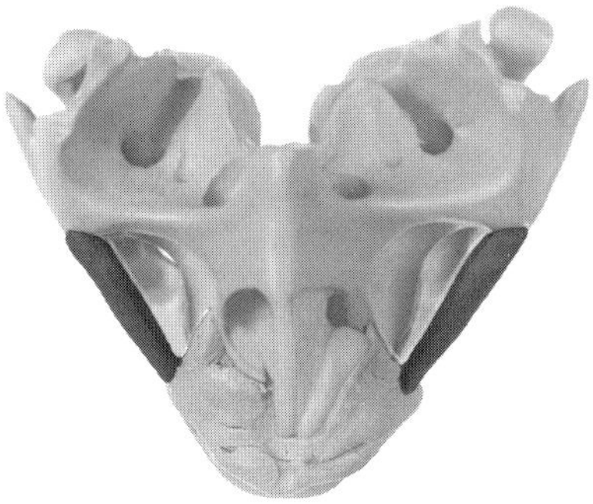

Caudoventral

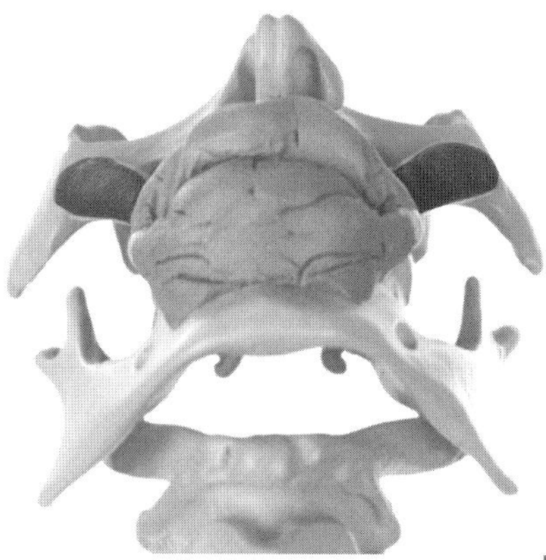

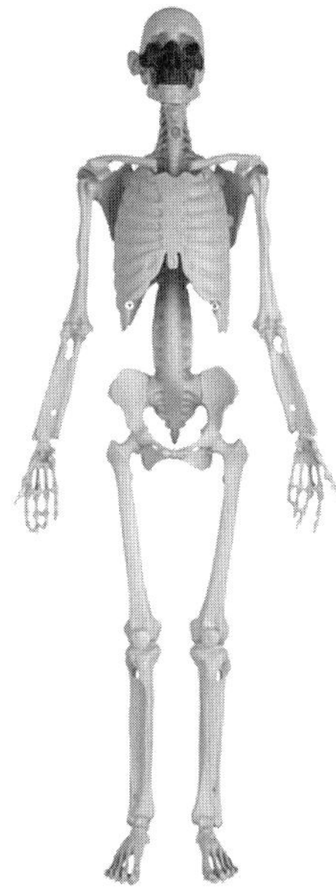

Actions

Elevates upper lip laterally.

Attachments

Cranial

Ventral surface of zygomatic bone.

Caudal

Skin of nasolabial groove and upper lip.

*zy•go•**mad**•i•kus*

From the Greek, "zygon"+"ikos," meaning related to the yoke or connector.

my•*nur*

Latin, meaning smaller.

Zygomaticus major

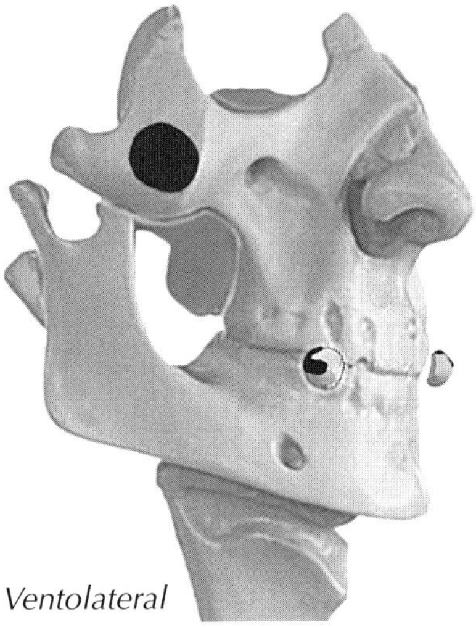

Ventolateral

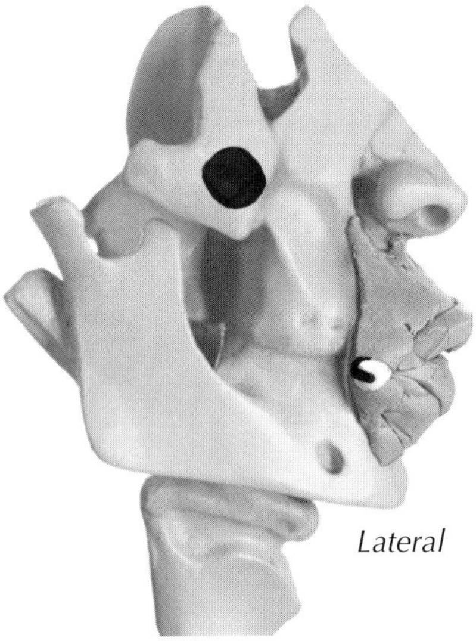

Lateral

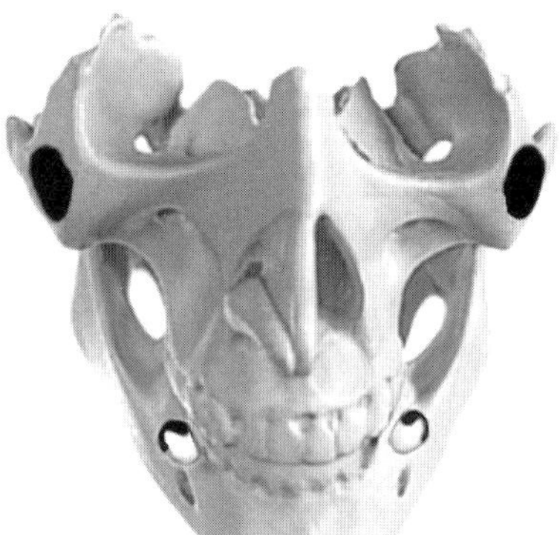

Ventral

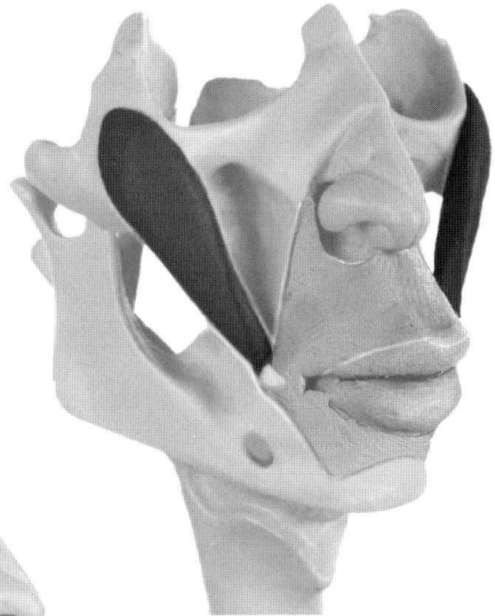

Ventrolateral

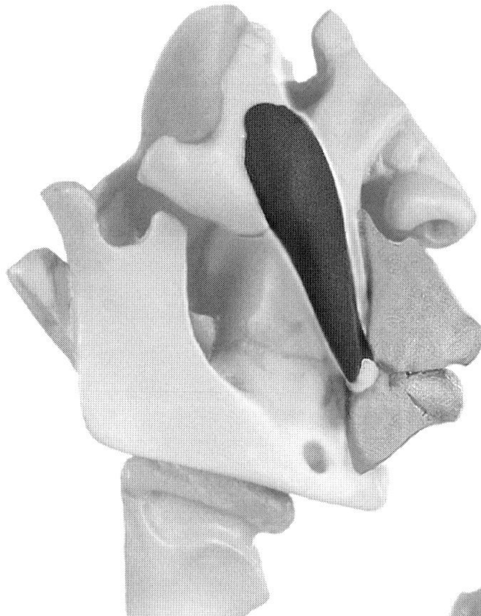

Lateral

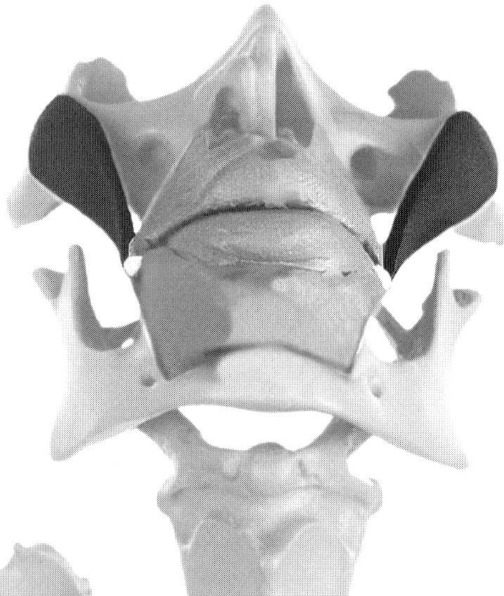

Caudoventral

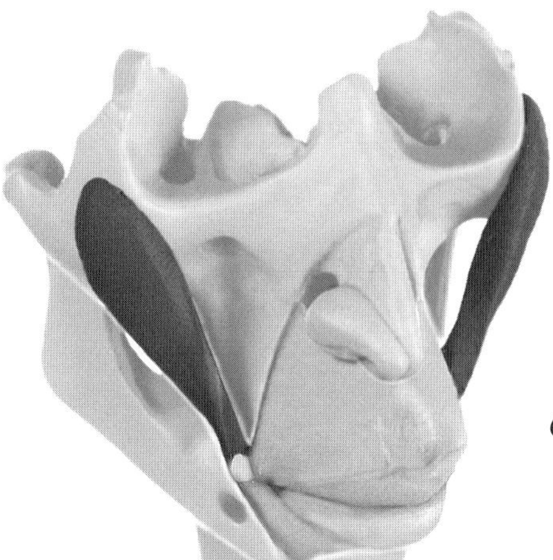

Cranioventrolateral

Zygomaticus
major

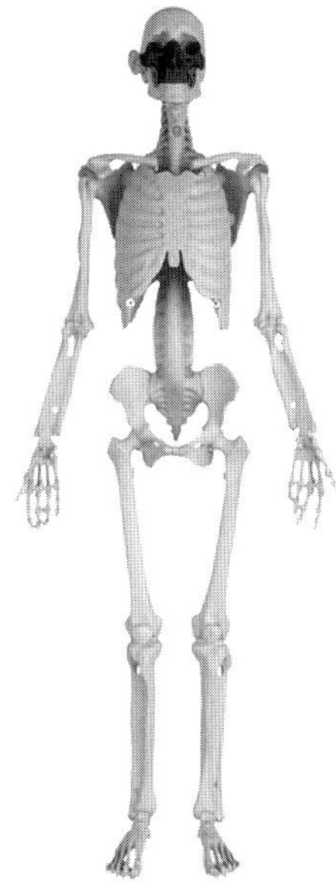

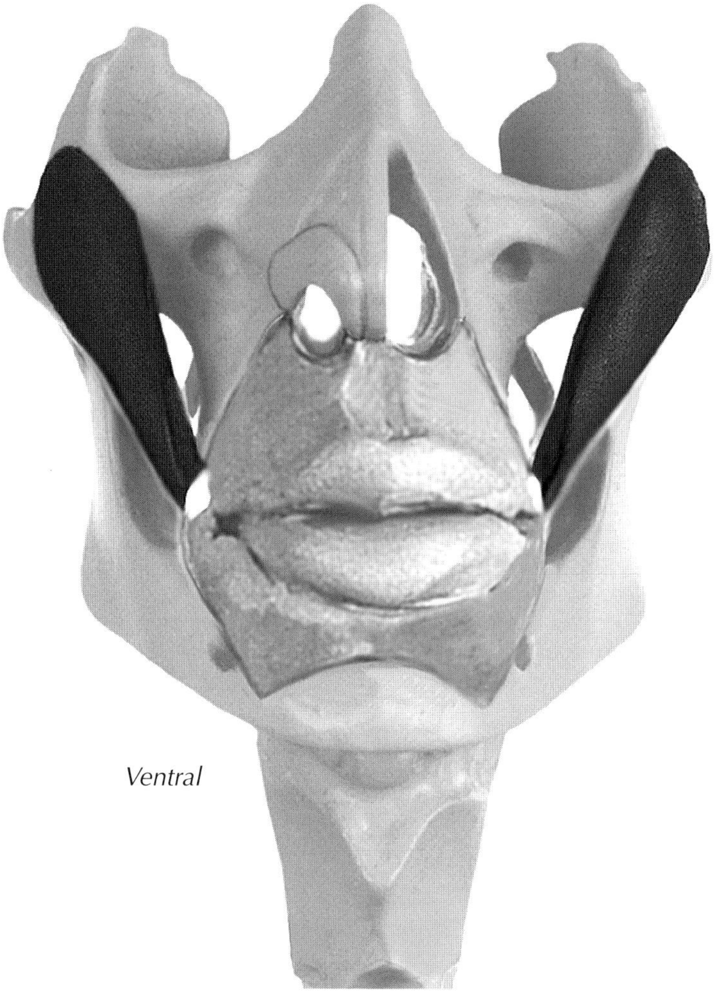

Ventral

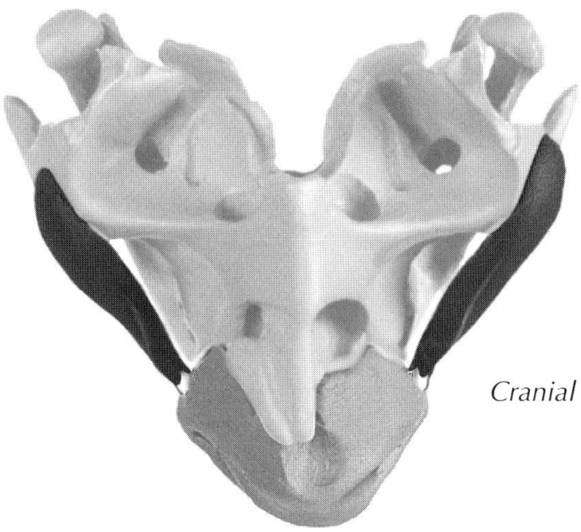

Cranial

Actions

Pulls upward and laterally on the modiolus at the corner of the mouth.

Attachments

Cranial

Ventral (zygomatic) part of the zygomatic arch.

Caudal

Modiolus at the corner of the mouth.

*zy•go•**mad**•i•kus*

From the Greek, "zygon"+"ikos," meaning related to the yoke or connector.

may•*jur*

Latin, meaning larger.

Depressor labii inferioris

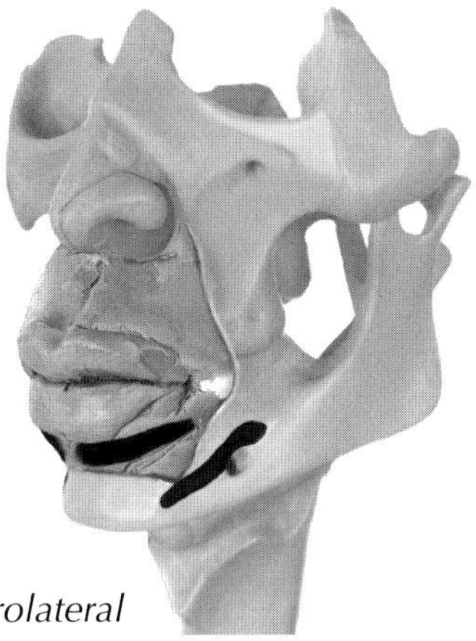

Ventrolateral

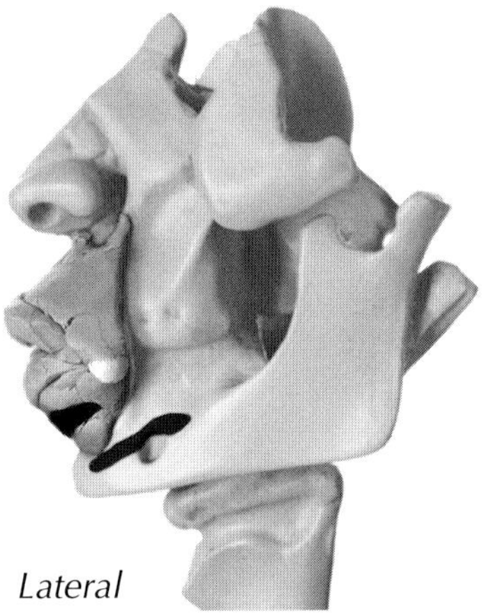

Lateral

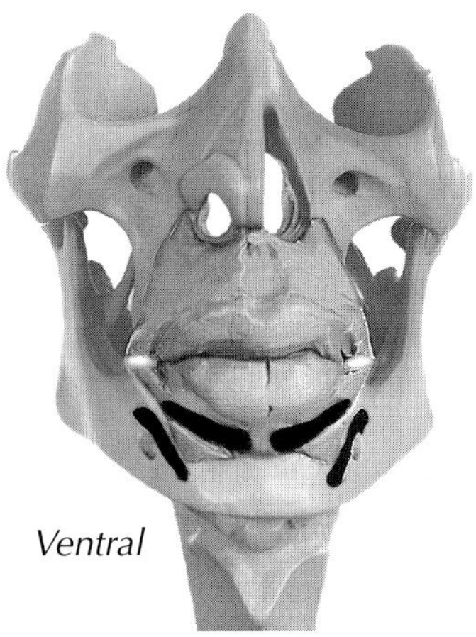

Ventral

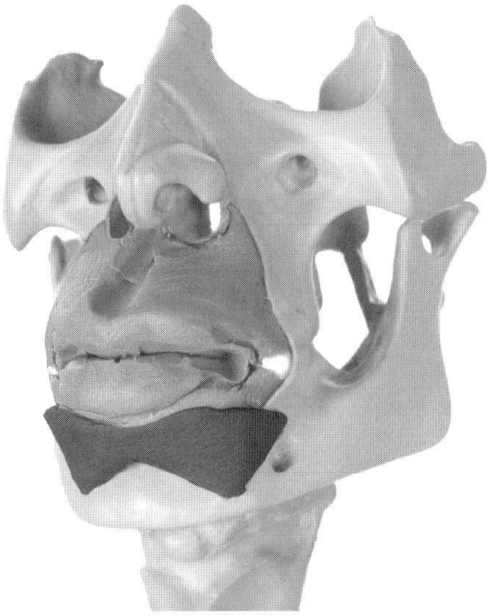

Ventrolateral

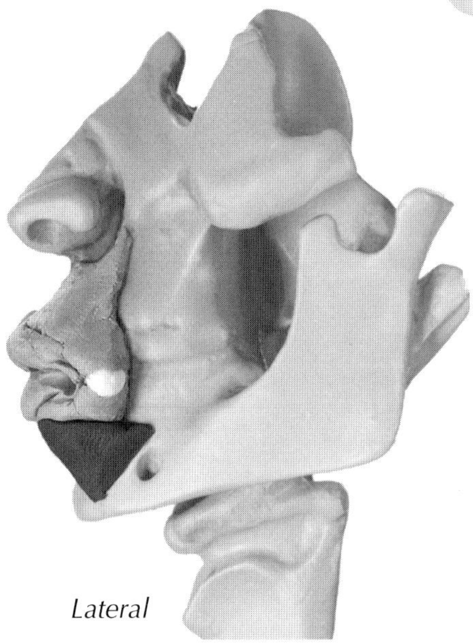

Lateral

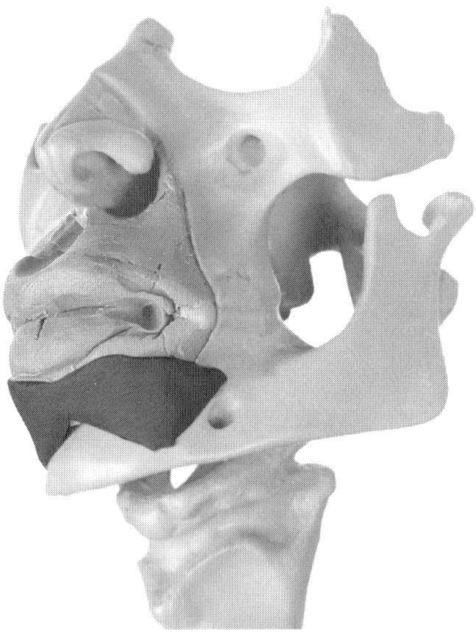

Caudoventrolateral

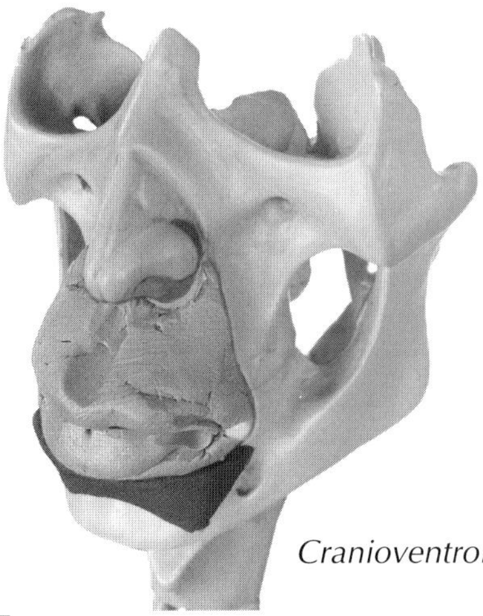

Cranioventrolateral

Depressor labii inferioris

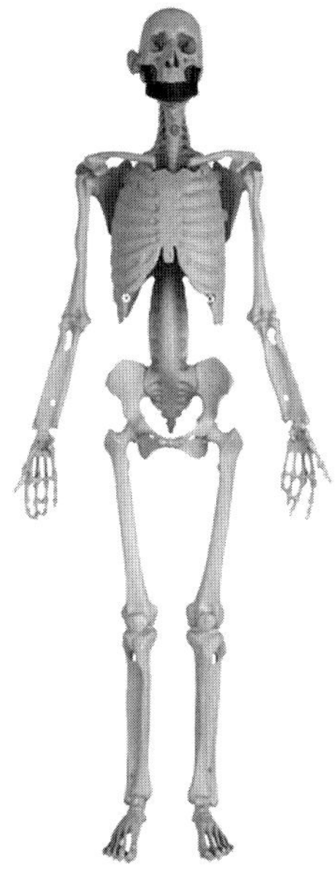

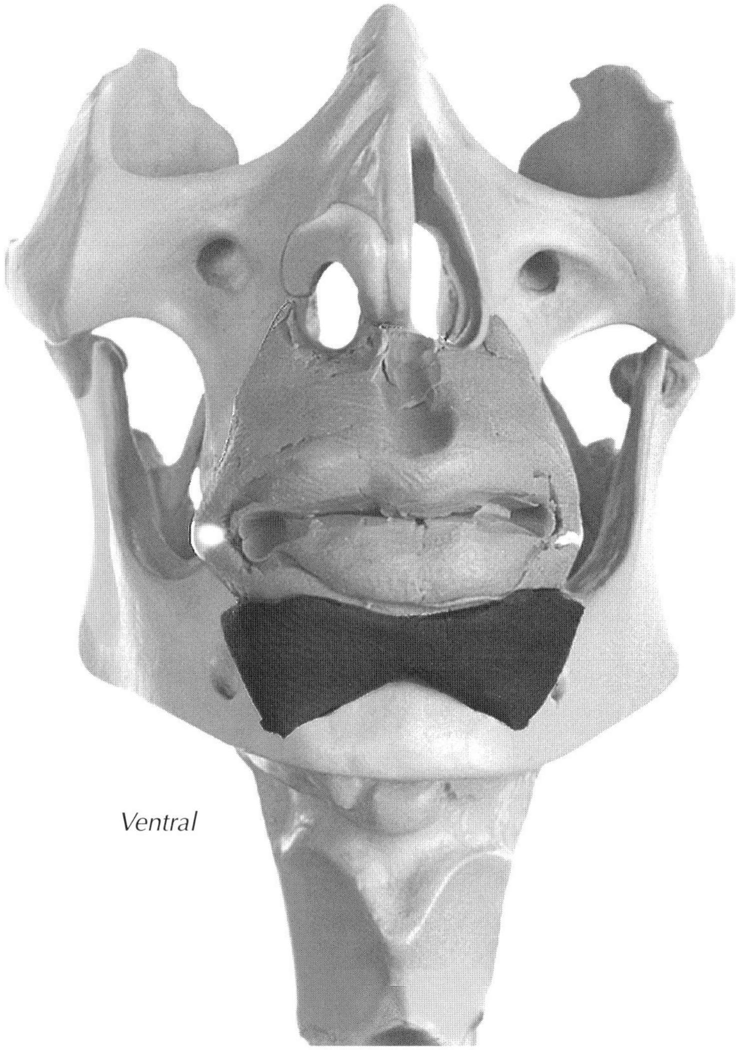

Ventral

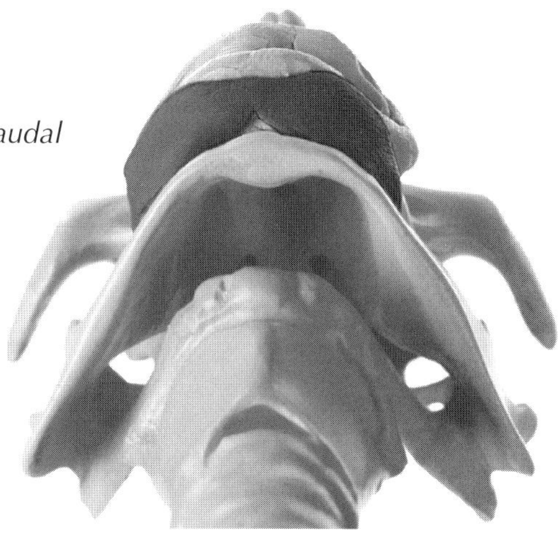

Caudal

Actions

Depresses the lower lip.

Attachments

Caudal

Lateral surface of mandible anterior to the mental foramen.

Cranial

Skin of lower lip.

*duh•**pres**•ur*

From the Latin, "depressus"+"or," meaning to one that presses down

lay•*bee•eye*

Latin, meaning belonging to the lip.

*in•**fee**•ree•or•is*

From the Latin, "inferior," meaning lower or beneath

Depressor anguli oris

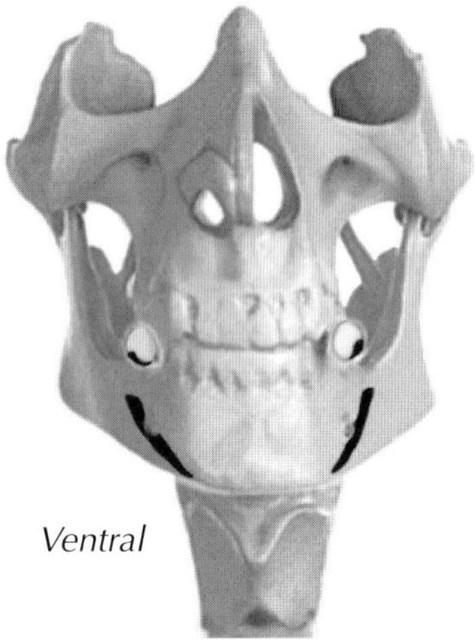

Ventral

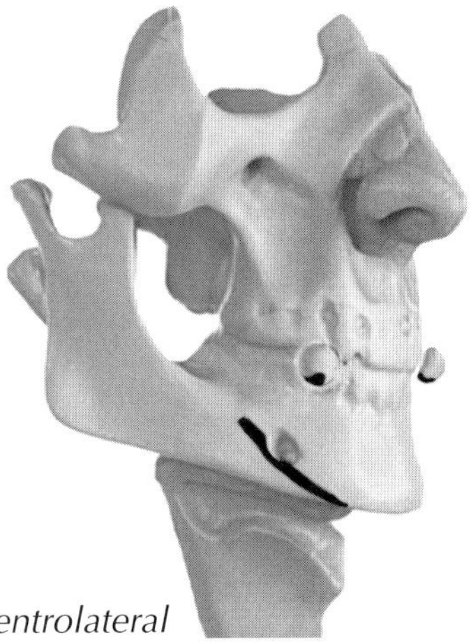

Ventrolateral

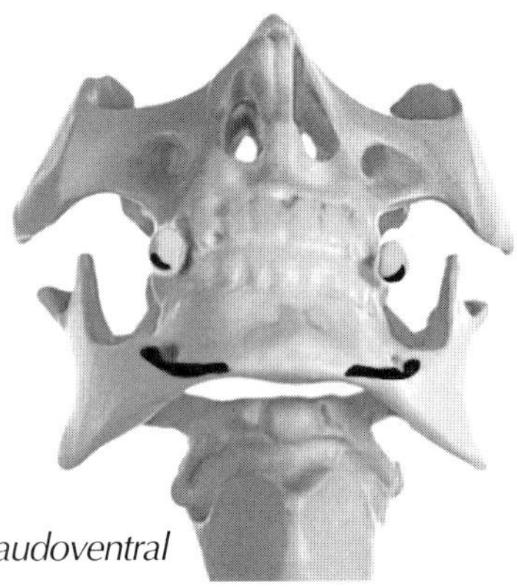

Caudoventral

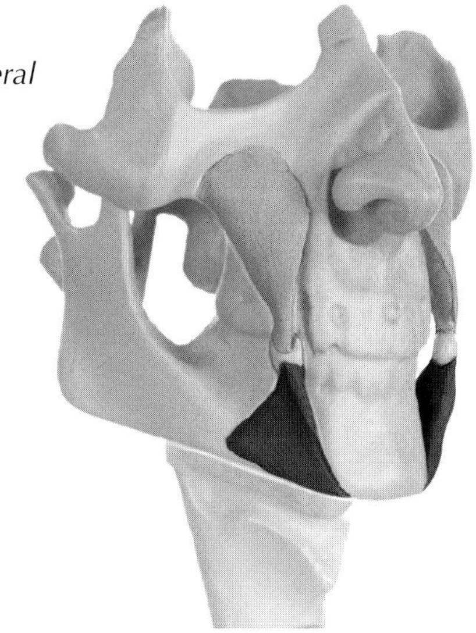

Ventrolateral

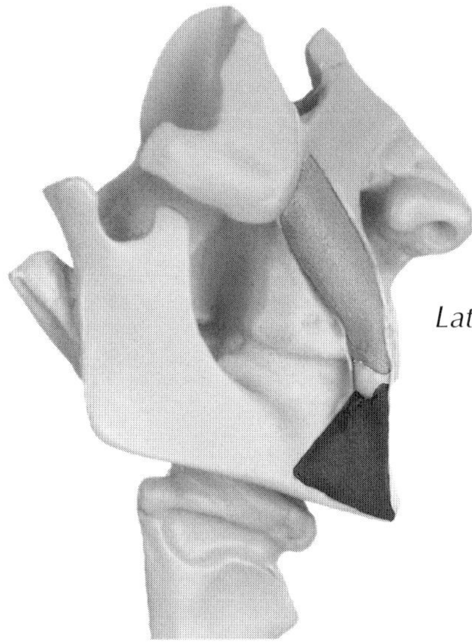

Lateral

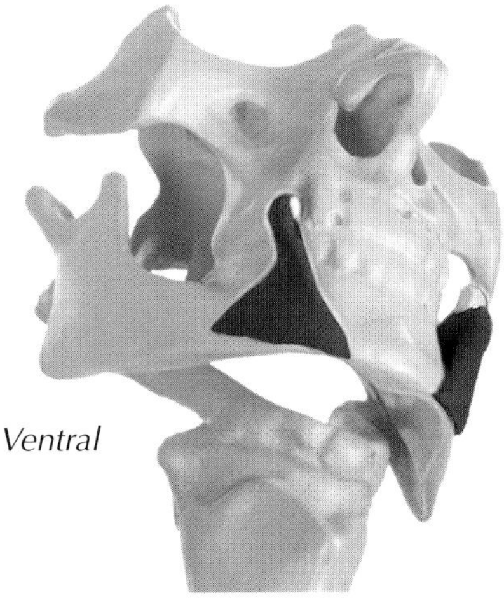

Ventral

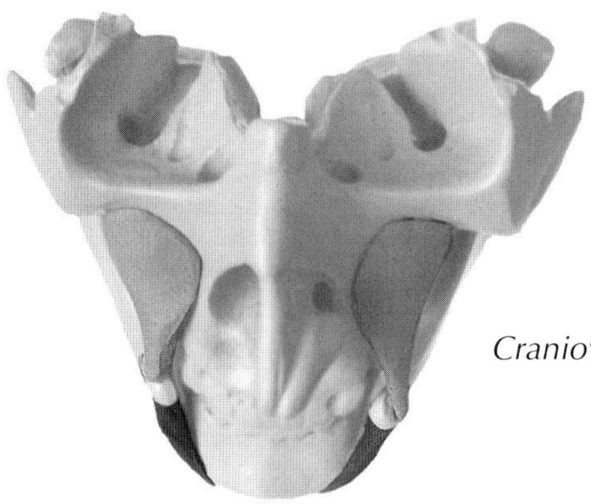

Cranioventral

Depressor anguli oris

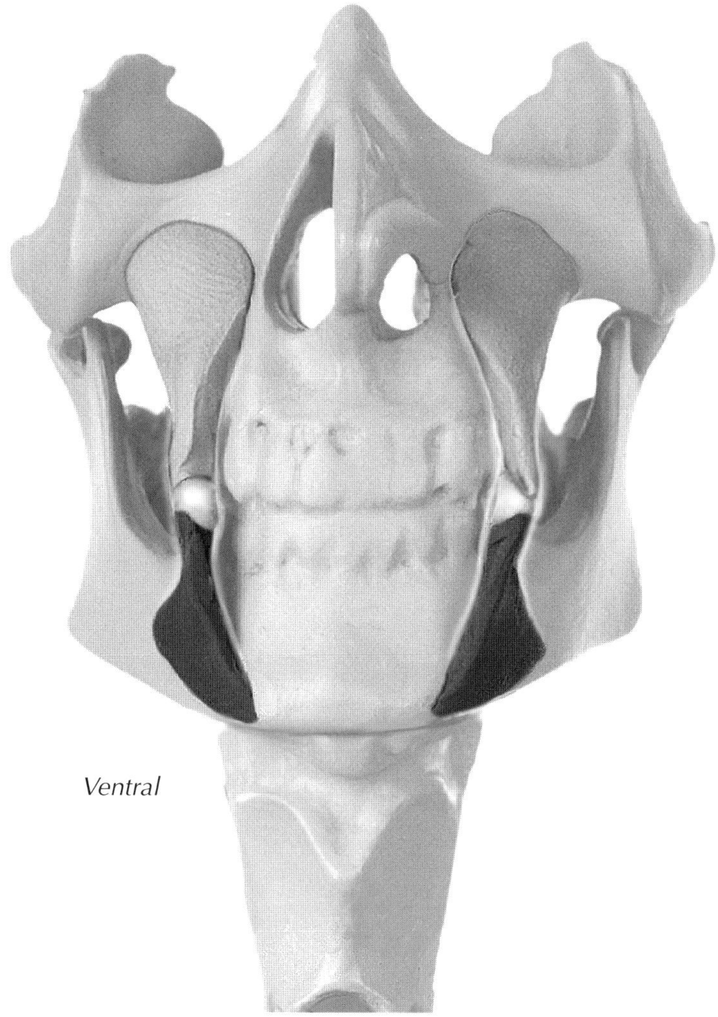

Ventral

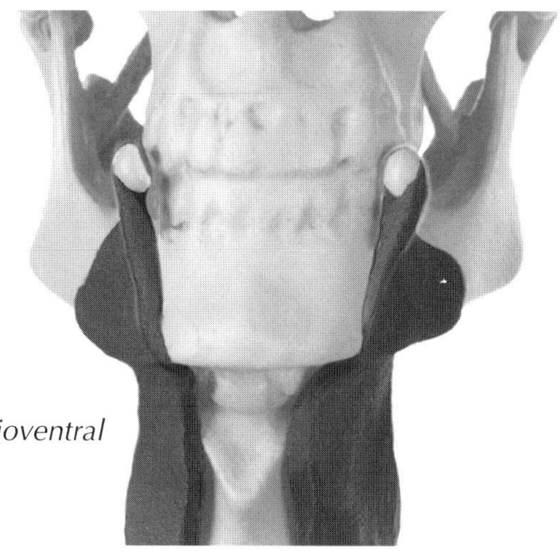

Cranioventral

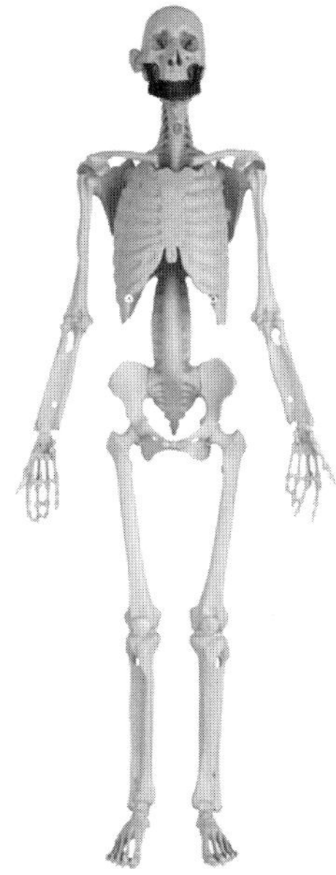

Actions

Depresses the corner of the mouth.

Attachments

Caudal

Oblique line of the mandible.

Cranial

Modiolus at the corner of the mouth.

Note

This muscle is blurs with platysma.

duh•**pres**•*ur*

From the Latin, "depressus"+"or," meaning to one that presses down.

ang•*you*•*lee*

Latin, meaning belonging to a corner or angle.

or•*us*

Latin, meaning belonging to the mouth.

Transversus menti

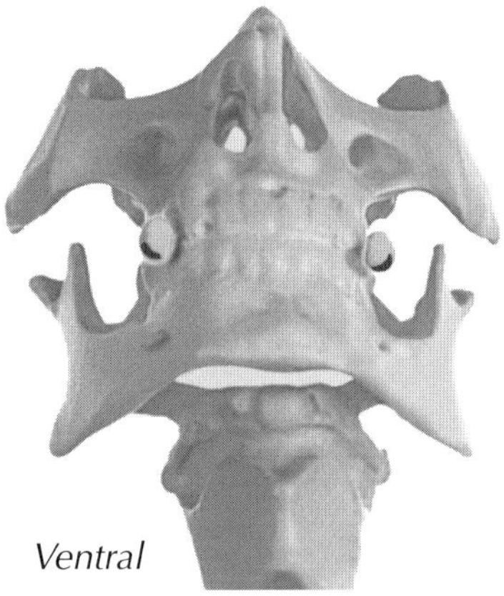

Ventral

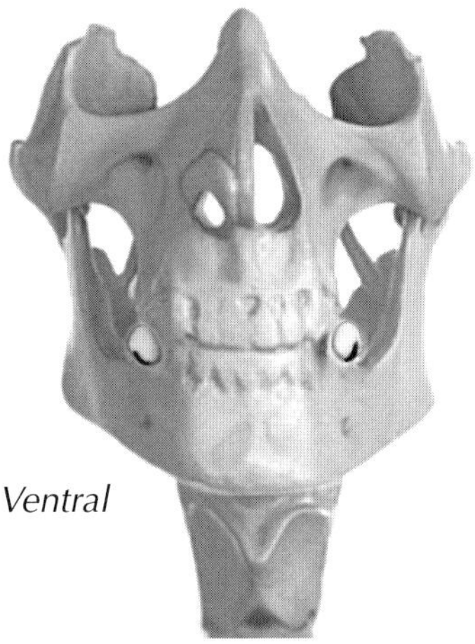

Ventral

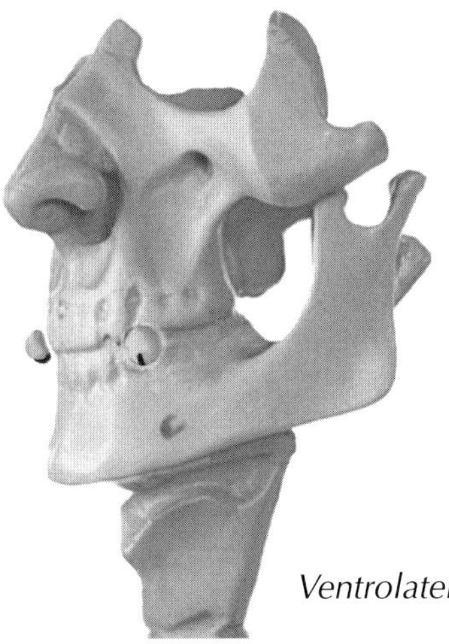

Ventrolateral

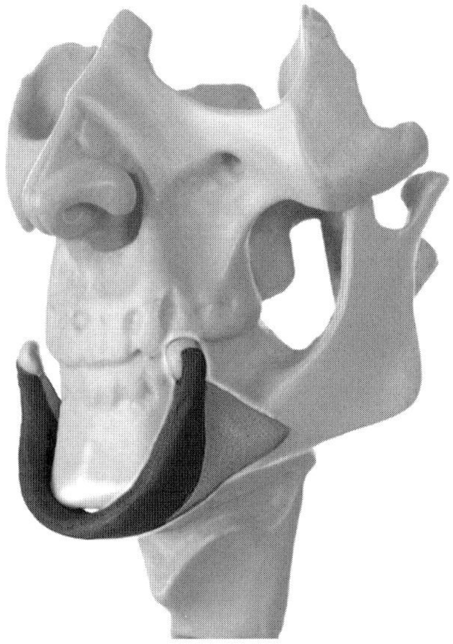

Ventrolateral

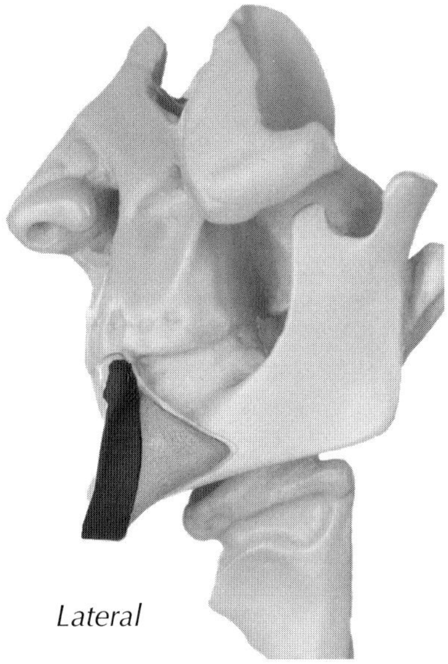

Lateral

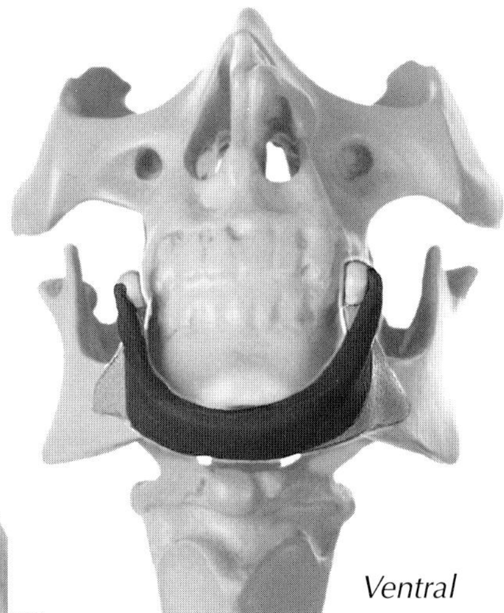

Ventral

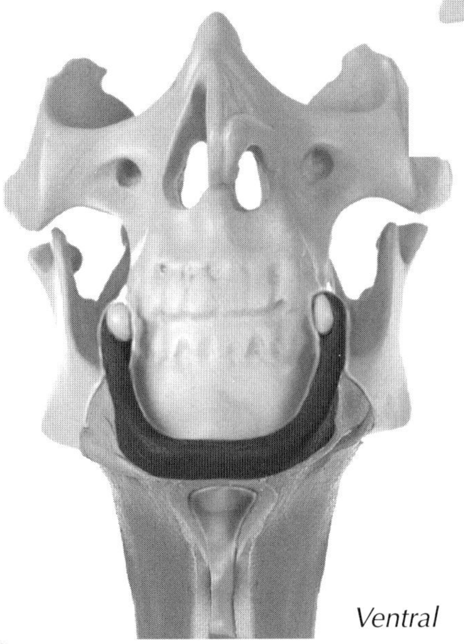

Ventral

Transversus menti

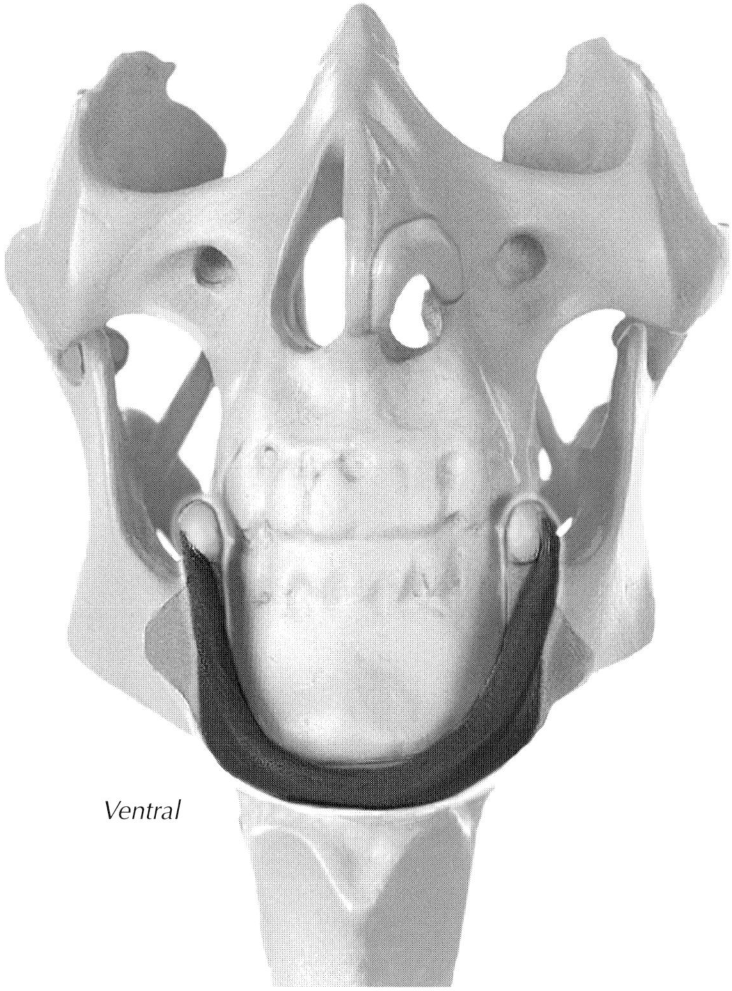

Ventral

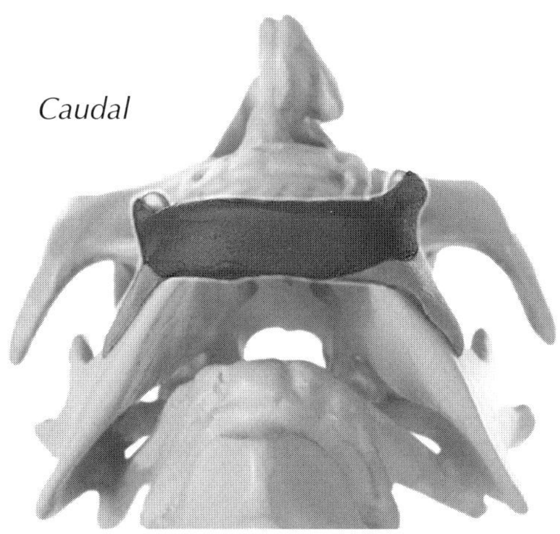

Caudal

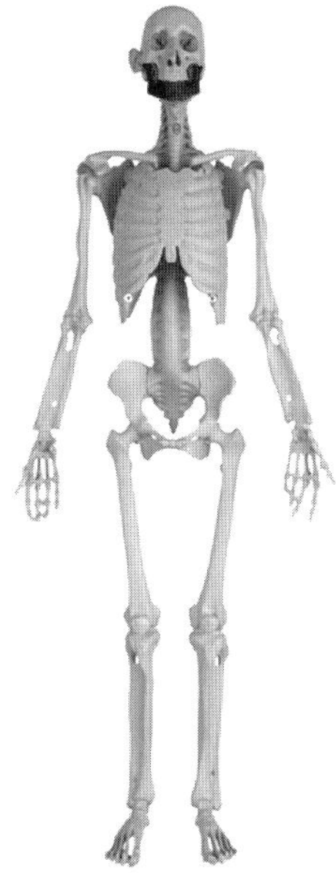

Actions

Helps depress corners of the mouth.

Attachments

Cranial

Modiolus.

Caudal

Blends at median with opposite muscle.

Note

When present, this muscle is thought of as a few slips of depressor anguli oris that by-pass the mental margin of the mandible to interlace with the slips from the other other side.

tranz•**vers**•*us*

From the Latin, "depressus"+"or," meaning to one that presses down.

men•*tee*

Latin, meaning belonging to the chin.

Risorius

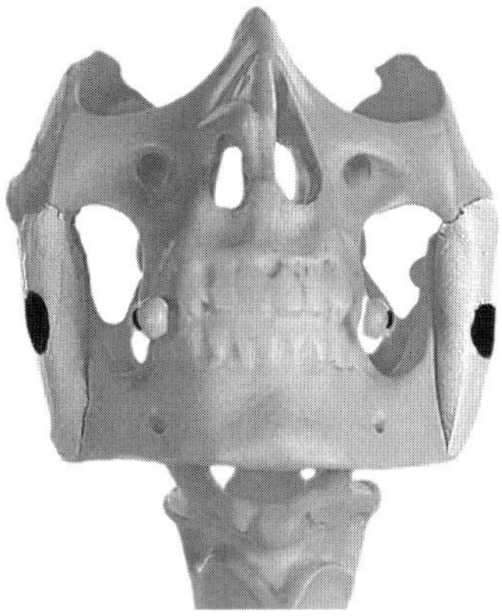

Ventral

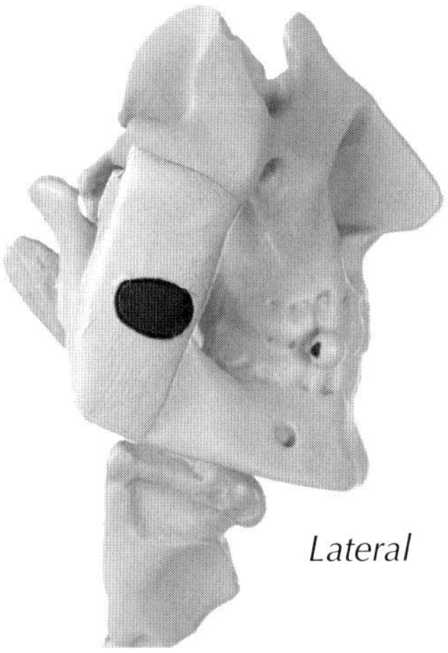

Lateral

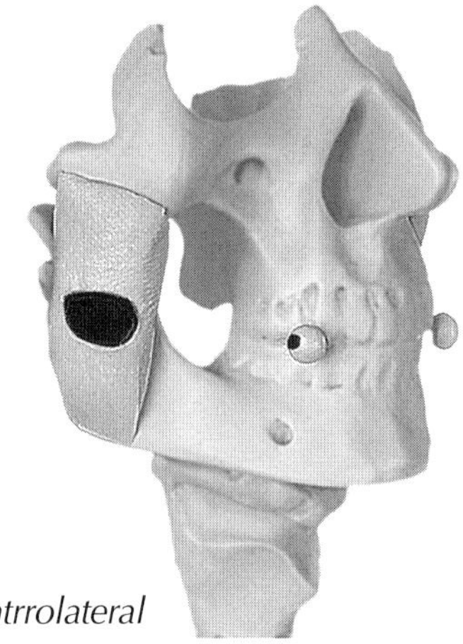

Ventrrolateral

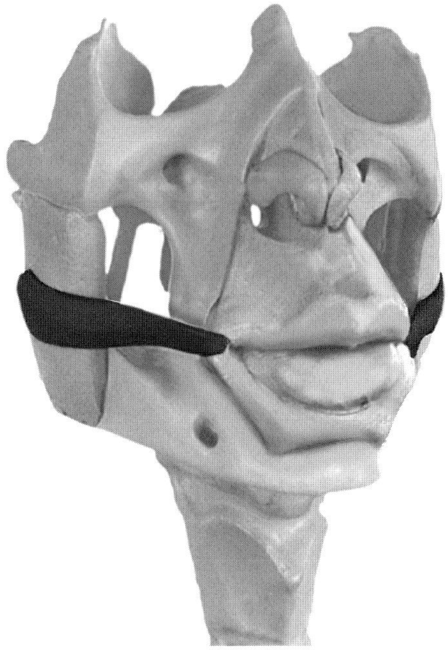

Ventrolateral

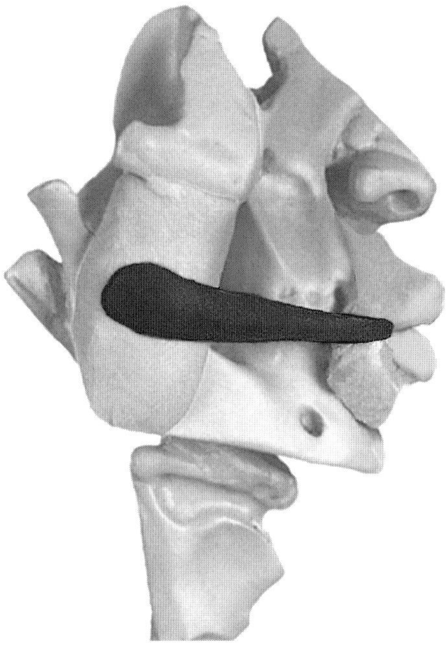

Lateral

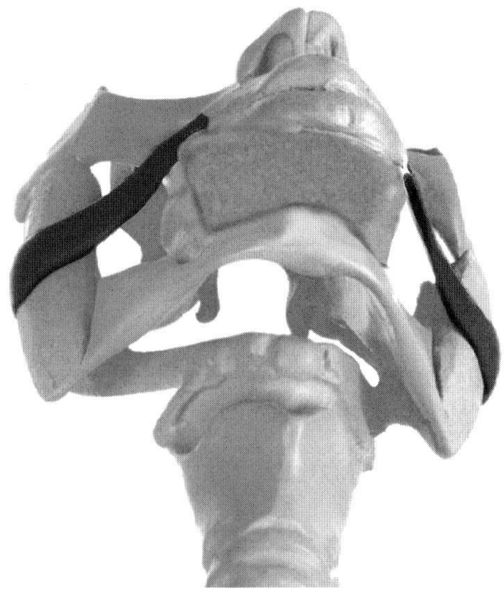

Caudoventolateral

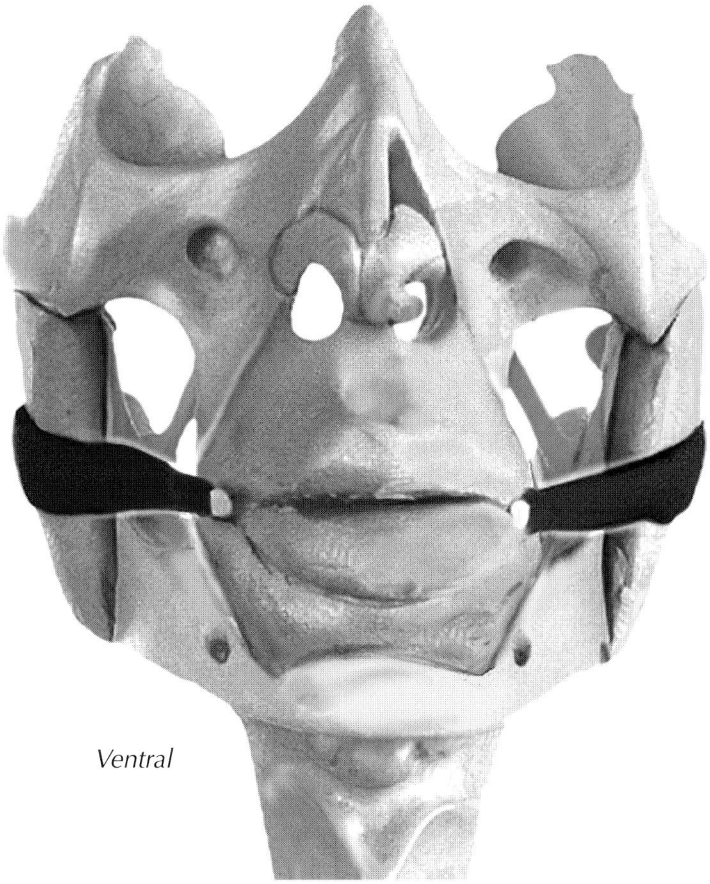

Ventral

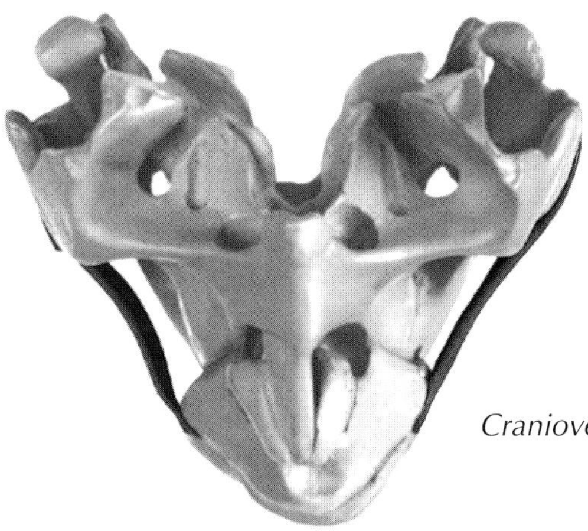

Cranioventral

Risorius

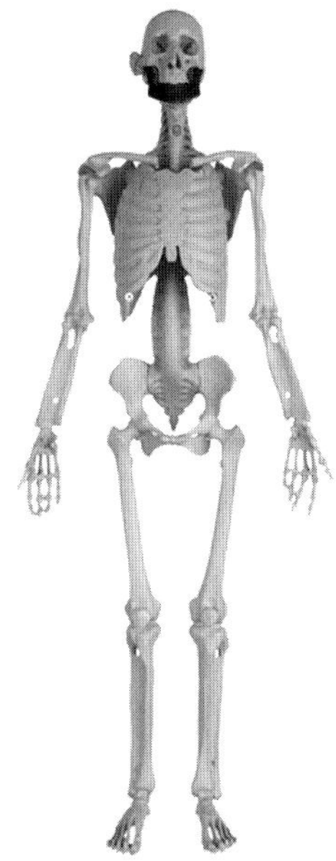

Actions

Retracts corner of the mouth.

Attachments

Dorsal

Fascia over the masseter.

Ventral

Modiolus at the corner of the mouth.

Note

The dorsal attachment is extremely variable, with possible attachments on the zygomatic arch, galea aponeurotica, the ear or the fascia of the parotid gland.

*ruh•**sore**•ee•us*

From the Latin, "risus"+"or," meaning one who causes one to laugh.

Index

Notes